Intellectual Property Management in Health and Agricultural Innovation

a handbook of best practices

EXECUTIVE GUIDE

Intellectual Property Management in Health and Agricultural Innovation

a handbook of best practices

EXECUTIVE GUIDE

ANATOLE KRATTIGER

WITH

RICHARD T. MAHONEY

LITA NELSEN

JENNIFER A. THOMSON

ALAN B. BENNETT

KANIKARAM SATYANARAYANA

GREGORY D. GRAFF

CARLOS FERNANDEZ

STANLEY P. KOWALSKI

MIHR
A STEP NEARER EQUITY

PIPRA
public intellectual property resource for agriculture

Ministério da Saúde
FIOCRUZ
Fundação Oswaldo Cruz

bioDevelopments
INTERNATIONAL INSTITUTE

Published by

MIHR (Centre for the Management of Intellectual Property in Health Research and Development), Oxford Centre for Innovation, Mill Street, Oxford, OX2 0JX, U.K. www.mihr.org, mihr@bioDevelopments.org

PIPRA (Public Intellectual Property Resource for Agriculture), University of California, Davis, Plant Reproductive Biology Bldg., Extension Center Circle, Davis, CA, 95616-8780, U.S.A. www.pipra.org, info@pipra.org

Oswaldo Cruz Foundation (Fiocruz), Av. Brasil 4365, Rio de Janeiro, RJ 21040-900, Brazil. www.fiocruz.br, ccs@fiocruz.br

*bio*Developments-International Institute, Cornell Business and Technology Park, PO Box 4235, Ithaca, NY 14852, U.S.A. www.bioDevelopments.org, info@bioDevelopments.org

www.ipHandbook.org, info@ipHandbook.org

Manuscript Editors
Paula Douglass (Dryden, New York, U.S.A.) and Faye Farmer (ASU, Tempe, U.S.A.)
with Jacqueline Stuhmiller (Ithaca, New York, U.S.A.) and Katy Dixon (Champagne, Illinois, U.S.A.)

Layout and Design
Linette Lao (Invisible Engines, Ypsilanti, Michigan, U.S.A.)

Copy Editor
Barry Hall (OmniGlyph, Atlanta, Georgia, U.S.A.)

Online version
Dynamic Diagram (Providence, Rhode Island, U.S.A.)

Citation
Krattiger A, RT Mahoney, L Nelsen, JA Thomson, AB Bennett, K Satyanarayana, GD Graff, C Fernandez and SP Kowalski. 2007. *Executive Guide to Intellectual Property Management in Health and Agricultural Innovation: A Handbook of Best Practices*. MIHR (Oxford, U.K.), PIPRA (Davis, U.S.A.), Oswaldo Cruz Foundation (Fiocruz, Rio de Janeiro, Brazil) and *bio*Developments-International Institute (Ithaca, U.S.A.). Available online at www.ipHandbook.org.

Translations
Requests for permission to reproduce the book in its entirety or for translations,
whether for sale or for non-commercial distribution, should be addressed to any of the publishers.

Library of Congress Preassigned Control Number
2007930234

(ISBN: 978-1-4243-2028-8)

Distributed free of charge to low- and middle-income countries
(subject to availability of funding and support for distribution).
www.ipHandbook.org, info@ipHandbook.org

PRINTED WITH SOY INK

green press INITIATIVE

MIHR and PIPRA are committed to preserving ancient forests and natural resources. We elected to print *Intellectual Property Management in Health and Agricultural Innovation* on 30% post consumer recycled paper, processed chlorine free. As a result, for this printing, we have saved:

14 Trees (40' tall and 6-8" diameter)
6,059 Gallons of Wastewater
2,437 Kilowatt Hours of Electricity
668 Pounds of Solid Waste
1,312 Pounds of Greenhouse Gases

MIHR and PIPRA made this paper choice because our printer, Thomson-Shore, Inc., is a member of Green Press Initiative, a nonprofit program dedicated to supporting authors, publishers, and suppliers in their efforts to reduce their use of fiber obtained from endangered forests.

For more information, visit www.greenpressinitiative.org

Anatole Krattiger
The Biodesign Institute at ASU, *bio*Developments-International Institute,
and Cornell University, U.S.A.

WITH

Richard T. Mahoney
International Vaccine Institute, Republic of Korea

Lita Nelsen
Massachusetts Institute of Technology, U.S.A.

Jennifer A. Thomson
University of Cape Town, South Africa

Alan B. Bennett
PIPRA and University of California, Davis, U.S.A.

Kanikaram Satyanarayana
Indian Council of Medical Research, India

Gregory D. Graff
PIPRA and Colorado State University, Fort Collins, U.S.A.

Carlos Fernandez
Foundation for Agriculture Innovation, Chile

Stanley P. Kowalski
The Franklin Pierce Law Center, U.S.A.

COMPREHENSIVE EDITORIAL CONSULTANT

David P. Alvarez
University of California, Davis, U.S.A.

THE ROCKEFELLER FOUNDATION

Judith Rodin
President

September 22, 2007

Dear Reader,

Over the last decade, the world has been paying increasing attention to the agricultural, health, and economic disparities between industrialized and developing countries. The Rockefeller Foundation is proud to have helped develop and launch some of the numerous initiatives to address these issues—initiatives such as the African Agricultural Technology Foundation, the Alliance for a Green Revolution in Africa, the International AIDS Vaccine Initiative, the Global Alliance for TB Drug Development, and others.

We believe, however, that launching the success of these and other similar initiatives requires that we both engage directly with research universities in the industrialized world and encourage the growing innovation capacity of developing countries. The Public Intellectual Property Resource for Agriculture (PIPRA) and the Centre for the Management of Intellectual Property in Health Research and Development (MIHR) were created for precisely these reasons. Their mission is to enhance the power of publicly funded research institutions to harness new technologies and to ensure that the benefits of globalization are shared more equitably. This *Executive Guide* and the companion *Handbook* and online version are a natural outcome of their efforts to contribute new solutions to this two-fold challenge. A follow-on interactive electronic version will reach an even wider audience and, we hope, provide even greater benefits.

The Rockefeller Foundation is delighted to have supported the creation of this unique resource. It holds lessons that are valuable (in many senses of the word) for policy makers, leaders of research institutions, researchers, and technology managers alike—in both industrialized and developing countries. Indeed, this *Executive Guide* and *Handbook*, a testament to the committed, excellent work of MIHR, PIPRA, and *bio*Developments-International Institute, might be the most thorough primer on intellectual property management for the public interest ever assembled. As such, it will be an indispensable tool for both planners and practitioners for years to come.

Judith Rodin

420 Fifth Avenue
New York, New York
10018-2702
Tel 212.852.8400
Fax 212.852.8277
president@rockfound.org
www.rockfound.org

Message from the Board of Patrons

Dear Reader,

Intellectual property can be a powerful tool. When effectively and ethically managed, it can both accelerate the development of lifesaving, poverty-alleviating innovations and secure access to them. Both development and access are urgently needed in health and agriculture to improve the lives of people in need—particularly those living in the developing world.

This *Executive Guide* and its companion *Handbook* constitute an authoritative, comprehensive, and practical reference on intellectual property management and best practices. These works will be invaluable for anyone seeking to use intellectual property strategically to enhance economic growth and equitably distribute innovative technologies. This *Executive Guide* and *Handbook* uniquely contributes to efforts in global health and food security. We are pleased to endorse its use. ■

Charles J. Arntzen, *U.S.A.*
Roger N. Beachy, *U.S.A.*
Norman E. Borlaug, *Mexico*
Zhang Liang Chen, *China*
Jacques Diouf, *Senegal and Italy*
Mahmoud F. Fathalla, *Egypt*
Maria Freire, *U.S.A.*
William Hennessey, *U.S.A.*
Kamil Idris, *Sudan/Switzerland*
Molly Jahn, *U.S.A.*
Gurdev Khush, *India/U.S.A.*
Mosibudi Mangena, *South Africa*
R. A. Mashelkar, *India*
Alex F. McCalla, *U.S.A.*
Carlos Morel, *Brazil*
Ingo Potrykus, *Switzerland*
Judith Rodin, *U.S.A.*
Emil Salim, *Indonesia*
Albert Sasson, *Morocco/France*
Akinori Seki, *Japan*
Ismail Serageldin, *Egypt*
Victor S. Shevelukha, *Russia*
M. S. Swaminathan, *India*
Baron Marc Van Montagu, *Belgium*
Vo-Tong Xuan, *Vietnam*
Yongyuth Yuthavong, *Thailand*

Sponsors and Distribution Supporters of the *Handbook*

THE ROCKEFELLER FOUNDATION

with

Ewing Marion
KAUFFMAN
Foundation

Distribution Supporters

ABIISP

AATF

AUTM

BURRILL & COMPANY
A LIFE SCIENCES MERCHANT BANK

Cornell University

GOBIERNO DE CHILE
FUNDACIÓN PARA LA INNOVACIÓN AGRARIA
INNOVA-CORFO

dcvmn
Developing Countries Vaccine Manufacturers Network

Law Offices of
DODDS AND ASSOCIATES

PIERCE LAW
FRANKLIN PIERCE LAW CENTER
AND GEROW D. BRILL, ESQ. '91

Global Forum for Health Research
HELPING CORRECT THE 10|90 GAP

innovation fund

ipn International Policy Network

IRRI
INTERNATIONAL RICE RESEARCH INSTITUTE

ISAAA
INTERNATIONAL SERVICE FOR THE ACQUISITION OF AGRI-BIOTECH APPLICATIONS

Kirkhouse Trust

MONSANTO
imagine®

National Institute of Allergy and Infectious Diseases
National Institutes of Health

PIIPA

SARIMA

Sathguru

An employee-owned company
www.tshore.com
Thomson-Shore
Book Manufacturing

THE UNIVERSITY OF CALIFORNIA 1868

VENABLE® LLP

WIPO

A number of organizations, after seeing sample chapters of the *Handbook*, came forward to purchase copies for wide distribution in developing countries. We are grateful for their generosity and extend our thanks in advance to others who will join their ranks. For an updated list of Distribution Supporters, please visit www.ipHandbook.org.

Indirect or in-kind support provided by AUTM, NIAID, PIIPA, and Venable LLP.

Individuals are also encouraged to purchase *Handbooks* on behalf of developing countries; we will distribute such orders at cost.

Table of Contents

Foreword by Dr. K. Anji Reddy

The instinctively appealing proposition that knowledge ought to be used for the public good—the common cause—is at the core of the debate surrounding the ethics of intellectual property. This notion of the role of knowledge in advancing the public good has deep roots in numerous cultures and communities. Rabindranath Tagore, the universally celebrated poet and Nobel laureate penned a stirring plea of hope:

Where knowledge is free
Where the world is not broken up into fragments by narrow domestic walls
... Into that heaven of freedom, my Father
Let my country awake!

The debate is particularly intense in relation to health and food, areas in which restrictions to access can profoundly affect the human condition. Jonas Salk, the inventor of the first polio vaccine, refused to patent it and famously said: "*Who owns my polio vaccine? The people! Could you patent the sun?*" But the debate over what constitutes patentable subject matter has actually intensified over the years, as concerns have deepened over access. Nowhere is this more pertinent than with antiretrovirals, with the patenting of live organisms, cell lines, and plant varieties.

A persistent disparity in the human condition has accompanied us into the 21st century. This disparity is mirrored by differences, within our global community, regarding access to knowledge and technology, between the rich and the poor. Despite grave incongruities, there is consensus that new knowledge and innovation can contribute significantly to rectifying the inequalities. The faster and more efficient the flow of knowledge "*through narrow domestic walls*," the faster global equity and prosperity will replace global disparity.

The ethical debate over intellectual property rights focuses on the flow of knowledge. But there must be innovation initially to trigger the flow. The great centers of education and research in the public sector have made enormous contributions to furthering knowledge that have significantly improved the human condition, directly or through inventive augmentation by others. Private corporations and individuals have contributed stunning advances as well, and have promoted productive and creative uses of knowledge through science and technology. The crucial question, therefore, is how to increase investment—in

Reddy KA. 2007. Foreword. In *Executive Guide to Intellectual Property Management in Health and Agricultural Innovation: A Handbook of Best Practices* (Krattiger A, RT Mahoney, L Nelsen et al.). MIHR (Oxford, UK), PIPRA (Davis, USA), Oswaldo Cruz Foundation (Fiocruz, Rio de Janeiro, Brazil), and *bio*Developments-International Institute (Ithaca, USA). Available online at www.ipHandbook.org.

both the public and private sectors—to accelerate innovation and facilitate its flow for the greatest possible benefit.

Though the question may be simple, the answer is not. Yet, clear is the message from the writers of the *Executive Guide to Intellectual Property Management in Health and Agricultural Innovation: A Handbook of Best Practices*—that innovation, which can be transformed into intellectual property and then owned and sold for profit in order to sustain investment for further innovation, can be managed to benefit all people, and particularly those who are poor and stand to benefit most from this process. Most experts acknowledge that the notion of intellectual property rights is a compromise between the twin imperatives of providing a reward system that will spur investment in innovation, on the one hand, and use it to *effect the greatest public good*, on the other. Like all compromises, this one too may be imperfect, but it is actionable and practical.

Winston Churchill foresaw the potential of the knowledge economy and asserted that "*the empires of the mind are the empires of the future*." The "haves" of the world own the lion's share of global intellectual property and it is no coincidence that they represent the lion's share of global wealth. The challenge for the "have nots" among nations, most of which have not had the benefit of experience with a robust intellectual property system, is to rapidly develop legislative and policy frameworks to foster a culture of innovation and to manage intellectual property for the greatest benefit. This *Executive Guide* includes many thought-provoking ideas that will support such transformations.

Remarkable in its scope, the *Handbook* not only provides discussion of broad, general, and theoretical issues, it also provides practical ideas that can help institutions and companies strategically manage their intellectual property. The *Handbook* is an epic compilation of over 2000 pages in two volumes, with contributions from nearly 200 authors. It is an invaluable and instructive sourcebook for scholars and students, policymakers and practitioners.

Now, the vast knowledge represented by the *Handbook* has been distilled in the companion *Executive Guide*, which summarizes the *Handbook's* 17 sections and presents key implications and best practices related to each of the major topics. I am delighted to recommend the *Executive Guide*. It will serve as a concise guide for experts and provide fascinating insight for all citizens of the Brave New World.

September 2007
Hyderabad, India

DR. K. ANJI REDDY, *Chairman, Dr. Reddy's Laboratories Ltd., 7-1-27 Ameerpet, Hyderabad 500 016, India.* drreddy@drreddys.com

Foreword by Sam Dryden

Intellectual property! The concept seems to some individuals to represent a vague abstraction. To others it represents a contradiction. But to many, without doubt, intellectual property is a powerful tool. Intellectual property refers to a set of global practices with undeniably real effects. Indeed, the fundamental cultural, intellectual, and commercial enterprises of any nation increasingly intersect with the implementation of intellectual property rights. This is especially true with respect to the life sciences, and specifically to health care and agriculture, because the way intellectual property is managed dramatically affects the pace of innovation, the dissemination of knowledge, and the delivery of new technologies.

These reasons alone are enough to warrant the publication of this *Executive Guide* and companion *Handbook* on intellectual property and innovation. But there are more ambitious and more practical reasons as well. These essays (together with the companion *Handbook*) offer a truly global snapshot of the emerging worldwide practice of IP management. Depending on your point of view (and these essays integrate many points of view)—or on how you practice it—IP management either retards or stimulates innovation and access to new technologies. The authors' views converge, however, on two main points: the growing reach of the emerging global IP rights system and the importance of IP management practice by the public sector. All agree that both the current IP rights systems and IP management practices, in general, are far from perfect. Yet it is evident that solid IP management can be a powerful tool for advancing the public interest. To manage intellectual property well, however, requires knowledge—not simply the knowledge necessary to navigate these systems, but an understanding of the concepts that govern the systems and the values that activate them. With informed use, the public benefits that IP rights regimes are capable of providing can be maximized, especially in developing countries.

Even to function minimally in the modern global economy requires a thorough understanding of how IP systems work. Governments that wish to be part of the global economy will need to radically adjust their approaches to intellectual property; indeed, both intellectual and real property should be reexamined and redefined in both the public and private realms. As new systems evolve, they will need to be understood, established, and enforced by countries' legislative and judicial branches. Innovations will have to be both protected and exploited. These demands have broad commercial implications for all parties involved

Dryden S. 2007. Foreword. In *Executive Guide to Intellectual Property Management in Health and Agricultural Innovation: A Handbook of Best Practices* (Krattiger A, RT Mahoney, L Nelsen et al.). MIHR (Oxford, UK), PIPRA (Davis, USA), Oswaldo Cruz Foundation (Fiocruz, Rio de Janeiro, Brazil), and *bio*Developments-International Institute (Ithaca, USA). Available online at www.ipHandbook.org.

in the development and use of intellectual property, including private inventors, academia, corporations, and, more recently, farmers.

To many people in the plant sciences, these evolving global systems appear to go against the grain of many time-honored precepts, such as breeders' rights, farmer-saved seed, and the free dissemination of germplasm. Many misconceptions exist. Some individuals believe IP rights function simply to enable the developed world to take advantage of the developing world (for example, by depriving farmers of seed that could support their livelihood or by depriving nations of the value of their indigenous germplasm). But the exercise of IP rights is far more complex than these suspicions represent and the provisions for adapting IP rights agreements are much more flexible than is commonly imagined.

This *Executive Guide* is important in many ways. It explains what makes IP systems work and how the public sector, in particular, can best use the system to achieve its mission and objectives. An authoritative undertaking written by world authorities on the subject, this *Executive Guide* is an exceedingly valuable—and timely—contribution to the fields of IP management and economic and social development. The *Guide* is a manual for understanding, not just the mechanics of IP rights, but also their conceptual foundations. A resource for translating IP rights into realistic deals and practical solutions, the *Executive Guide* demystifies intellectual property, making the subject accessible to all. We hope that this *Guide* will level the playing field with respect to developed and developing nations, open up new avenues of collaboration between the public and private sectors, and move us all in the direction of a healthier and more equitable world.

September 2007
Boulder, Colorado, U.S.A.

SAM DRYDEN, *1900 9th Street, Boulder, CO 80302, U.S.A.* sdryden@wolfensohn.com

Foreword by Ismail Serageldin

Today, about two-thirds of global research and development (R&D) is being done by the private sector. That enormous investment in science, and the technology that ensues from it, would not be possible without a system of intellectual property protection that rewards innovators and allows the investors to recoup their investments. No one would want to jeopardize this enormous investment that advances human welfare so much.

However, it is increasingly apparent that in many quarters, such as in research and academia in developing countries, it is difficult to undertake the basic research needed to generate new knowledge rather than simply to be users of technology developed elsewhere. Obstacles include costs as well as a lack of strategic patented technologies as inputs into the research process. True, most of the time a research exemption is included in national law, but it is usually accompanied by a "reach back" clause that comes into play if the research yields some useful product.

To navigate the shoals of this new legal terrain, many researchers and decisions makers need help in setting priorities, developing effective research strategies, and drawing on new knowledge and on the experiences of others to make better and more informed decisions. It is here that this marvelous book becomes absolutely essential.

This *Executive Guide*, a summary of a massive two-volume work, is both readable and authoritative. It represents the fruits of enormous effort and much deep thinking by the people who are most knowledgeable about these complex subjects. The authors are to be congratulated for having put at the fingertips of executives and researchers this most valuable guide and to the authoritative resource that they have produced. I recommend this *Guide* to anyone interested in the practice of science in the twenty-first century or in the promotion of R&D in this new age of globalization of knowledge and trade, where the agenda is increasingly driven by the enormous resources of the private sector.

September 2007
Alexandria, Egypt

ISMAIL SERAGELDIN, *Librarian of Alexandria, Director of the Bibliotheca Alexandrina, P.O. Box 138, Chatby, Alexandria 21526, Egypt.* secretariat@bibalex.org

Serageldin I. 2007. Foreword. In *Executive Guide to Intellectual Property Management in Health and Agricultural Innovation: A Handbook of Best Practices* (Krattiger A, RT Mahoney, L Nelsen et al.). MIHR (Oxford, UK), PIPRA (Davis, USA), Oswaldo Cruz Foundation (Fiocruz, Rio de Janeiro, Brazil), and *bio*Developments-International Institute (Ithaca, USA). Available online at www.ipHandbook.org.

Preface

"No man is an island, entire of itself," wrote John Donne. The concept was apropos in the 17th century and remains so today, perhaps more so. Today, our actions affect everyone, our lives knit ever tighter, and the world is now our community. The strength of myriad environmental, economic, media, technological, and knowledge connections is propelling globalization forward in a complex network of products, people, images, ideals, and aspirations. Intellectual property is perhaps the most scrutinized component of the process of globalization, being a process of ever-increasing interconnectedness. Respecting each other's property (intellectual, cultural, or physical), and sharing one's own properties, is a fundamental principle of building useful and strong connections between people, institutions, cultures, and continents. This is why the *Executive Guide* is so timely. It lays out the major ideas and concerns of the companion *Handbook* in compact, lucid examinations of the full range of intellectual property (IP) issues. It is a virtual map of the basic contours of the IP management response to the increasing interconnectedness. This *Guide* offers readers quick access to current information about IP essentials, particularly as they relate to the public sector, to developing countries, and to making the world a better place.

Even a cursory glance at the *Guide* will reveal the momentous changes that have already taken place in the field of intellectual property and innovation management and the enormous potential that has been generated. The public sector has begun to recognize the value of sound IP management. The result: new, creative relationships between the public and private sectors that are helping to address the urgent health and agricultural needs of people in developing, as well as developed, countries.

Pragmatic IP management is *building bridges* between the world's islands, be they economic, institutional, or geographic. The choice of this metaphor is not accidental. It affirms a key claim that reverberates within the pages of this book: the global IP system and innovation management is not about changing islands. Rather, it is about building bridges between them.

The *Executive Guide* mirrors the principle of building bridges. Approaching topics from various expert perspectives, the *Guide* allows the reader to bridge the conceptual structures, case studies, and IP models contained within it to see new connections and craft new approaches.

Krattiger A. 2007. Preface. In *Executive Guide to Intellectual Property Management in Health and Agricultural Innovation: A Handbook of Best Practices* (Krattiger A, RT Mahoney, L Nelsen et al.). MIHR (Oxford, UK), PIPRA (Davis, USA), Oswaldo Cruz Foundation (Fiocruz, Rio de Janeiro, Brazil), and *bio*Developments-International Institute (Ithaca, USA). Available online at www.ipHandbook.org.

Through the *Guide* and *Handbook*, we hope to liberate the potential of the global IP system, and we believe that such liberation will come *through* the readers as they react to the chapters, draw their own conclusions, imagine their own solutions, and create a positive outcome.

This is clearly not a time for complacent isolation. In fact, it is impossible, as interconnectedness continues to increase. The task at hand is to engage in the process of connecting, with awareness and intention, to ensure that the needs and wishes of the poor in developing countries are not (once again) overlooked or disregarded. It is therefore time to join *your* voices, *your* values, *your* knowledge, and your actions to pursue networks of partnerships that will nurture global access to innovation. Solo efforts at innovation, as I wrote in the Prelude to the *Handbook*, will be ineffectual without collaboration; the components of a partnership create a dynamic entity that achieves more than any one party could achieve. Imagining and cultivating such partnerships, however, requires an understanding and orchestration of numerous components.

This *Executive Guide* opens the door to such an understanding and, I hope, will not only empower its readers to envision a more equitable world but will also inspire them to realize that vision.

Anatole Krattiger
September 2007
Fréjus/St. Raphaël, France

ANATOLE KRATTIGER, *Research Professor, the Biodesign Institute at Arizona State University; Chair, bioDevelopments-International Institute; and Adjunct Professor, Cornell University, P.O. Box 26, Interlaken, NY, 14847, U.S.A. afk3@cornell.edu*

Acknowledgments

Like the progress of any invention from bench to bedside, or from sowing to harvesting, the road leading to the creation of this *Executive Guide*, companion to *Intellectual Property Management in Health and Agricultural Innovation: A Handbook of Best Practices*, has required a great deal of effort and more than a little good fortune. Together with my colleagues, I have discovered that introducing 17 topics and distilling the major points of more than 150 chapters, written by some 200 authors, not only presents unique challenges but also yields exceptional rewards.

The acknowledgments section of the *Handbook* contains a most detailed list of all the wonderful people and institutions who contributed to the total endeavor, which includes this *Executive Guide*. For the *Guide* specifically, I would like to thank:

- The Rockefeller Foundation, which provided substantial funding for the *Handbook*, the online version of the *Handbook*, and this *Executive Guide*
- The Ewing Marion Kauffman Foundation and a large and generous group of supporters, which purchased substantial quantities of the *Handbook* and *Executive Guide* for distribution in low- and middle-income countries
- The Swiss Agency for Development and Cooperation (SDC), Berne, Switzerland, for substantial distribution support for low- and middle-income countries
- My colleagues at MIHR in Oxford, U.K., particularly Junko Chapman, Rachelle Harris, Board Chair Pramilla Senanayake, and Jerry Keusch
- My colleagues at PIPRA, particularly Sara Boettiger, Cecilia Chi-Ham, and Kathleen Bess
- The editorial team of David Alvarez, Paula Douglass, Faye Farmer, and Barry Hall
- The design and page layout team composed of Linette Lao and Mary Penn
- Colleagues, including Charles Arntzen, John Dodds, and others too numerous to mention here

Particular thanks go to my fellow Editorial Board members, Richard Mahoney, Lita Nelsen, Jennifer Thomson, Alan Bennett, Kanikaram Satyanarayana, Gregory Graff, Carlos Fernandez, and the ever-diligent Stan Kowalski.

The dedication and concentrated effort of all of these individuals has made it possible to condense and bring into clear focus the essential points, key implications, and best practices found in more than 2000 pages of the *Handbook*—and beyond. All of us hope that our readers will find this a useful service.

Anatole Krattiger

Executive Summary of Key Implications and Best Practices

EDITORIAL

Abridged Message from the *Handbook's* Editorial Board

The road from sound principles to best practices must include a broad, clear vision of a more equitable world, because when vision is limited, action is circumscribed and constrained. One way to achieve this vision is by expanding and accelerating access—especially in developing countries—to life-saving and poverty-alleviating innovations in health and agriculture. A belief in this path to equity is shared by all of the contributors to the *Executive Guide* and the *Handbook*. The Message from the Editorial Board included in the *Handbook* elaborates on this path as well as on the relationship of intellectual property (IP) to innovation management as this critical interface is perceived by the board members. The key points are:

First, **intellectual property is a tool to foster innovation**. Intellectual property is here—and here to stay—because of its undisputable value as a business asset and an instrument to achieve humanitarian objectives. Since inventions can become property and can therefore be owned and sold, many individuals have been encouraged to invest in innovation, based on the profit potential from resulting technologies. But because IP protections by definition—or by design—exclude competitors and encourage higher pricing, they limit and, in some cases, can altogether prevent access by some individuals and populations. There are many ways, however, for intellectual property to be distributed and utilized **and put to work for the public interest**. Hence we agree that intellectual property should be neither feared, nor blindly embraced; rather, it should be managed to maximize the benefits of innovation for all of society, especially the poor.

Second, **IP rights are a compromise and an imperfect solution**, representing the search for balance between *public domain* and granting *ownership*. Historically, we have seen that this balance encourages investment—and reinvestment—in innovation, although the innovation too infrequently is directed toward the needs of the poor. Fortunately, as numerous case studies have shown, the public sector can craft effective solutions that can approach or even achieve a suitable balance. This can be accomplished by using the existing IP system, especially as it addresses situations in which companies agree to donate or otherwise share their intellectual property.

Krattiger A, RT Mahoney, L Nelsen, JA Thomson, AB Bennett, K Satyanarayana, GD Graff, C Fernandez, and SP Kowalski. 2007. Abridged Message from the *Handbook's* Editorial Board. In *Executive Guide to Intellectual Property Management in Health and Agricultural Innovation: A Handbook of Best Practices* (Krattiger A, RT Mahoney, L Nelsen et al.). MIHR (Oxford, UK), PIPRA (Davis, USA), Oswaldo Cruz Foundation (Fiocruz, Rio de Janeiro, Brazil), and *bio*Developments-International Institute (Ithaca, USA). Available online at www.ipHandbook.org.

EDITORIAL

Third, **genius can flourish anywhere, and the emerging global systems of innovation in health and agriculture open up new prospects for innovation everywhere.** This notion has profound implications for the management of innovation, technology transfer, market competition, and economic development in every country. Irrespective of whether inventions are home grown or originate abroad, authoritative IP management will play a crucial role in enabling and preserving access to the resulting technologies.

Fourth, **policies to promote the creation and management of intellectual property by public sector institutions should give first priority to advancing the mission of those institutions.** Put differently, technology transfer should support the larger mission and not merely be seen as potential revenues.

Fifth, intellectual property has historically benefited mostly the affluent. This is due, in part, to the fact that **insufficient attention has been paid by the public sector to managing intellectual property. This lack of focused attention must be corrected.** Fortunately, there is growing interest, within both the public and private sectors, in putting intellectual property to work for public benefit, although concurrently, there is a lack of knowledge and capacity to use IP appropriately and responsibly.

This *Executive Guide* and companion *Handbook* are designed to help address these complex needs. We hope the pages that follow (particularly the Key Implications and Best Practices) will encourage all parties to take greater advantage of the unprecedented opportunity to strategically manage intellectual property to especially benefit those who have been unable to benefit from technology. Seizing this opportunity will lead, in turn, to a healthier and more equitable world. ■

ANATOLE KRATTIGER, *Research Professor, the Biodesign Institute at Arizona State University; Chair,* bio*Developments-International Institute; and Adjunct Professor, Cornell University, P.O. Box 26, Interlaken, NY, 14847, U.S.A. afk3@cornell.edu;* **RICHARD T. MAHONEY**, *Director, Vaccine Access, Pediatric Dengue Vaccine Initiative, International Vaccine Institute, San Bongcheon-7dong, Kwanak-ku, Seoul 151-818, Republic of Korea. rmahoney@pdvi.org;* **LITA NELSEN**, *Director, M.I.T. Technology Licensing Office, Massachusetts Institute of Technology, Five Cambridge Center, Kendall Square, Room NE25-230, Cambridge, MA, 02142, U.S.A. lita@MIT.edu;* **JENNIFER A. THOMSON**, *Professor, Dept. of Molecular and Cell Biology, University of Cape Town, Private Bag, Rondebosch 7701, South Africa. Jennifer.Thomson@uct.ac.za;* **ALAN B. BENNETT**, *Associate Vice Chancellor and Executive Director, PIPRA, Office of Research, University of California, Davis, 1850 Research Park Drive, Davis, CA, 95616, U.S.A. abbennett@ucdavis.edu;* **KANIKARAM SATYANARAYANA**, *Chief, IP Rights Unit, Indian Council of Medical Research, Ansari Nagar, New Delhi 110029, India. kanikaram_s@yahoo.com;* **GREGORY D. GRAFF**, *Agricultural & Resource Economics, B306 Clark Hall, Colorado State University, Fort Collins, CO 80523, U.S.A. Gregory.Graff@colostate.edu;* **CARLOS FERNANDEZ**, *Director, Strategic Studies, Foundation for Agriculture Innovation (FIA), Loreley 1582, La Reina, Santiago, Chile. carlos.fernandez@fia.com;* **STANLEY P. KOWALSKI**, *The Franklin Pierce Law Center, 2 White Street, Concord, NH, 03301, U.S.A. spk3@cornell.edu and skowalski@piercelaw.edu*

Pictured above, from left to right: Carlos Fernandez, Alan Bennett, Anatole Krattiger, Lita Nelsen, Jennifer Thomson, Kanikaram Satyanarayana, and Richard Mahoney Right: Gregory Graff and Stanley Kowalski

Executive Summary of Key Implications and Best Practices

Throughout this *Guide*, best practices refers to the strategies and approaches that the public sector in particular can employ to achieve its goals within an evolving IP framework. These proven ways that arguably represent the best or most innovative ideas in IP management can help the public sector better mobilize the resources to take products through the process of innovation, and collaborate with the private sector throughout that process. Best practices, therefore, include:

- enactment of comprehensive national laws and policies
- formulation of institutional IP policies and effective IP management strategies
- application of creative licensing practices that ensure global access and affordability
- building institutional IP management capabilities
- the creation of functioning national IP systems that include efficient patent offices and transparent IP court systems

It is important to note that the Key Implications and Best Practices throughout this *Guide* are intended as starting points to be adapted to specific institutional contexts and national circumstances. Some practices are evolving and will depend on context; but most are applicable across countries and continents, and within many institutional contexts.[1]

The pages that follow present in a highly condensed manner the Key Implications and Best Practices distilled from this *Guide*. They are presented in four parts and are, in language and content, specifically aimed at four different constituencies (who need to act in concert to make innovation work). These are:

- Government Policymakers (see page 7)
- Senior Management (university presidents, R&D managers, etc.) (see page 10)
- Scientists (see page 14)
- Technology Transfer Officers (see page 18)

Krattiger A and SP Kowalski. 2007. Executive Summary of Key Implications and Best Practices. In *Executive Guide to Intellectual Property Management in Health and Agricultural Innovation: A Handbook of Best Practices* (Krattiger A, RT Mahoney, L Nelsen et al.). MIHR (Oxford, UK), PIPRA (Davis, USA), Oswaldo Cruz Foundation (Fiocruz, Rio de Janeiro, Brazil), and *bio*Developments-International Institute (Ithaca, USA). Available online at www.ipHandbook.org.

We look forward to feedback on all of the best practices given in this *Guide*,[2] and we encourage all parties to take greater advantage of the unprecedented opportunity to strategically manage intellectual property to benefit especially those who have, so far, been left behind. Seizing this opportunity will lead, in turn, to a healthier and more equitable world. ■

ANATOLE KRATTIGER, *Research Professor, the Biodesign Institute at Arizona State University; Chair, bioDevelopments-International Institute; and Adjunct Professor, Cornell University, P.O. Box 26, Interlaken, NY, 14847, U.S.A. afk3@cornell.edu*

STANLEY P. KOWALSKI, *The Franklin Pierce Law Center, 2 White Street, Concord, NH, 03301, U.S.A. spk3@cornell.edu and skowalski@piercelaw.edu*

---

1 The publishers, editors, and authors have given their best efforts in preparing this publication, and, while we believe the *Executive Guide* (including the *Handbook* and the online version) will all be useful resources relating to intellectual property and the management thereof, the *Executive Guide* and the *Handbook* are not intended to serve as the sole source of information on the topic. Readers are advised to seek independent legal counsel and/or other professional advice for all intellectual property and contractual matters with regard to appropriate practices for specific situations and countries. No warranties or representations of any kind are made as to the accuracy, usefulness, or completeness of any suggestions or information provided in the *Executive Guide*. Neither *bio*Developments-International Institute, MIHR, PIPRA, Oswaldo Cruz Foundation, nor any of the contributors to the *Executive Guide*, nor the editors, funding agencies, or sponsors will be liable for any loss or damage arising out of the use of any information or suggestions in the *Executive Guide* and *Handbook*. This comprehensive limitation of liability applies to all damages of any kind, including (without limitation) compensatory, direct, or consequential damages; loss of data, income, or profit; loss of or damage to property; and claims of third parties. All Web pages have last been accessed between 10 and 21 September 2007.

2 The online version (www.ipHandbook.org) contains, for each of the 153 chapters of the *Handbook*, a detailed Editor's Summary, Implications, and Best Practices. These are more detailed than the summaries provided in this *Guide*. The Web site will also, in future, contain a blog and discussion forum where users can discuss the best practices presented here and share additional ones.

GOVERNMENT POLICYMAKERS

KEY IMPLICATIONS AND BEST PRACTICES
FOR GOVERNMENT POLICYMAKERS

The page numbers refer to the page in this Executive Guide *where the specific concept is discussed in detail.*

The Intellectual Property–Innovation Nexus

- ✔ Whether viewed as a legal concept, a social construct, a business asset, or an instrument to achieve humanitarian objectives, **intellectual property is an important driver of innovation**. (p. 25)
- ✔ The use of the IP system, via balanced patenting and sound licensing strategies, can **serve the public interest through private rights.** This has profound implications for the management of innovation, technology transfer, market competition, and economic development in every country. (p. 25)
- ✔ **Innovation is a complex process**, stimulated by coordinated and structured policies and programs. An IP management system is one of six important factors in determining a country's, or institution's, ability to innovate. Others are R&D capacity of the public and private sectors; safe and effective regulatory systems; the ability to produce new products to high standards of quality; a national distribution system in both the public and private sectors; and international distribution systems and trade in technologies. (p. 25)
- ✔ Policies promoting the management of intellectual property by public sector institutions can **advance the missions of these institutions, fostering creativity and innovation**. (p. 25)

Public Sector Institutions and Intellectual Property

- ✔ When **public research institutions own IP rights,** they can use licensing to control how technology is deployed, meeting both commercial and noncommercial goals. (p. 35)
- ✔ Best practices in IP management will facilitate **global access**, provided the entire **innovation process** is considered from the outset. This includes a mission-driven mindset to establish optimum goals for the public sector. (p. 35)
- ✔ An important best practice by the public sector is incorporating **humanitarian use reservation** provisions in commercial licenses whenever possible. (p. 35)
- ✔ Public sector institutions should have **IP policies and institutional capacity** for implementing best practices in IP management. Licensors are more likely to grant licenses to institutions that respect and protect third party IP rights. (p. 35)
- ✔ Public sector institutions need **strategies that balance the public domain and IP rights**. Commercial and humanitarian objectives are not in conflict, but rather are complementary, indeed mutually reinforcing, aspects of best practices in IP management. (p. 111)

Building IP Management and Technology Transfer Capacities

- ✔ Technology transfer converts scientific findings into useful products or services for society. In the increasingly global economy, **technology transfer collaborations are particularly effective when spanning geographic boundaries**. (p. 73)
- ✔ Strategies for establishing and operating **a TTO must be grounded in realistic economic expectations**. Technology transfer will not make an institution rich. It takes time (ten years or more) to build an IP portfolio, establish contacts, and develop skills in technology transfer. Furthermore, **a critical mass of R&D activity is necessary to justify the costs of a fully functioning TTO**. (p. 73)
- ✔ Alternative models for an institutional TTO are possible. **For example, costs can be shared among a consortium of universities or research institutions.** Such **hub-and-spokes configurations** allow essential policy decisions and scalable functions to be centralized, while keeping essential context-specific functions localized. (p. 73)
- ✔ **Technology transfer is a talent-based business.** Building networks is essential for success. Governments should encourage the creation and operation of national technology transfer associations that concurrently build international linkages. (p. 73)

GOVERNMENT POLICYMAKERS

- **Putting pressure on TTO officers to break even or to generate revenues can constitute a perverse incentive**, forcing a TTO to go with up-front payments. (p. 101)

Statutory IP Considerations

- The statutory **tools of IP** (patents, copyright, trademarks, trade secrets, plant variety protection) are per se neutral; what matters is how these tools are used. (p. 57)
- **Membership in the Patent Cooperation Treaty (PCT)** will provide national institutions (public and private) with greater opportunities for international technology transfer, licensing, product development and penetration of global markets. Membership in the **PCT can also provide** for cost-effective examination of patent applications. (p. 111)
- Providing for legislation, or for amendments to current statutes, that **facilitates patent filing by foreign entities** can be an important component of technology transfer and development. (p. 111)
- Patent licenses are most valuable when coupled with access to associated know-how. Trade secret laws are thus conducive to the transfer of know-how through licensing. **Patents and trade secrets** are thus complementary forms of IP protection. (p. 121)
- Successful licensing of crop varieties (and of accessing improved varieties from other countries) increasingly depends **on the strength of plant variety protection legislation**. Such legislation can support the interests of the variety owner and the farmer and also facilitate the transfer of technology and provides incentives for further investments. (p. 121)
- **Trademarks allow public and private institutions to capture value.** To benefit from trademark strategies, internationally harmonized legislation is important, as is the maintenance of high quality standards and stewardship, since trademarks (and geographical indications) serve as indicators of source and quality. (p. 121)
- Although IP rights are governed by national statutory protection, **contract law is also important**, as contracts are the legal mechanisms that structure the orderly exchange of intellectual property. (p. 121)
- A **functioning court system is essential to encouraging partnerships** and to accelerate national and foreign investments. Indeed, suppliers of biological materials (and of confidential information and intellectual property) will be encouraged to enter into agreements if the suppliers are confident that their property rights will be protected and that agreements will be enforced. (p. 85)
- Pursuant to the TRIPS Agreement and the Doha Declaration provisions on parallel trade, countries can **implement patent rights exhaustion regimes** that either permit or restrict parallel importation. Despite the evident benefits of parallel trade, there are also disadvantages. (p. 163)

Public and Private Sector Intersections

- **Product development partnerships (PDPs)**, essentially **alliances between the public and private sectors**, facilitate and accelerate the flow of public and philanthropic investment through the innovation pipeline. The ultimate measure of success should not be maximum profit but maximum social benefit. (p. 133)
- PDPs enable industry to invest and apply its expertise to address the needs of the poor. In many contexts **PDPs are now driving the drug-development pipeline in neglected-disease R&D**. National institutions in developing countries should be encouraged to participate in PDPs. (p. 45)
- Negotiating **agreements between public and private sectors** is an opportunity to forge a long-lasting and mutually beneficial relationship. But every partnership is different. National institutions need much **latitude and flexibility to craft appropriate deals**. (p. 133)
- **Negotiation and technology marketing skills** are fundamental for successful licensing and technology transfer. People working in the public sector generally need better negotiating skills, thereby enabling institutions to take advantage of their own R&D efforts and to realize broad public sector and commercial goals. (p. 133)

Commercialization of Public Sector R&D

- ✔ Rather than venture capital driving the creation of new companies, **it is usually the creation of new companies that attracts venture capital**. (p. 143)
- ✔ While a government cannot *legislate* entrepreneurship, it can *encourage* entrepreneurship by providing a **favorable environment for creating and growing new companies**. (p. 143, 153)
- ✔ Policies and legislation that benefit **biotechnology companies and start-ups** can accelerate the pace of innovation, particularly when it comes to commercializing public-sector-generated inventions. (p. 133)
- ✔ Much of **the success of a spinout or start-up will depend on the entrepreneurial spirit** at the institution. The more entrepreneurial, the more likely it will be that someone wants to set up a new company. **Spinouts** can create jobs, enhance economic development and create international opportunities. (p. 143)
- ✔ **Cluster formation** fosters the movement of new technologies into existing industry. This requires a commitment to science education and research, a strategically situated anchor institution with a technology transfer office, and reliance on market forces as the engine for technology transfer. (p. 45)
- ✔ Governments can encourage regional economic development by fostering and **financing business incubators**. (p. 143)
- ✔ As intellectual property becomes more prevalent in health and agricultural research, public sector institutions need to consider the intellectual property of **third parties**. Knowledge of "who owns what" is needed. That is what a freedom to operate (FTO) analysis provides, facilitating the handling of products for further development or commercialization, even if the goal is to address the needs of the poor. (p. 153)

IP Dispute Resolution

- ✔ A country's statutory code, combined with a reliable system of **adjudication and enforcement**, is the basis for enforcing IP rights. (p. 163)
- ✔ **Alternative dispute resolution procedures** for settling differences between parties to an agreement can be an effective strategy for public sector institutions. These procedures are particularly important in international contract dispute resolution. (p. 163)
- ✔ Governments and public institutions can help make **arbitration or mediation procedures** accessible by identifying and supporting neutral institutions that can provide cost-efficient, timely dispute-resolution services. The World Intellectual Property Organization offers such services through the **WIPO Arbitration and Mediation Center**. (p. 163)
- ✔ IP protection mechanisms depend upon effective and **equitable enforcement** by national governments. This requires effective, transparent, and enforceable **contract law**. (p. 57)

Biodiversity, Traditional Knowledge and Bioprospecting

- ✔ **Indigenous communities** often play a significant role as gatekeepers to a country's biodiversity wealth. They are the regional specialists with respect to the flora and fauna. Their knowledge can often exceed that of leading scientists. (p. 173)
- ✔ Patent laws per se do not "create" biopiracy. Rather, **biopiracy is a form of misappropriation**, unfair acquisition, and inequitable sharing of benefits with respect to biological resources. (p. 173)
- ✔ Formulate procedures for **equitable access to traditional knowledge** held by indigenous societies. However, this requires balance: access should be granted only via authorized permission, yet the price that is assessed for permission to bioprospect should not dissuade potential development. (p. 173)

SENIOR MANAGEMENT

KEY IMPLICATIONS AND BEST PRACTICES
FOR SENIOR MANAGEMENT
(UNIVERSITY PRESIDENT, R&D MANAGER, ETC.)

The page numbers refer to the page in this Executive Guide *where the specific concept is discussed in detail.*

Intellectual Property in the Innovation Context

- ✔ **Benefits from public sector health and agriculture investments have tremendous global potential.** Government-sponsored research can thus make a big difference in meeting the needs of the poor, nationally, regionally and globally. (p. 25)
- ✔ **Intellectual property, as a tool to foster innovation, is a compromise**, balancing the public domain with limited grants of ownership. An appropriate balance requires sound IP management and capacity to do so in the public sector. (p. 35).
- ✔ **Balancing public benefit and economic returns can be influenced through government policy.** (p. 1). But decisions to place inventions in the public domain can only meaningfully be determined on a case-by-case basis. (p. 45)
- ✔ In order to fulfill both commercial and noncommercial goals, **public research institutions** can, through IP management and licensing strategies, control how their patented inventions are deployed and developed. (p. 25, p. 45)
- ✔ **By working with their respective governments to implement national policies**, public sector institutions can help establish a national IP system that fosters best practices in IP management in the public sector. (p. 93)
- ✔ Well-crafted contracts based on best practices can be instrumental in achieving **global access**, provided the entire innovation process is considered from the outset. Such an approach requires preparation, detailed knowledge, and a **public sector mission-driven mindset**. (p. 45)
- ✔ The use of IP rights is not a panacea for the management of innovation, nor is the public domain. Both public and private goods have utility and limitations. The **art of innovation management is in using both public and private goods** and to manage the interface between them. (p. 111)

Institutional Mission and Policy

- ✔ A sound **institutional IP policy** typically addresses ownership, conflicts of interest and commitment, confidential information, broad IP licensing approaches, and IP generated revenues. (p. 65) Other important considerations include **ethical guidelines** for IP management (p. 35) and the **reservation of humanitarian rights** (also called philanthropic-use provisions) on IP and the **retention of research and teaching rights** of all inventions and technologies. (p. 45, 121)
- ✔ An effective **institutional IP strategy** describes long-term goals and the allocation of resources necessary for their realization. Public sector institutions have much to gain by articulating how their IP management strategies foster global access to innovations. (p. 65)
- ✔ **IP audits** can be useful mechanisms that form the basis for an internal review and revision of an institution's IP strategy and IP policy. (p. 65)
- ✔ Government policies ought to be **flexible and enable research institutions to customize technology transfer strategies** that align with the institutions' missions. Different approaches will serve different types of research and academic organizations working within various disciplines and cultures. (p. 101)

IP Management Principles and Technology Transfer

- ✔ Technology transfer invariably brings **conflicts of interest**. These should not be viewed as negative. They can usually be managed in a transparent and consistent manner. (p. 65)
- ✔ **Technology transfer and licensing are context-specific.** (p. 35). Licensors, public or private, are more willing to license to institutions that consistently protect third-party property. This builds confidence and thereby promotes licensing and technology transfer.

(p. 45) A successful approach is to allow **maximum flexibility whereby institutions can set, or negotiate, the terms that best fit the mission and goals of the institution** and the purpose of the partnership. (p. 85)

- ✔ The **core element for successful technology transfer is people**. The technology transfer office (TTO) should be led by an individual who understands the details of running a business. (p. 73).
- ✔ Successfully establishing and operating a TTO will require **visible and sustained support, financial and otherwise, from senior administration**. Clear mandates will help technology transfer professionals choose among competing priorities. (p. 73). This includes the implementation of rigorous IP-related policies and procedures. (p. 57) Above all, **it requires TTO officers' ability to assume risks knowing that they have backing from senior management**. (p. 73)
- ✔ A critical mass of R&D activity is necessary (typically $100-$500 million) to justify the costs of a fully functioning TTO. **Consortium approaches to technology transfer** can ameliorate this imposing initial requirement by pooling resources and expertise. (p. 73)
- ✔ For companies, the ultimate purpose of IP management is to **enhance competitiveness and reduce risk**. For public sector institutions, the purpose of IP management should be to **serve the greater public interest**. These are not mutually exclusive goals, and they can be reconciled through sound technology marketing and licensing practices. (p. 133)
- ✔ Scientists should be encouraged to **use public domain technologies** as research inputs whenever feasible to reduce possible future constraints in the downstream commercialization of innovations. In many circumstances, however, **relying on patented technologies** may be the more effective way to go, particularly when the goal is to develop products. (p. 111)
- ✔ Building strong **institutional capacity in IP management** will enable technology managers and scientists alike to understand the complex array of options that should be considered before publishing research results or filing patent applications. (p. 111)
- ✔ An important element during the **development of an IP strategy** is to document the technologies that already exist in the organization, plus those technologies in development. This can be achieved through an **IP audit**, among other approaches. (p. 111)
- ✔ Management should encourage good laboratory practices and diligent **record keeping** of data to ensure that research can later be used in possible regulatory filings. Doing so could lower costs and reduce the time to market. (p. 111)
- ✔ To **benefit from trademark strategies**, the maintenance of high quality standards is important, since the trademark (brand) indicates the source and quality of the product. (p. 121)

Supporting Entrepreneurship

- ✔ **The creation of business incubators as a tool for stimulating local economic development** should not be underestimated. Incubated companies have a dramatically higher rate of survival than the average spinouts. (p. 143)
- ✔ **A spinout often creates enhanced opportunities for its faculty.** If spinouts remain in the region, faculty inventors can remain active as consultants. Also, a university's success with spinouts can attract new talent. But much of **the success will depend on the entrepreneurial spirit** at the institution. The more entrepreneurial, the more likely it will be that someone wants to set up a new company. (p. 143)
- ✔ If public sector institutions found ways to reduce the risk of investing in agricultural projects, more **venture capital would be attracted**. (p. 143)
- ✔ Rather than venture capital driving the creation of new companies, **it is usually the creation of new companies that attracts venture capital**. (p. 143)
- ✔ **Robust innovation clusters** require a commitment to science and entrepreneurship, a strategically situated anchor institution, and reliance on market forces as the engine for technology transfer. **Cluster formation** will strengthen the ability of local and national economies to absorb new technologies into existing industry or entrepreneurial sectors. (p. 25)

SENIOR MANAGEMENT

Licensing

- ✔ **Nonexclusive licensing** can be a strategy to maximize the utilization of research tools. **Exclusive licensing**, on the other hand, can be quite effective for broad dissemination of patented products, particularly when coupled with milestone clauses. Complementary strategies are **market segmentation, field-of-use licensing,** and the negotiation of **tiered pricing clauses**. (p. 45)
- ✔ The dual goals of **economic growth and social/humanitarian benefits** through licensing are not mutually exclusive. Indeed, they are often complementary. Much will depend on a sound institutional licensing strategy and on good relationships with licensees. (p. 121)
- ✔ **Business decisions, more than legal aspects, should determine licensing terms.** Nevertheless, lawyers should ensure that the contracts comply with prevailing law. (p. 121)
- ✔ **Confidentiality agreements rely on a culture of trust, not a culture of secrecy.** They protect sensitive information transferred between parties, and, when well managed, and are not inconsistent with public sector missions or the publication of research results. (p. 85)
- ✔ **Patent licenses are most valuable when coupled with access to associated know-how,** which is often shared under confidentiality provisions. Comprehensive staff training in the handling of confidential information from third parties is therefore critical. (p. 121)
- ✔ Specific **best practices in licensing terms that allow public sector entities** to meet public sector goals (ensuring broad access to innovation) include area of use, territory, price, labeling, white-knight conditions, and royalties. (p. 133)
- ✔ The key to successful negotiation is having a clear understanding of the value each party brings to a relationship. **Value may be monetary or non-monetary.** (p. 133)
- ✔ Negotiating between public and private sectors ought **not be confrontational**, but as a first step in a long-lasting and mutually beneficial relationship. **Negotiating a fair licensing agreement** should thus not be seen as a process of "bargaining" toward a win-win outcome. (p. 133)
- ✔ For the public sector, a well-tested and successful approach to negotiating an agreement is to **offer initial terms to a company that the public sector organization itself would be willing to agree to** if it were on the other side of the negotiating table. (p. 133)
- ✔ Senior management can **set a positive tone for negotiation** that will ensure that deals made with others are a vehicle for building strong relations and trust between parties. (p. 133)
- ✔ **Networking** is essential for successful technology marketing. Technology transfer officers and scientists particularly should be encouraged and given opportunities to network. (p. 133)

Managing Risk, Maximizing Benefits

- ✔ As public sector and nonprofit institutions pursue **product development, freedom to operate (FTO)** will become a strategic component in an organization's **risk-management strategy**. Implementing the strategy requires clear pathways of communication and dialogue between science managers, product development, licensing personnel, counsel, and senior management. (p. 153)
- ✔ FTO is an **interdisciplinary endeavor** and considered within the context of the institution's mission, business development, research and technology transfer, and tolerance for risk. (p. 153)
- ✔ The more downstream a research-based institution operates, the more important FTO considerations become. A system should be in place to help decide **whether, when, and how** a public sector institution should conduct or commission a legal FTO opinion. (p. 153)

Maintenance of IP Rights

- ✔ It is important to have the flexibility to opt for **legal action** if this seems to be in their best interests. But legal action is often complex due to cost, length of procedure, uncertainty as to outcome, confidentiality/publicity, the difficulty of seeking action in foreign jurisdictions, and the negative impact on existing business relationships. (p. 163)

- Encouraging **alternative dispute-resolution procedures** can be a viable strategy and, indeed, often a preferred one, for settling differences between parties to an agreement. These are particularly important in international contract dispute resolution. (p. 163)

Biodiversity, Traditional Knowledge: Access and Equity

- There is a strong interaction between bioprospecting activity and national scientific capabilities. In countries with strong scientific capability, bioprospecting is robust. Moreover, such **capacity increases the negotiating strengths and benefit-sharing opportunities.** (p. 101)
- The technology transfer office should work with senior management to establish **policies and systems for accessing indigenous or traditional knowledge** (TK), bioprospecting activities, and benefit sharing in an equitable manner. (p. 173)

SCIENTISTS

KEY IMPLICATIONS AND BEST PRACTICES
FOR SCIENTISTS

The page numbers refer to the page in this Executive Guide *where the specific concept is discussed in detail.*

Inventions, Inventors and Innovations

- ✔ **Research is one of the very foundations of innovation.** Research leads to discovery; discovery fosters invention; inventions nourish innovation. Your work is part of a larger **innovation process** that spans R&D across the public and private sectors. (p. 35) As the creator of inventions and technologies, **your role in technology transfer is critical**. (p. 65) **So please read on!**
- ✔ As a scientist, you recognize the **interconnected web of science, R&D, technological advance, and commercial investment**. Take the time to share these insights with your institution's technology transfer office (TTO) and its senior managers. (p. 25)
- ✔ **Determining how to translate an invention into an innovation** that makes a difference in people's lives (economically or socially) is one of the principal reasons TTOs exist. (p. 101)
- ✔ **The emerging global system of innovation in health and agriculture creates opportunities worldwide.** This key concept, that public interest can be served through private rights, has profound implications for the management of innovation, technology transfer, market competition, and economic development in every country, regardless of its economic status. (p. 35)
- ✔ Countries engaged in reforming their R&D and technology transfer efforts often include royalty-sharing provisions for scientists in publicly funded research institutions. This often requires assignment of ownership rights to the institution and a duty to disclose inventions. This should be seen as an **incentive to turn inventions into innovations that benefit society.** (p. 25)
- ✔ While access to foreign technology is integral to development, it is also important to **focus on capturing the national innovation potential** of any country. Through the activities of your research program, you may be positioned to facilitate the development of indigenous innovation and traditional knowledge. (p. 25)
- ✔ **Your continued interest in your invention's development is important.** This will help it reach the marketplace, and **especially** benefit those who most need it, yet can least afford it. (p. 35) Hence, as the inventor, you can **significantly influence how your technology is used**. For example, you might request that licenses reserve your right to continue research using your inventions or reserve rights for humanitarian uses of your technology. (p. 45)
- ✔ **Collaboration with private sector entities** can significantly contribute to your institution's broader participation in innovative initiatives, particularly product development. (p. 45)

Networks

- ✔ Collaborations create contacts. Contacts build networks. **Networks provide opportunities.** (p. 133)
- ✔ Collaboration is often based on establishing personal contacts and **building strong professional networks**; these foster the formation of collaborative research projects and are fundamental for effective sharing of know-how and show-how. (p. 25, 73) Accessing other's intellectual property can be facilitated through networks of committed professionals. (p. 45)
- ✔ Keep your TTO informed about your networking activities, particularly if there is a possibility of **shared research endeavors.** These collaborative research projects, and your network in general, can be starting points for technology transfer and licensing. (p. 73, 173)

IP Management

- ✔ Your role can best be carried out if you have good relations with your TTO. But fulfilling your role also requires an understanding of your institution's **IP policy**. The policy will likely articulate ownership of intellectual property, conflict of interest, the handling of confidential information, and more. (p. 65)

SCIENTISTS

- ✔ The purpose of such a policy, and more importantly of your institution's **IP strategy**, is not just to protect your inventions, but also to control technologies and IP assets such as to determine how these can be managed to spur economic growth and contribute to the greater public good. **If your institution does not "own" anything, how can it place conditions upon its use?** (p. 65)
- ✔ As your institution implements IP policies and patenting strategies, **your right to publish is not jeopardized**. IP protection and licensing are but one form of knowledge transfer that, if well undertaken, can very much be in the public interest. (p. 25)
- ✔ **Philanthropic donors** increasingly expect to find IP management components in **grant applications** and to understand how intellectual property will be used to achieve global access and humanitarian benefits. This is one reason why a close relationship with your TTO is important since your colleagues at TTOs may increasingly be required to prepare access strategies as part of your grant applications. (p. 65)
- ✔ When your institution conducts or commissions an **IP audit**, view this as an opportunity to better identify the intellectual property generated in your research program, to improve and streamline the management of third-party intellectual property (allowing you to concentrate more on research), and to contribute to the formulation and execution of an IP strategy that benefits your program and its global impact. (p. 65)
- ✔ It is your responsibility to disclose any **potential conflict of interest.** Know your institutional **conflict of interest policy**. Most conflict of interest issues arise when procedures are not properly followed. (p. 73) You are not guilty of anything if you have a potential, perceived, or even real conflict of interest. It is only a matter of "managing" these conflicts. (p. 65)
- ✔ Everyone in your group or laboratory should know the **obligations entered into through any agreement that affects your program**. (p. 85)
- ✔ Increasingly contracts will include **milestones**. Research schedules and goals may be directly linked to specific milestones, and you need to know how such milestones might impact your program. (p. 45)
- ✔ Published information, or research tools provided by a colleague, may be covered by IP rights. This should neither deter nor distract you from good science. An **awareness of basic IP management best practices** will help you to understand and identify potential IP issues. (p. 111) Encourage your TTO to organize **occasional seminars on the basics of IP management**. This will facilitate communication with your TTO staff and answer your questions about IP management. (p. 93)

Laboratory Notebooks and Records

- ✔ Good **data management and accurate record keeping** through comprehensive laboratory notebooks (p. 163), is the foundation for building a portfolio of IP assets. Essentially, best practices in scientific record keeping should be precisely the same as best practices in record keeping for purposes of IP management. (p. 57, 93)
- ✔ The **confidentiality of your data** may be critical in ensuring global access. Data is a valuable form of intellectual property that can be used in licensing negotiations. (p. 57)
- ✔ **Confidentiality agreements** are meant to protect sensitive information exchanged between parties and are not inconsistent with public sector missions or research publication. **Confidentiality agreements rely on a culture of trust, not a culture of secrecy.** (p. 85)
- ✔ Make an especially strong effort to document the **origin of biological and other materials** you use in your research, and keep a comprehensive record. (p. 121)

Invention Disclosures and Patenting

- ✔ **Recognize when you actually have an invention** (it is generally much, much earlier than most scientists think)! Invention disclosures are the first step in protecting intellectual property. **Disclose early and often.** But expect only a small portion of your invention disclosures to lead to patent applications. (p. 93)

SCIENTISTS

- By **filing an invention disclosure** with your TTO, you are initiating a dialogue. Even if the TTO does not immediately file a patent based on your first invention disclosure, it is a process that has started, and follow-up will be much easier. (p. 93)
- **Invite your TTO liaison to visit your laboratory occasionally** and discuss with you and your research team what you have been doing. Discussions with technology transfer experts, especially patent attorneys, can help you to identify inventions. (p. 93)
- **If patenting and public disclosure are your goals**, consult with your institution's technology transfer manager prior to disclosure. Your institution should have a mechanism to determine whether or not a patent should be filed without significantly delaying publication. Just be aware that premature publication can lead to a loss of IP rights. (p. 111)
- **Patents** often disclose more technical and scientific information than academic publications. Read published patent applications or issued patents in your field. You can access this information for free on the Internet. (p. 57)
- Your institution's technology **transfer managers will need your input** in order to make strategic decisions about where to pursue **foreign patent applications**. You likely know where different competitors are located and where products arising from your research are needed. (p. 111)
- When you disclose an invention to your TTO officers, inform them of any ideas you may have on the various fields of endeavor in which your invention could find applicability. This will help the TTO plan patent applications and later design license agreements under different **field-of-use licenses**. (p. 121)

Licensing Inventions and Marketing Technologies

- The "**unique selling proposition**" **of your invention or technology** (the features, advantages, or benefits it offers) may not be the science behind the technology, but your invention's use. (p. 133)
- **Technology marketing** is a process by which owners of a technology create relationships, between themselves and potential users, which will drive technology development and availability, through commercialization or other methods. (p. 133)
- When speaking to potential licensees or investors, it is often best to, in extremely simple language, **stress the potential applications of your invention** rather than the superb science. (p. 133)
- **Your role in field-of-use licensing is essential.** You can provide your TTO with valuable information on licensable components for different applications and entities. (p. 121)
- **In agreement negotiations**, your role may be to share relevant information, advice and insights. In some cases, especially with collaborative research agreements, you may be an integral member of a team that will address issues such as research plans. (p. 85)
- Detailed aspects of **negotiations**, such as collaboration or license agreements, are conducted by the relevant offices of your institution. However, do **participate in the internal discussions** prior to licensing negotiations. Your input will be important and should be valued. (p. 121)
- **Material transfer agreements are tools for gaining greater access to tangible materials from a number of sources** (scientists from the public and private sectors, both in your own country and abroad). (p. 85)
- In most institutions, you will not be authorized to sign most agreements without review by counsel or by your TTO. Know **whether or not you are authorized to sign a given agreement**. (p. 85)
- Understand the obligations that are attached to different funding sources. The **impact of joint public and private financial support** can be complex but will increase, particularly as your institution positions itself strongly within an innovation cluster and engages in product development. (p. 25)

SCIENTISTS

Scientists and Entrepreneurs

- ✓ Not all university inventors are **entrepreneurs** interested in being company founders, and not all spinout company founders from a university are the technology's inventors. **Involvement as a company founder entails a greater degree of risk and commitment** to move an invention to commercialization. (p. 143)
- ✓ Much of **the success of a spinout or start-up will depend on the entrepreneurial spirit** at the institution. The more entrepreneurial, the more likely it will be that someone wants to set up a new company. (p. 143)
- ✓ **Venture capital investors** combine a broad view of the market with solid technical expertise. Venture capital investors can be great allies, but will impose, for good reasons, distinct conditions on the project. **Be open, patient, and willing to work with investors.** (p. 143)

Freedom to Operate

- ✓ Collaboration among scientists and the professionals who conduct freedom to operate (FTO) analyses is essential. **The scientist can explain the science and technology** to help others understand the materials and methods. A scientist is the expert in the area of research and can provide important leads to other scientific groups and publications. (p. 153)
- ✓ Teams conducting FTO analyses will also need to understand precisely **what the product is, how it was developed, what materials were used**, and what reports were prepared. The purpose is to ascertain that all relevant information has been considered in the FTO analysis. (p. 153)
- ✓ The results of an FTO analysis may **allow you to make better use of technologies in the public domain** and inform your choice of research tools or vector constructs. The analysis also may alert you to scientific discoveries and inventions related to your work. (p. 153)
- ✓ Knowledge of how to access, manipulate, and mine patent and publication search tools for valuable information will serve you and your program well. Hence, become **versed in Internet database search skills**, and ask your TTO to organize short patent search workshops. (p. 153)

Maintaining IP Rights and Obligations

- ✓ As a scientist, you should **regularly review** all of the agreements that relate to your projects. This specifically includes ensuring that milestones are met, royalties paid, and that any other obligations are taken care of. (p. 163)
- ✓ Your institution should continuously monitor **patent infringements** through various surveillance protocols. A lack of patent enforcement can lead to a loss of patent rights. Your role in this is important, since you are well connected in the area of your research and can indicate to the TTO which companies might be practicing your inventions. (p. 163)
- ✓ If your institution conducts alternative dispute-resolution procedures such as **mediation or arbitration**, you might be called upon to participate, particularly if aspects of your research program are involved in the ongoing discussions. (p. 163)

Biodiversity, Bioprospecting, and Traditional Knowledge

- ✓ When working with colleagues from foreign countries, be aware that those colleagues may be authorized to make **collections of biological materials** only under specified circumstances, ensuring **fair and equitable terms with prior informed consent** (p. 101.) Before proceeding with joint activities, check with your institution's TTO to make sure that all the requirements have been met. (p. 173)
- ✓ **Scientists and anyone else accessing biodiversity must ask, and answer, the following questions prior to initiating collecting activities**: Under which conditions may I *enter* another sovereign state's territory in my scientific capacity? Under which conditions may I *collect* biological material and related information? Under which conditions may I *carry out or export* biological material and related information from that sovereign state's territory? Under which conditions may I make *further use* of collected biological material and related information? (p. 173)

TECHNOLOGY TRANSFER OFFICERS

KEY IMPLICATIONS AND BEST PRACTICES

FOR TECHNOLOGY TRANSFER OFFICERS

The page numbers refer to the page in this Executive Guide *where the specific concept is discussed in detail.*

The Innovation Landscape and Intellectual Property

- ✔ **The emerging global systems of innovation in health and agriculture open up new prospects for innovation everywhere.** This has profound implications for the management of innovation, technology transfer, market competition, and economic development in every country. (p. 25)
- ✔ **Innovation is complex and integral to all six components of innovation:** IP management, R&D in the public and private sectors, safe and effective regulatory systems, the ability to produce new products to high standards of quality, a national distribution system in both the public and private sectors, and international distribution systems and trade in technologies. **Consider this entire innovation process** when making patenting and licensing decisions. (p. 25)
- ✔ The use of IP rights is not a panacea for the management of innovation, nor is the public domain. Both public and private goods have utility and limitations. The **art of innovation management is in using both public and private goods** and to manage the interface between them. (p. 111)

The Role of the Technology Transfer Office

- ✔ The traditional mission of technology transfer offices (to bring university-generated intellectual property to benefit the public) is broadening, reaching the global community. **Technology transfer also enhances the reputation of academic institutions and facilitates their missions** of education, research, and community outreach, ensuring social impact. (p. 25, 45)
- ✔ A TTO is **responsible for creating incentives to move discoveries toward product development** by motivating public sector researchers, not by a promise of revenue streams, but by the satisfaction of seeing their work applied to serve the public good. (p. 45)
- ✔ **The primary role of a technology transfer office should not be the generation of financial returns**; they can take years to come. Be realistic when making forecasts about expected income; a positive return can take eight to ten years to achieve. (p. 45)
- ✔ **Your role in communicating** the use of IP tools and the benefits of good IP management is critical, thereby cultivates an IP management "culture" throughout the organization. Such communication should be directed to senior management, your institution's board, and to scientists. (p. 57, 153)

IP Policy and IP Strategy

- ✔ An **IP policy** should address, at a minimum, ownership of intellectual property, conflicts of interest and conflicts of commitment, the handling of confidential information, the principles of IP licensing approaches, the sharing of income derived from intellectual property, and any rights the institution will retain (such as for research and for humanitarian uses). (p. 65)
- ✔ An institutional **IP strategy** addresses how IP management will be used to achieve global access/humanitarian benefits of the inventions and products developed at an institution. It should include how the institution deals with **incoming** third-party intellectual property, how it deals with internally generated intellectual property, and how it will **out-license** its intellectual property to third parties. (p. 65)

Agreements and their Uses

- ✔ A public sector institution can use a variety of agreements to both manage and protect intellectual property, regardless of whether that intellectual property is owned by the public sector institution or by licensing partners in the private sector. The key issue is to allow for

TECHNOLOGY TRANSFER OFFICERS

maximum flexibility whereby institutions can set, or negotiate, the terms that best fit the mission and goals of the institution and the purpose of the partnership. (p. 85)

- ✔ **A template agreement** should be used only as a starting point for discussions. (p. 85)
- ✔ Contracts should be tailored to fit local customs and business practices. Be **sensitive to cultural and linguistic differences between parties** to a contract. (p. 85)
- ✔ Your office ought to be the **official repository** of all agreements dealing with incoming and outgoing biological materials. (p. 85)
- ✔ Legal jargon in agreements should be avoided. Instead, **use short, clear sentences that are free of vague adjectives and are written in the active voice.** (p. 85)
- ✔ Confidentiality agreements rely on a culture of trust, not a culture of secrecy. Make sure that **confidentiality agreements contain the necessary exceptions** appropriate for the mandate of your institution. (p. 85)
- ✔ When negotiating **collaborative research agreements**, involve scientists. Their input will be critical at various stages of the process. (p. 85)

Some IP Management Nuts and Bolts

- ✔ Conduct occasionally **comprehensive IP audits** to determine where your IP assets are, when IP protection is needed, whether there are potential IP liability issues, whether there are licensing needs and/or opportunities, and whether there are inventions to be harvested. (p. 157) IP audits can be useful mechanisms that form the basis for an internal review and revision of an institution's IP strategy and IP policy. (p. 65)
- ✔ Technology transfer invariably brings **conflicts of interest**. The challenge is to manage them in a transparent and consistent manner. Most problems arise when **potential** conflicts are not disclosed. (p. 65)
- ✔ All employees (and visitors in some cases) should be required to sign an **invention assignment agreement** on their date of arrival. (p. 65)
- ✔ Any TTO will have a wide range of legal matters to be addressed, and procedures for **working with external patent and general counsel** should be well established. (p. 73)
- ✔ **Many technology valuation approaches exist. None is perfect.** Considering that each deal is highly context specific, each technology transfer office should be able to select the best approach and adapt it to the specific circumstances. (p. 101)
- ✔ When devising a **patenting strategy**, you will need to make three decisions: First, should you seek patent protection? Second, what is the best patent-marketing approach? Third, what license fees and/or royalties ought to be levied? (p. 101)

Modes of IP Protection

- ✔ **Trademarks** are a critical, and often overlooked, option for IP protection. They can be used as stand-alone IP protection, or they can be integrated into an overall strategy for integrated IP protection. (p. 57)
- ✔ Because public-domain technologies play an important role in publicly funded research, **defensive publishing** can increase accessibility of technologies in the public domain. Scientists need your help to ensure that such disclosure truly places the invention into the public domain. (p. 111)
- ✔ There are **advantages in filing provisional patent applications** (such as controlling costs and providing additional time for weighing options as to whether it is worthwhile to pursue a full patent application) but beware of the downsides. (p. 111)
- ✔ For any invention, evaluate whether **foreign patent rights** are truly required. Keep in mind possible applications in developing countries; a patent may be critical to ensure access. This will require a combination of business, marketing, and legal analyses. (p. 111)

Licensing Inventions and Technology

- ✔ **Both nonexclusive and exclusive licenses can be applicable to meeting socio-economic goals.** Within exclusive licensing, there are many options, such as exclusivity limited to a certain field of use, or geography, or for limited periods of time. (p. 35)

TECHNOLOGY TRANSFER OFFICERS

- ✔ **Reserving rights for humanitarian use** may require additional work and will likely not generate licensing revenue; conversely, such provisions, if used in a strategic way, are unlikely to lead to loss of revenues. (p. 35)
- ✔ Though potentially useful, IP managers should **be cautious of imitating open licensing procedures** in the field of biotechnology. It is still unclear to what extent the software models of open source can be adapted to other technological fields. (p. 35)
- ✔ Any organization engaged in high-volume licensing will find it useful to develop its own **internal template agreements** that are then modified and adapted to suit each special circumstance. **Checklists** for different types of recurring licensing negotiations should be reviewed prior to and during negotiations. (p. 121) For the licensing of plant varieties, certain software may be useful (p. 123)
- ✔ **Field-of-use licensing** should be adopted as the preferred method of licensing whenever possible. It allows you to gain greater control while maximizing the use and value of your licensed technology. (p. 121)
- ✔ In a license agreement, the **rights (or prohibitions) to sublicense and assign a license** ought to be explicitly articulated. (p. 121)
- ✔ Licensee agreements are contracts. Hence, a practical understanding of contract law will be fundamental to negotiating and drafting good license agreements. TTOs can ask counsel to **ensure that agreements are compliant with national law**. (p. 121)

Milestones

- ✔ **Creative licensing strategies** will help your institution gain the greatest benefits from the research it conducts. Such strategies include, at a minimum, the balancing of exclusive and nonexclusive rights, defining field of use, setting **appropriate milestones**, requiring the delivery of products to developing country markets, and exercising control over pricing. (p. 25)
- ✔ The **public sector must specify in writing exactly what it wants to accomplish with a commercial partner**, detailing when and how this will be achieved by articulating milestone obligations. (p. 35)
- ✔ Avoid "best effort" clauses in agreements. Instead, draft **comprehensive contracts with articulated milestones**. This up-front investment will pay off later if a problem arises. (p. 35)
- ✔ Developing meaningful **milestones that provide the appropriate balance of incentives, rewards, and penalties** requires detailed preparations, a sound understanding of the processes related to developing and marketing the product, realistic forecasting of product potential, and a mission-driven mindset. (p. 35)

"Moving" Technologies to the Market

- ✔ One of your responsibilities will be to bring together individuals with different backgrounds and experiences before negotiating agreements. Ideally, a team should include **business strategy, marketing, legal, scientific, regulatory, production, and finance expertise**. (p. 133)
- ✔ Marketing inventions should not simply be a push of technologies; rather, it should be an approach that allows the needs of **buyers to *pull* inventions**. (p. 133)

IP Management and Entrepreneurship

- ✔ One of the most important factors for a successful TTO is **the institution's entrepreneurial culture**. This is strongly influenced by the attitude and degree of support from senior management. (p. 73)
- ✔ **Spinouts carry a number of risks**, but with certain factors in place they can represent the best opportunity for developing early-stage technology. (p. 143)
- ✔ **Potential investors** in a spinout will ask two major IP questions. Could previously existing intellectual property block the technology? Could your intellectual property dominate the market and prevent entry by others? (p. 143)

- ✔ When licensing to or creating new ventures, several key attributes are essential for **attracting venture capital investment**: a strong management team, a viable technology, a strong IP position, a large potential market, and location in an environment favorable for entrepreneurship. (p. 143)
- ✔ New ventures in developing countries have much to gain by **attracting and building on international investor networks**. They have the potential to open new markets and bring in new alliances. (p. 143)
- ✔ It is often appropriate to strike a balance between reliance on **licensing-out to existing companies** and investing time and resources in **creating new companies**. (p. 143)

Risk Management and Freedom to Operate (FTO)

- ✔ The role of the technology transfer officer, and that of attorneys who may produce legal FTO opinions, is generally **to advise senior management on risks**. It is a manager's purview, based on your input, to decide how to deal with the risks identified in your FTO analysis. (p. 153)
- ✔ A freedom to operate (FTO) analysis is an interdisciplinary endeavor best executed through FTO teams. These teams, made up of legal, business, and scientific professionals, are in themselves useful for **strengthening intra-institutional dialogue** and communications. (p. 153)
- ✔ For an academic or public institution, legal **FTO opinions are unlikely to be needed** for the majority of technology transfer functions. They would be relevant only if the institution is engaged in downstream product development and commercialization. (p. 153)

Monitoring and Protecting Intellectual Property

- ✔ Potential **patent infringements** should be monitored continuously through sound surveillance protocols, and action taken to remedy infringement is an essential part of IP asset management. The lack of patent enforcement can lead to a loss of patent rights. (p. 163)
- ✔ Early communication with potential infringers and **good license and licensee diligence**, are the foundations for policing and maintaining intellectual property, irrespective of whether the intellectual property is owned by a public or a private entity. (p. 163)
- ✔ Essential to contract management is a **well-organized electronic filing system**. A TTO should establish such a system as early as possible and before the number of agreements and licenses becomes large. An agreement management system (donated by the Whitehead Institute for Biomedical Research) is available on the *Handbook's* Web site for download. (p. 163)
- ✔ Most **IP disputes should not end up in litigation**, as there are many options and strategies for resolving disputes. Good contracts and good licensing practices anticipate that disputes arise with partnerships and licenses. (p. 163)
- ✔ **Mediation and arbitration** can be effective dispute-settlement procedures, provided they have been agreed upon and established in contract clauses at a time when a license or partnership is being negotiated—and before any problems arise. (p. 163)
- ✔ A technology transfer office must have systematic procedures to **administer, monitor, and enforce** its technology licenses. This includes compliance with royalty payments and reporting obligations in a non-confrontational manner. (p. 163)

IP Training and Capacity Building

- ✔ When **scientists learn the basics of IP management**, communications with the technology transfer office will improve. Public sector institutions should offer training to every scientist, student researcher, and technicianwhen they join an institution. (p. 93)
- ✔ Part of the aim of IP management training is **team building** that encourages communication between your office and the scientists in your institution. It is part of creating a culture of IP awareness and particularly useful to encourage invention disclosures. (p. 93)

- ✔ It is good practice to **include senior management** as participants in the training sessions. This is especially useful when the training program includes case studies. (p. 93)

Intellectual Property, Bioprospecting, and Traditional Knowledge

- ✔ The technology transfer office should work with senior management to establish **policies and systems for accessing indigenous or traditional knowledge** (TK), bioprospecting activities, and benefit sharing in an equitable manner. (p. 173)

Executive Guide to the *Handbook of Best Practices*

Innovation and IP Management: A Contextual Overview

The pressures on innovative organizations and countries are many and varied—economic, technological, organizational, environmental, political, and societal. Leaders of public sector institutions and private companies have limited control in most of these areas. But a leader of any entity, public or private, can exercise a high degree of control over the entity's own intellectual property (IP) through sound IP management and critical tools to accelerate innovation. Section 1 of the *Handbook*[1] presents an overview of innovation and IP management, which are the focal points of the entire *Handbook*, and logically, the *Executive Guide*.

IP rights are a critical tool for fostering innovation. Managed judiciously, they balance private rights and public necessity in a manner that, overall, encourages innovation. The first chapter of the *Handbook*, by Mahoney and Krattiger,[2] directly addresses the paradox that underlies the *Handbook*: the **pursuit of the public interest through private rights.** Focusing on the life sciences, the chapter asks how IP rights can best be managed to promote public welfare. It finds an answer, not in the system of IP rights, but in the judicious and skillful *use* of proprietary science. In short, the authors argue that creative management of IP rights, especially by public sector institutions, is essential for achieving public benefits.

Understanding how intellectual property fits into the much broader context of innovation and product development is important for any public sector entity, whether in a developing or developed country, for addressing neglected diseases, for alleviating poverty, for development in agriculture, and for eliminating chronic malnutrition. Mahoney and Krattiger discuss how, within the rapidly evolving global IP landscape, public sector institutions can better mobilize resources in order to accelerate products through the innovation pipeline through best practices in IP management. These **best practices include creative licensing practices that ensure global access and affordability, improved institutional IP management capabilities, the formulation of comprehensive national IP policies, and the strengthening of IP court systems and patent offices.**

Recent national and international changes in IP treaties, legislation, and frameworks are having profound effects on innovation systems and on how public and private research and development institutions implement their missions and how health and agricultural innovations reach the poor. Seen within this broader context, **intellectual property is one of six interrelated components of innovation management**[3] that focuses on developing a variety of issues:

- R&D capability by the public and private sectors
- safe and effective regulatory system that covers drugs, vaccines, and agricultural products
- manufacturing capability for health products and for the inputs into and outputs of agricultural production

---

Krattiger A, RT Mahoney, L Nelsen, JA Thomson, AB Bennett, K Satyanarayana, GD Graff, C Fernandez and SP Kowalski. 2007. 1: Innovation and IP Management: A Contextual Overview. In *Executive Guide to Intellectual Property Management in Health and Agricultural Innovation: A Handbook of Best Practices* (Krattiger A, RT Mahoney, L Nelsen et al.). MIHR (Oxford, UK), PIPRA (Davis, USA), Oswaldo Cruz Foundation (Fiocruz, Rio de Janeiro, Brazil), and *bio*Developments-International Institute (Ithaca, USA). Available online at www.ipHandbook.org.

- expansion of national health or agricultural delivery systems, including an attractive, private sector domestic market for health or agricultural products and services
- international trade systems for health products (including global procurement funds) and agricultural inputs and outputs; and an IP system (legal framework, judiciary to enforce it)
- institutional management capabilities

The authors raise important questions and make significant proposals, framing IP management within a global context. This global approach is inexorably spreading and expanding at an accelerating pace. Because of the increasing interaction between developed and developing economies and the increased number and complexity of relationships between the public and private sectors, understanding the best ways to forge and maximize partnerships has a high priority. Such **partnerships will be the engines that drive global innovation.**

The chapter further reviews recent dramatic developments in the institutional aspects of intellectual property, as well as global policy shifts and international case studies. In the field of health, changes have been particularly pronounced with the founding of a novel form of institution for innovation: product-development partnerships (PDPs). Mahoney and Krattiger make the case for a fundamental shift in the way IP management in health and agricultural innovation is viewed and conducted: **the public sector can employ new ways to achieve its goals within the evolving IP framework.** In response to rapid global evolution, nongovernmental organizations (NGOs) and PDPs will have important roles to play in the global IP environment, particularly for developing countries.

It is well established that intellectual property advances product development because **intellectual property provides incentives for R&D, commercialization, and product distribution.** Investors in biotechnological R&D naturally want to protect their investments, and must, therefore, secure IP rights for their inventions. Before the creation of IP rights protection, the private sector had little incentive to develop safe and efficacious pharmaceuticals. Historically, the public sector had neither the funds nor the capability to develop products. Yet, the world is changing. As the public sector devotes more of its efforts to humanitarian missions, and engages in more development partnerships (such as PDPs) in the fields of health and agriculture, it will also have to consider the critical role of intellectual property in a broader innovation context.

The permanence of intellectual property is evident. **If the public sector does not effectively utilize the IP system, it will neither be serving its own interests nor the interests of those it has promised to serve.** Without effective IP management skills, the public sector risks squandering the rights, powers, and opportunities that the IP system provides. Intellectual property is a tool, and, like all tools, its impact depends on how it is used, who uses it, and for what purpose. IP strategies can serve to either restrict or expand access to innovations; it's all a matter of *capacity, management,* and *context.* These three aspects are addressed in the creation of a best practices document.

For IP management to efficiently function within a larger framework of innovation, **best practices need to be documented.** That is what the *Handbook* and this *Executive Guide* seek to do: to provide a teaching and capacity building resource for IP management, with a focus on health and agricultural biotechnologies. When it comes to increasing developing countries' access to fundamental innovations in health and agriculture, success requires knowledge, capacity and active engagement. This is what best practices in IP management strive to bring about and what the *Handbook* promotes.

To illustrate how best practices may be used, Mahoney[4] reviews the **impact of the Agreement on Trade-Related Aspects of Intellectual Property Rights (TRIPS) on the development of the pharmaceutical industry** in Korea. Korea was able to implement a wide range of initiatives—including an upgrade of its IP regulatory systems—that benefited its pharmaceutical industries. It is likely that Korea's success was largely achieved because its IP systems aligned with those of developed countries. The empirical data gleaned from the experience of Korea is promising, suggesting that TRIPS compliance will improve biotechnological capabilities in developing countries.

Building innovation capabilities in developing countries often relies less on the creation of IP systems and more on the creation of markets and support for R&D. Broadly speaking, the evolution of IP rights in developing countries involves three basic stages (before reaching stage 4):

- **Stage 1**: In the early stage of development, little intellectual property is generated domestically and few foreign companies are interested in introducing their technologies to these countries due to inadequate IP rights protection.
- **Stage 2**: At a later stage of development, the country's innovative capabilities improve, but due to the same inadequate IP rights protection, there is limited foreign investment in technology.
- **Stage 3**: Eventually, when domestic companies are able to generate their own intellectual property, they demand more-effective IP protection. With more-effective protection, foreign investment in technology increases along with the presence of foreign technologies.

Within the pharmaceutical industry (and many parallels exist with the agri-biotechnology industry), companies in developing countries tend to move through four stages described in Table 1. This shows that technological and IP capacities tend to develop in tandem. However, such a measured-pace "natural coevolution" is no longer likely with TRIPS requirements in effect. These state that signatory countries' pharmaceutical industries *must* rapidly progress from either Stage 1 or Stage 2 on to Stage 3. It remains to be seen whether this mandatory accelerated development will hinder or help developing-country pharmaceutical industries in the longer term and to what extent TRIPS will facilitate the transfer of technology to and within developing countries.

However, elsewhere, especially in the United States, Europe, Japan, and parts of Australasia, **technology transfer from universities works and works well.** Examples of this are when we get into a car and buckle up, when we sweeten our coffee with saccharin, when we search the Internet using Google™, or when we take advantage of the innumerable medical and agricultural advances of the last quarter-century. In essence, we are reaping the benefits of technology transfer. Universities do not only educate the next generation and create new knowledge, but also create knowledge that enhances the quality of life, increases economic productivity, and even saves lives.

Fraser[6] points out that university–industry collaborations and licensing have soared, for example, in the United States, ever since the Bayh-Dole Act of 1980, which forces the moving of inventions from laboratories to store shelves more quickly. The law allowed U.S. universities and public research institutions to patent inventions that were based on federally funded research, then to license those inventions to the private sector.

Some people continue to question the fairness of the global IP system, but others are using new opportunities created by this system to improve lives in the developing world. Technology transfer is thus changing rapidly. Traditionally, the mission of technology transfer offices has been to make university-generated innovation available to the public as rapidly as possible. However, **technology transfer offices now have a broader purpose: to enhance the reputation of academic institutions and to help them to achieve their missions of education and outreach by assisting in forming relationships with the private sector.** Technology transfer has the potential to benefit the entire world. As technology transfer develops, it will undoubtedly evolve again, in response to new conditions.

University technology transfer professionals are already becoming increasingly aware of their obligation to ensure that the underserved communities of the world have access to medicines and agricultural biotechnology that have been developed from basic research conducted in their universities. But certain conditions need to be remediated, for example: many university administrators, technology transfer officers, and businesspeople are unaware of the need for new health technologies in developing countries; few

Table 1: The Four Stages in the Development of Biotechnology

	Development of Manufacturing	Development of Distribution Systems		Development of R&D Capability		IP Systems	Drug and Vaccine Regulation
		National	International	Private Sector	Public Sector		
Stage 1. Establishing the Foundation	Importation of finished goods or assembly of parts into finished products	Small domestic market	Very little, except as toll manufacturer	Very little	Very little	Initial development allowing patents for local inventors; no interest from foreign inventors	Very limited
Stage 2. Capacity Building	Production on license or by copy	Growing local market of increasing interest to foreign companies; import substitution	Growing companies learning how to establish export markets	R&D to understand technology either to produce on license or to copy	Development of university and independent research centers; capacity building	Interest growing among foreign inventors; local inventors starting to file more patents	Limited services but without enforcement capabilities
Stage 3. Maturation IDCs	Manufacture of domestically developed, high technology products	Rapidly growing domestic market of interest to foreign companies	Increasing exports that account for a growing share of GNP	Small-scale, advanced R&D effort capable of creating new products for domestic and export market	Vast acceleration of funding for R&D; development of major research centers; linking with private sector	Advanced IP system but with certain limitations such as lack of enforcement	Advanced capabilities but not at highest level because of lack of enforcement capabilities
Stage 4. The Most-Developed Countries, with a Drug or Vaccine Industry	Highest capabilities to produce high technology drugs and vaccines	Highly profitable market in both the public and private sectors, generating profits to support, in part, advanced research	Global companies	Generous support for health research from basic to applied; large research investment by private companies, including large pharmaceutical manufacturers and biotechnology companies		Sophisticated system of IP management operating according to the requirements of the TRIPS Agreement	Sophisticated agency overseeing regulatory approvals of drugs and vaccines; government oversees clinical trials and production facilities and enforces regulations

Source: Mahoney[5]

people know how to incorporate patenting and licensing practices into global access strategies; and best practices for global access strategies have not yet been defined.

Nelsen and Krattiger[7] describe **several possible strategies for ensuring both that everyone has access to technologies and that for-profit companies have incentives to develop those technologies.** The Association of University Technology Managers (AUTM) has formed Technology Managers for Global Health (TMGH), the purpose of which is to draw attention to global health issues and compile and promote a collection of best practices, policies, and licensing terms. The National Institutes of Health (NIH) issued guidelines on the patenting and licensing of research tools as a way to increase global access to health innovations. And this *Executive Guide* serves similar and perhaps broader purpose.

Careful **patent-filing strategies** can help ensure that developing countries have access to the technologies they need. One example of this is when using a *prohibition of filing strategy*, one does not file for patent protection in developing countries if there is a very large market for the product in developed countries. In fact, it may be a good idea to file a patent in developing countries if those countries create a substantial demand for the drug or vaccine in question. These patents can provide incentives for private sector development, and also provide powerful tools for the consolidation of resources via aggregation of developing-world markets (in addition to enabling or at least strengthening technology transfer through the licensing of know-how associated with patents).

Questions to consider include: whether it is better to prohibit or to require filing patents in developing countries, whether patents encourage private sector investment by aggregating the developing world market, what kinds of licenses should be granted, and what requirements for development milestones, product delivery in developing countries, pricing, or sublicensing options work best for a particular situation.

There are a number of **licensing strategies:**

- Licenses can be exclusive, nonexclusive, or a combination thereof.
- Licenses can specify milestones, such a requirement that the licensee has to contribute a minimum toward the development of a product earmarked for developing countries.
- Establish pricing controls in developing countries.
- Insist upon sublicensing, which ensures that the licensee finds partners who can move the product to developing countries.

However, as the chapter points out, there are no clear answers as to how best to increase global access to necessary technologies. Each of the above strategies has been tried, but they are all relatively new, besides which each situation will require a tailored solution. There is no one-size-fits-all approach nor are there boilerplate strategies that can be applied in a suite of different contexts. For this reason, everyone has to build upon his or her own experiences and find creative solutions. Universities can take the lead. Indeed, their public sector missions compel them to do so. When universities implement consistent and effective licensing strategies, they not only stimulate investment in R&D but also ensure that the products of that research are affordable and widely available in developing countries. **Where there is a will, there is a way.**

Some individuals and organizations have denounced, on ethical grounds, any patents for biotechnological applications, genes, or living organisms, especially patents on pharmaceuticals. However, there is neither a single articulation of such concerns nor is there a branch of any government that can address the concerns. For example, patent offices are ill-equipped to address ethical questions. In addition, any blanket prohibition on patenting genes or other biological materials would *generally* be inconsistent with TRIPS, which requires countries to allow IP protection for *most* biotechnology products. Still, under TRIPS, there are exceptions to this general prohibition. TRIPS also contains a provision that certain unethical inventions or innovations may be denied patents; a similar provision is found in the European Union's *ordre public* clause. U.S. patent law, however, does not have this kind of morality provision.

But are some or all patents for genes and cells unethical per se? How should tissue samples be collected? And how exactly are patents used to restrict access to medical and ag-biotech inventions? These are **some of the fundamental ethical questions** addressed by Marchant.[8]

With respect to obtaining biological samples, the ethical norm is that people who donate tissue for research purposes relinquish property rights to the donated cells, genes, and other biological material. Problems can arise, however, when human, animal, and plant materials and specimens are collected in developing countries and subsequently used to create biotechnological inventions that are patented in developed countries; some people have questioned the ethics of what they call biopiracy. Marchant explains that ethically questionable situations can usually be avoided if the following principles are followed:

- **Prior informed consent** should be obtained from the relevant entities before taking any samples.
- An organization agrees to share with a developing country any economic benefits that result from patented inventions based on biological materials collected in that country, called **benefit sharing.**

The above methods help to minimize, but do not eliminate, the ethical quandaries related to the collection of biological materials. Many questions remain: Who is authorized to give prior consent? Should more than one authority give consent (for example, both tribal and government officials)? How much must be disclosed about the proposed research to ensure that the authorities are adequately informed?

Benefit sharing can also be difficult to implement. First, many scientific researchers do not have funding to properly compensate indigenous peoples for their assistance. Second, who decides how benefits will be allocated? Third, if benefits are offered to people who assist researchers, some individuals may "assist" researchers purely for the money they will receive, not unlike some blood donors.

There are few laws that address the ethics of patenting. In the absence of any clear consensus, **ethical decisions concerning biotechnological patents will need to be made on a case-by-case basis.** The ethics of patenting is an evolving field that currently is more gray than black-and-white.

This is just one reason why it is hoped that this *Executive Guide* and *Handbook* will encourage all parties to take greater advantage of the unprecedented opportunity to benefit from the strategic management of intellectual property aimed at promoting the public welfare—especially those people who have, until now, been unable to benefit from today's technology—and that this will contribute to building a healthier and more equitable world. ■

All chapters refer to: *Intellectual Property Management in Health and Agricultural Innovation: A Handbook of Best Practices*. 2007. A Krattiger, RT Mahoney, L Nelsen, JA Thomson, AB Bennett, K Satyanarayana, GD Graff, C Fernandez and SP Kowalski (eds.). MIHR: Oxford, U.K., and PIPRA: Davis, U.S.A. Available online at www.ipHandbook.org. The online version contains for each chapter a detailed Editor's Summary, Implications and Best Practices.

1 Krattiger A, RT Mahoney, L Nelsen, JA Thomson, AB Bennett, K Satyanarayana, GD Graff, C Fernandez, and SP Kowalski. (eds.). 2007. *Intellectual Property Management in Health and Agricultural Innovation: A Handbook of Best Practices*. MIHR: Oxford, U.K., and PIPRA: Davis, U.S.A. Available online at www.ipHandbook.org.

2 Chapter 1.1 by RT Mahoney and A Krattiger titled The Role of IP Management in Health and Agricultural Innovation, p. 3.

3 Chapter 1.2 by RT Mahoney titled Building Product Innovation Capability in Health, p. 13.

4 Ibid.

5 Ibid.

6 Chapter 1.3 by J Fraser titled IP Management and Deal Making for Global Health Outcomes: The New "Return on Imagination" (ROI), p. 19.

7 Chapter 1.4 by L Nelsen and A Krattiger titled Ensuring Developing Country Access to New Inventions: The Role of Patents and the Power of Public Sector Research Institutions, p. 23.

8 Chapter 1.5 by GE Marchant titled Genomics, Ethics, and Intellectual Property, p. 29.

FOR GOVERNMENT POLICYMAKERS

- **Intellectual property (IP) is a tool to foster innovation.** Whether viewed as a legal concept, a social construct, a business asset, or an instrument to achieve humanitarian objectives, the value of intellectual property cannot be disputed.

- **IP rights are a compromise and an imperfect solution**, representing the search for balance between making all knowledge available within the *public domain* and granting *ownership* of valuable discoveries to the inventors. Reaching an appropriate balance requires continuous, sound IP management.

- The use of the existing IP system, especially coupled with sound patenting and licensing strategies, resolves the **apparent paradox: the pursuit of the public interest through private rights**.

- **The emerging global systems of innovation in health and agriculture open up new prospects for innovation everywhere.** This notion, that the public interest can be served through private rights, has profound implications for the management of innovation, technology transfer, market competition, and economic development in every country, regardless of its economic status.

- **Innovation is a complex process.** It is stimulated by coordinated and structured policies and programs. The IP management system is an important factor, but it is only one of six factors that determine a country's or institution's ability to innovate.

- **Intellectual property is integral to all six components of innovation** that are, in addition to IP management: R&D in the public and private sectors; safe and effective regulatory systems; the ability to produce new products to high standards of quality; a national distribution system in both the public and private sectors; and international distribution systems and trade in technologies.

- Policies to promote the creation and management of intellectual property by public sector institutions should give first priority to **advancing the missions of those institutions**.

- There are few laws that address the ethics of patenting. In the absence of a clear consensus, **ethical decisions concerning biotechnology patents will need to be made on a case-by-case basis**.

- **Protection and licensing go hand in hand.** Public research institutions have much to gain if they are permitted to protect their inventions. A system that allows technologies to be patented and that encourages institutions to license them will both help countries to reach their economic goals and better serve the poor.

- Policymakers should encourage and fund **national technology transfer managers' associations** to the extent that doing so is feasible. Such associations are working to determine best practices in technology transfer and licensing.

Given that IP management is heavily context specific, these Key Implications and Best Practices are intended as starting points to be adapted to specific needs and circumstances.

FOR SENIOR MANAGEMENT

(UNIVERSITY PRESIDENT, R&D MANAGER, ETC.)

- ✔ **Intellectual property is a tool to foster innovation.** Whether viewed as a legal concept, a social construct, a business asset, or an instrument to achieve humanitarian objectives, the value of intellectual property cannot be disputed.

- ✔ **IP rights are a compromise and an imperfect solution,** representing the search for balance between making all knowledge available within the *public domain* and granting *ownership* of valuable discoveries to the inventors. Reaching an appropriate balance requires continuous, sound IP management.

- ✔ The use of the existing IP system, coupled with sound patenting/licensing strategies, resolves the **apparent paradox: the pursuit of the public interest through private rights**.

- ✔ **The emerging global systems of innovation in health and agriculture open up new prospects for innovation everywhere.** This notion, that public interest can be served through private rights, has profound implications for the management of innovation, technology transfer, market competition, and economic development in every country, regardless of its economic status.

- ✔ **Innovation is a complex process**. It is stimulated by coordinated and structured policies and programs. The IP management system is an important factor, but it is only one of six factors that determine a country or institution's ability to innovate.

- ✔ Public sector institutions that **optimize their IP management capacity** and those that have capacity in any of the additional areas, such as regulatory systems, will be better equipped to actively participate in innovation.

- ✔ Often the most innovative organizations are those with the most **dynamic networks**, and those that reach out to other entities and potential partners.

- ✔ The case studies in the insert of the *Executive Guide* demonstrate how **public sector technology transfer can make a difference** in the developing world and elsewhere.

- ✔ Technology transfer officers should have ample **opportunities for professional development and networking**. Technology transfer is a field in which much information is shared informally.

- ✔ **Technology transfer and licensing are heavily context-specific.** A one-size-fits-all patenting and licensing policy and strategy is rarely effective for an institution.

- ✔ Public sector institutions ought to have **ethical guidelines** for IP management that are consistent with national laws and an institution's mission.

Given that IP management is heavily context specific, these Key Implications and Best Practices are intended as starting points to be adapted to specific needs and circumstances.

FOR SCIENTISTS

- **Intellectual property is a tool to foster innovation.** Whether viewed as a legal concept, a social construct, a business asset, or an instrument to achieve humanitarian objectives, the value of intellectual property cannot be disputed.

- **IP rights are a compromise and an imperfect solution,** representing the search for balance between making all knowledge available within the *public domain* and granting *ownership* of valuable discoveries to the inventors. Reaching an appropriate balance requires continuous, sound IP management.

- The use of the existing IP system, coupled with sound patenting and licensing strategies, resolves the **apparent paradox: the pursuit of the public interest through private rights**.

- **The emerging global systems of innovation in health and agriculture open up new prospects for innovation everywhere**. This notion, that public interest can be served through private rights, has profound implications for the management of innovation, technology transfer, market competition, and economic development in every country, regardless of its economic status.

- **Innovation is a complex process**. It is stimulated by coordinated and structured policies and programs. The IP management system is an important factor, but it is only one of six factors that determine a country's or institution's ability to innovate.

- Your work is part of a larger **innovation process** that spans R&D across the public and private sectors, using regulatory systems, enabling the ability to produce new products to high standards of quality, allowing for the national distribution of new products through the public and private sectors, accessing foreign technologies, and managing intellectual property in a way that fosters partnerships.

- **Research is the very foundation of innovation.** Research leads to discovery; discovery fosters invention; inventions nourish innovation.

- **Your sustained interest in your invention is important** if it is to reach the marketplace, especially if it is to benefit those who most need it.

- It will be wise to consider the **ethical implications of your research**.

- You should **always obtain prior informed consent** when you access other people's materials or samples irrespective where they originate.

Given that IP management is heavily context specific, these Key Implications and Best Practices are intended as starting points to be adapted to specific needs and circumstances.

KEY IMPLICATIONS AND BEST PRACTICES: SECTION 1

FOR TECHNOLOGY TRANSFER OFFICERS

- ✔ **Intellectual property is a tool to foster innovation.** Whether viewed as a legal concept, a social construct, a business asset, or an instrument to achieve humanitarian objectives, the value of intellectual property cannot be disputed.

- ✔ **IP rights are a compromise and an imperfect solution,** representing the search for balance between making all knowledge available within the *public domain* and granting *ownership* of valuable discoveries to the inventors. Reaching an appropriate balance requires continuous, sound IP management.

- ✔ **The emerging global systems of innovation in health and agriculture open up new prospects for innovation everywhere.** This notion has profound implications for the management of innovation, technology transfer, market competition, and economic development in every country, regardless of its economic status.

- ✔ **Innovation is a complex process** and coordinated and structured policies and programs stimulate it. The IP management system is an important factor, but it is only one of six factors that determine a country's or institution's ability to innovate.

- ✔ **Intellectual property is integral to all six components of innovation** which are, in addition to IP management: R&D in the public and private sectors; safe and effective regulatory systems; the ability to produce new products to high standards of quality; a national distribution system in both the public and private sectors; and international distribution systems and trade in technologies.

- ✔ An IP manager should **consider the entire innovation process** when making patenting and licensing decisions.

- ✔ The traditional mission of technology transfer offices (to bring university-generated intellectual property to the public as rapidly as possible) is broadening. **Technology transfer enhances the reputation of academic institutions and helps them achieve their missions**, both at home and abroad.

- ✔ IP managers should join **professional national and international licensing and technology transfer societies** whenever possible.

- ✔ **Creative licensing strategies** will help your institution gain the greatest benefits from the research it conducts. Such strategies include, at a minimum, the balancing of exclusive and nonexclusive rights, defining field of use, setting appropriate milestones, requiring the delivery of products to developing country markets, and exercising control over pricing.

- ✔ In **benefit sharing**, an organization agrees to share with a developing country any economic benefits that result from patented inventions based on biological materials collected in that country. Make sure the individuals in your organization who collect biological resources are aware of this and obtain **prior informed consent**.

Given that IP management is heavily context specific, these Key Implications and Best Practices are intended as starting points to be adapted to specific needs and circumstances.

Specific Strategies and Mechanisms for Facilitating Access to Innovation

Innovation is a wonderful thing. Innovation occurs when any new knowledge is introduced into and utilized in an economic and/or social setting. People and institutions (agents) orchestrate this process. Simple economic theory states that such agents act in a rational way, responding to price signals as a way of maximizing investments. But modern innovation research indicates that this is not so, particularly when noncommercial or humanitarian goals are being pursued. Agents, both public (governments, universities, extension services) and private (small, medium, and large companies, as well as farmers, individual consumers, and communities) are essentially strategists who respond to other agents' behaviors. For example, a governmental "behavior" in this context would be changing a range of policies, such as research, regulatory, trade, and IP (intellectual property) policies. Each of these agents directly or indirectly engages in the production, processing, marketing, or distribution of products and services. Simultaneously, agents engage in the processes of knowledge creation or dissemination and the application of knowledge through both market and nonmarket relationships.

Using this definition of innovation, public research institutions and universities may be invention creators. They are not necessarily innovators per se but are important actors in the innovation process. Their roles are strengthened if they are well connected. That is, they are stronger if they function in partnerships that extend beyond their primary missions and include others who can turn inventions and knowledge into products and services that become economically successful or that have major social and humanitarian impacts. This impact can be measured through three key conditions that jointly determine whether an innovation is adopted. These are:

- availability
- affordability
- acceptability[1]

In addressing the needs of developing countries, achieving these three conditions concurrently can conveniently constitute **global access**. But translating the three into an effective innovation management plan or operational strategy is more challenging. To manage these goals, a strategy that consists of six thrusts should be considered:[2]

1. Development of R&D capability by the public and private sectors
2. Development of a safe and effective regulatory system that covers drugs, vaccines, and agricultural inputs and outputs
3. Development of manufacturing capability for health products, seed production systems, and value-added processing
4. Development of an IP system (legal framework, judiciary to enforce it, and institutional management capabilities)

Krattiger A, RT Mahoney, L Nelsen, JA Thomson, AB Bennett, K Satyanarayana, GD Graff, C Fernandez and SP Kowalski. 2007. 2: Specific Strategies and Mechanisms for Facilitating Access to Innovation. In *Executive Guide to Intellectual Property Management in Health and Agricultural Innovation: A Handbook of Best Practices* (Krattiger A, RT Mahoney, L Nelsen et al.). MIHR (Oxford, UK), PIPRA (Davis, USA), Oswaldo Cruz Foundation (Fiocruz, Rio de Janeiro, Brazil), and *bio*Developments-International Institute (Ithaca, USA). Available online at www.ipHandbook.org.

5. Development and expansion of national health or agricultural delivery systems, including an attractive, private-sector domestic market for health or agricultural products and services
6. Development of international trade systems for health products (including global procurement funds) and agricultural inputs and outputs

Although IP management is but one of six components, it can be viewed as the thread that runs through the innovation process or as the glue that holds partnerships together. Partnerships are based on the mutual interests of two or more parties and agreed upon in contracts. The terms and provisions of a contract can be almost limitless (provided they are both legal and agreed to by the parties) and are discussed elsewhere in great detail.[3] The present section focuses on the components that are most specific to global access. And one of the first and foremost sets of licensing terms, at least for universities and other public sector research institutions in the licensing of invention, is **the reservation of certain rights**. Indeed, Bennett[4] urges universities and public sector institutions to ensure that they preserve the right to use licensed technologies for educational, research, and humanitarian goals—including distribution rights in developing countries.

In essence, the terms of a license can subdivide the rights with respect to a technology. **Rights can be segmented** and apportioned across:

- *technological fields, markets,* or *economic contexts* in which the technology is used (such as farm size, farm income, or income derived from a particular crop, in the case of agriculture; or certain drugs to combat neglected diseases in the case of health)
- *income levels* (for example, per capita gross domestic product)
- *geographic regions* (by country, or by lowland or highland agricultural systems) or by customer (public procurement or private hospitals and pharmacies)[5]

This practice can be used by any technology owner to maximize the application of its inventions. Most importantly, subdividing and field-of-use licensing can allow both commercial and noncommercial uses of the technology to proceed in parallel and, thus, constitutes a central element in a global access strategy. Bennett provides suggestions for creating explicit reservation of rights in a commercial technology license. This will ensure that institutional objectives to support humanitarian applications of technologies that have applications to the needs of the poor are not inadvertently compromised.

The global **importance of humanitarian licensing** is also discussed in detail by Brewster, Hansen, and Chapman[6] with pragmatic answers about the why, who, and how of the process. The authors encourage IP managers, in both private and public sectors, to adopt such strategies, noting in particular that they are not incompatible with commercially driven businesses. The *why* for humanitarian-use licensing is obvious, with the vast unmet health and agricultural needs of developing countries. The *who* is increasing as more and more organizations have been using IP management practices to promote the health and food security of poor people (and the chapter offers many recent examples). The *how* is discussed through a number of case studies[7] on humanitarian licensing. The authors conclude, with others, that much can be achieved using best practices, many of which are enumerated at the end of the chapter.

Eiss, Hanna, and Mahoney[8] look at the same topic but on the basis of how various **product-development partnerships (PDPs)** seized upon public and private sector strengths and how they are **leveraging existing infrastructure and research** in developed and developing countries. Although each of the PDPs reviewed are very different, they nonetheless share some common strategies for maximizing IP management for global health. These include:

- defining a discrete territorial market
- establishing different structural incentives in public sector and private sector markets
- extending field of use to make the product applicable to diseases in developed countries
- using royalties to benefit the party that needs the most incentive

- providing access to the developed technology, should the private sector not follow through on the project

The authors conclude that whereas these issues can be complex, they should be addressed as early as possible in the formation of partnerships. But most importantly, the chapter concludes with the observation—grounded in much experience—that an approach that takes into account the six components of innovation discussed at the beginning of this chapter will have a much better chance of success than those efforts that take a piecemeal approach to product development and distribution. Such a comprehensive effort should not be considered daunting, but rather an opportunity for creativity.

It is widely acknowledged that IP rights are important drivers of innovation, and this applies equally to the private sector and the public sector. This is especially the case for product patents. However, for **research tools**[9] **in both medical and agricultural innovation**, Clift[10] demonstrates that patents related to research tools can have negative implications and, hence, should be balanced carefully with disclosure to place inventions in the public domain. Developing countries need to think about how to implement patent legislation (that is consistent with the Agreement on Trade-Related Aspects of Intellectual Property Rights, best known by its acronym TRIPS) while meeting their own objectives, particularly those related to genetic discoveries. For example, legislation should be formulated to set forth appropriate research exemptions, and patenting and licensing policies should be aimed at maximizing the availability of innovations in order to aid the development of urgently needed products.

For **product patents**, the decision to patent or not to patent is much simpler. For many products, even those aimed at developing countries' needs, patenting provides a mechanism to extend licenses, with specific provisions for licensees to meet these needs according to well-understood and well-defined market conditions. But this only works if best licensing practices are adopted (such as those defined and spelled out throughout the *Handbook*). A compelling reason for best practices in IP management is well illustrated by Stevens[11] who discusses how student activists ensured that Yale University and Bristol-Meyers Squibb quickly adopted humanitarian licensing (also called fair-access licensing).[12] The objectives for such **humanitarian licensing provisions** should be to maximize the possibility that the patented invention will be produced and to structure arrangements for tiered pricing, which will allow the poorest countries access to the drug at the lowest cost.

Best practices adopted by PDPs have also had the following important effects:

- Large companies have been motivated to contribute their drug-discovery skills and resources because they are assured that others are responsible for funding late-stage clinical development.
- Small companies have secured funding to develop technologies with dual-market uses, with the PDPs securing license rights for developing countries at zero or low royalty rates and the small company retaining rights for use in developed countries.
- Academic institutions have had a new channel to advance their neglected-disease discoveries.
- Developing country pharmaceutical companies have found their production and distribution skills in demand.

For products needed in developing countries, it could be argued that the question of patenting should not even arise. Some argue that not to patent would be the best strategy. Others believe that **open source licensing** may be an effective option for managing intellectual property. Indeed, the chapter by Hope[13] discusses the possibilities of open source licensing. Open source does not mean to place something squarely into the public domain nor does the decision to pursue open source licensing follow only from altruism: it *can* be based on commercial self-interest. Open source licensing has been most successful in the computer software field (Linux being the most prominent example), where anyone, anywhere, and for any purpose is essentially allowed to copy, modify, and distribute the company's software (either for free or for a nominal fee) and, therefore, anyone is allowed full access to the software's source code. The only condition is that everyone has to share its improvements with everyone else,

Open source is thus a form of IP licensing, and so it differs significantly from placing technology in the public domain via publication.[14] An open source strategy may indeed be a viable approach to encourage the widespread adoption and development of an innovation. Some innovations (such as software) are automatically protected by copyright, so a license clarifies the terms of its use. An open source license gives the innovator the right to set terms of use and exclude users who will not abide by those terms. But most of the incentives for open source licensing are indirect: cost savings, productivity gains, the capital provided by a good reputation, and, most importantly, an expanded user base, which correspondingly expands complementary goods and services. As the market expands, revenues from sales, one-time licenses, dual licensing, and complementary products and services may be enough to offset the opportunity cost of open source licensing. IBM has successfully used this approach. Importantly, open source licensing can place an institution in a network of innovation with enormous collaborative potential. Finally, open source licensing can encourage the development of alternatives to proprietary technologies. The greater the number of nonproprietary tools in a given tool kit, the greater the incentive for everyone in the field to invest in developing substitutes for the remaining proprietary technologies, since it will allow freedom to operate for the whole tool kit.

All of the foregoing relates to software licensing. Despite several attempts to adapt open source approaches to the biological sciences in both health and agriculture, none has been successful. Given the possible benefits associated with open source licensing, pursuing it further is warranted but will require additional work. Hope discusses the various unresolved challenges in the biotechnological field. She distinguishes between "copyleft" and "academic" forms of open source licensing. Using a proposed five-step decision-making process, Hope shows how open source technologies, for example, can be tailored to serve small agricultural and pharmaceutical markets in developing countries. And given the growing reliance on computer technologies in the life sciences—bioinformatics software programs, for example—the possibilities for open source licensing in this field have great potential.

Returning to the PDPs and their goal of moving candidate products through various stages up to clinical development and eventual distribution, many PDPs face a series of highly practical challenges. The central challenges are ensuring high-quality and low-cost production, sustained supply, affordable pricing, and effective delivery of their products. Indeed, the World Health Organization's Commission on Intellectual Property and Innovation in Health has found that many PDP research-and-development contracts defer issues related to manufacturing and distribution.[15] As a result, PDPs now increasingly face both negotiating and operational challenges regarding manufacturing sites, pricing to the public sector, market segmentation, market sizing, ensuring the lowest sustainable cost of production while guaranteeing sustainable supplies, quality control, and post-launch issues, such as pharmacovigilance and product liability.

Elaborating the requirements and **approaches for global access** early on is critical; otherwise, plans are developed incrementally or even after the product is developed. This leads to delays and creates large inefficiencies for the crucial last steps of distribution in developing countries. In addition, negotiating power may be diminished after development. In such cases, product uptake can be sluggish or stalled due to a variety of downstream considerations. A useful way to approach these situations is to **apply milestones** in the initial licensing agreements to ensure that key goals are met along the product development pathway. The kinds of milestones will vary based on product profiles and target markets, but in all cases, as Oehler[16] states, milestones require the following:

- intensive preparations
- detailed knowledge of the processes related to developing and marketing the product
- realistic forecasting of product potential

- persistence in quantitative forecasting and establishing a master plan for the entire product rollout
- a mission-driven mindset to establish optimum goals for the public sector

When public-private partnerships manage intellectual property, they are trying to **balance the private sector's commercial interests with the public sector's goal** of obtaining access to pharmaceutical products at the lowest possible cost. Although the goals and paradigms of the public and private sectors may appear too far apart, this chapter provides the tools and materials required to build the contractual architecture to span that divide. According to Oehler, a large part of the problem is simply a failure to communicate between the public and private sectors. Discussions between the public sector and industry are cross-cultural, no matter how well public sector players think they understand industry. In such a cross-cultural environment, there is nothing more dangerous and conducive to misunderstandings than to *assume the obvious*, since what is obvious for one person with a public sector background will not necessarily be the same for the other partner. Obligations and contractual performance cannot be left to vague *best efforts* and *common sense*.

The experiences of the Concept Foundation, which are the platform for the chapter by Oehler, demonstrate the successful use of milestones as a tool. The Foundation's business model initially considers downstream issues such as product delivery, and it utilises contractual milestones to achieve its principal goal of providing developing countries access to new medicines.

Milestones and open source are but one way for institutions to have easier access to inventions from third parties; there are many other mechanisms to assemble intellectual property. Krattiger and Kowalski[17] provide a brief overview of different **IP assembly options** (royalty collection agencies, information clearinghouses, technology clearinghouses, open source innovation clearinghouses, brokers and other kinds of facilitators, IP management services, IP commercialization agents, integrated commercial services, company-to-company arrangements, and other public technology transfer and financing mechanisms). The authors focus on the pros and cons of patent pools, which are receiving more and more attention as possible tools for improving technology transfer to developing countries.

There are many forms of **patent pools**: essentially they all allow for the interchange (cross-licensing) of rights to essential patents by a number of entities. They also include an agreed-upon framework for out-licensing the pooled intellectual property to third parties. A patent pool offers several benefits, a major one being that it cuts through patent thickets. But patent pools are also risky: the agreement to share technologies may run into problems based on antitrust legislation. Other considerations vis-à-vis patent pools include:

- They allow for the transfer of intellectual property. Know-how and trade secrets may also be required to use the intellectual property.
- They have generally flourished when all companies in a sector are stymied by restrictions on access to intellectual property. This makes them willing to compromise. It is unclear whether or not pharmaceutical companies feel similar inclinations. In agriculture, the interests of companies are not aligned to make patent pools feasible at the moment but this could be different for public sector organizations.
- They have been most successful in the electronics industry, since they facilitate industry-wide standards that create larger markets. Again, this may not apply to drug or agricultural biotechnology companies.
- They are typically expensive to create and maintain.

Despite these reservations, the benefits of patent pools are strong. They create an efficient "one-stop shop" for intellectual property, eliminate stacking licenses, avert litigation, decrease research and administrative costs, and can greatly improve the speed and efficiency of technological development. It is worth remembering, however, that patent pools are not the only ways to achieve these benefits. To help policymakers determine the appropriateness of patent pools for their unique situations, Krattiger and Kowalski provide a ten-step checklist for deciding whether or not to set up a patent pool and a ten-step procedure for setting one up. The authors point out legal pitfalls associated with patent pools and general suggestions are offered for identifying and avoiding them.

In sum, the reservation of rights, open source licensing, milestones, and different forms of IP assembly are all part of the toolbox of best practices that facilitate access to innovation. But to determine which specific rights should be retained, which elements should be licensed on an open-source basis, which milestones lead to the best results, and which assembly option works best, will always be difficult to determine because the answers depend heavily on the context. The context will include the actors involved in the innovative process, their relationships, and their connectivity within a global innovation network. Irrespective of which options are selected for greater global access, all strategies will require highly intensive preparations, detailed knowledge of processes related to the development and marketing of the product, detailed knowledge of markets, realistic anticipation and forecasting of product potential, and persistence in quantitative forecasting, as well as a master plan for the entire product rollout. The tools described in this section can be powerful for achieving public sector goals, and they are certainly worth the effort. ■

All chapters refer to: *Intellectual Property Management in Health and Agricultural Innovation: A Handbook of Best Practices*. 2007. A Krattiger, RT Mahoney, L Nelsen, JA Thomson, AB Bennett, K Satyanarayana, GD Graff, C Fernandez, and SP Kowalski (eds.). MIHR: Oxford, U.K., and PIPRA: Davis, U.S.A. Available online at www.ipHandbook.org. The online version contains for each chapter a detailed Editor's Summary, Implications, and Best Practices.

1 In the context of modern technologies, specifically food biotechnology, acceptability is particularly relevant.

2 The choice of the six strategic components stems from work in innovation theory and management and has been adapted for health by RT Mahoney (see Chapter 1.2 titled Building Product Innovation Capability in Health, p. 13).

3 See part 11 on Technology and Product Licensing, and part 12.

4 Chapter 2.1 by AB Bennett titled Reservation of Rights for Humanitarian Uses, p. 41.

5 See Chapter 11.8 by SL Shotwell titled Field-of-Use Licensing, p. 1113.

6 Chapter 2.2 by AL Brewster, SA Hansen and AR Chapman titled Facilitating Humanitarian Access to Pharmaceutical and Agricultural Innovation, p. 47.

7 The special insert in this *Executive Guide* offers 24 successful case studies.

8 Chapter 2.3 by R Eiss, KE Hanna and RT Mahoney titled Ensuring Global Access through Effective IP Management: Strategies of Product-Development Partnerships, p. 63.

9 These encompass a wide range of resources, including genes and gene fragments, cell lines, monoclonal antibodies, reagents, animal models, growth factors, combinatorial chemistry and DNA libraries, clones and cloning tools (such as the polymerase chain reaction or PCR), methods, laboratory equipment and machines, databases, and computer software.

10 Chapter 2.4 by C Clift titled Patenting and Licensing Research Tools, p. 79.

11 Chapter 2.5 by AJ Stevens titled Valuation and Licensing in Global Health, p. 89.

12 Yale University had granted an exclusive license for an anti-HIV compound to Bristol-Meyers Squibb, which also gave the company the right to file for patent protection in foreign countries. Fatefully, the company filed in South Africa, Mexico, and Egypt, among other countries. South Africa (and the rest of Africa) needed this anti-HIV drug (D4T), but it was far too expensive for all but a very few. Pressure was put on Yale University to make this drug available generically in South Africa. Yale resisted, but student activists brought the issue to the attention of the national press, and Bristol-Myers Squibb quickly acted to make the drug available at no cost to treat AIDS in South Africa.

13 Chapter 2.6 by J Hope titled Open Source Licensing, p. 107.

14 For a detailed discussion on patenting versus the public domain, see chapter 10.1 by S Boettiger and C Chi-Ham titled Defensive Publishing and the Public Domain.

15 WHO. 2006. Public Health Innovation and Intellectual Property Rights. Report of the Commission on Intellectual Property Rights, Innovation and Public Health. WHO: Geneva. www.who.int/intellectualproperty/en/.

16 Chapter 2.7 by J Oehler titled Using Milestones in Healthcare Product Licensing Deals to Ensure Access in Developing Countries, p. 119.

17 Chapter 2.8 by A Krattiger and SP Kowalski titled Facilitating Assembly of and Access to Intellectual Property: Focus on Patent Pools and a Review of Other Mechanisms, p. 131.

FOR GOVERNMENT POLICYMAKERS

- ✔ One of the **benefits of enabling public research institutions to own IP rights** is that institutions can control how technology is deployed through the terms of licensing contracts, thus meeting both commercial and noncommercial goals.
- ✔ Well-crafted contracts, based on best practices, can be instrumental in achieving **global access**, provided the entire innovation process is given due consideration from the outset. This includes consideration of R&D capabilities, regulatory environment, manufacturing capabilities, IP management, access to markets, and trade-related concerns. Such an approach requires a lot of preparation and detailed knowledge of the processes related to developing and marketing the invention; realistic forecasting of product potential; persistence in quantitative forecasting and establishing a master plan for the entire product rollout; and a **mission-driven mindset** to establish optimum goals for the public sector.
- ✔ One of many components of best practices by the public sector is incorporating **humanitarian-use reservation** provisions in commercial licensing contracts. This is becoming increasingly common with certain universities around the world, particularly with respect to agricultural inventions. There is conceptually no reason why this should not become common practice globally.
- ✔ Public sector institutions should have explicit IP policies and **demonstrated institutional capacity** to implement best practices in IP management. Any licensor, public or private, is more willing to give licenses to institutions that proactively protect third-party-property, which leads to confidence building and a higher degree of motivation to proceed with more licensing and technology transfer arrangements.
- ✔ **Open source** may offer an alternative mechanism for facilitating access to innovations in health and agriculture, provided the open-source approaches that are so popular and effective in the software area can be successfully adapted to the biological sciences. More conceptual research is needed to make open source an effective way to accelerate innovation in health and agriculture.
- ✔ Other policies and laws can foster and enable efficient IP assembly (or in-licensing by national institutions to obtain freedom to operate and the freedom to license bundles of technologies to manufacturers). These may include **patent pools** and other mechanisms.

Given that IP management is heavily context specific, these Key Implications and Best Practices are intended as starting points to be adapted to specific needs and circumstances.

FOR SENIOR MANAGEMENT

(UNIVERSITY PRESIDENT, R&D MANAGER, ETC.)

- ✔ One of **the benefits of enabling public research institutions to own IP rights** is that institutions can control how technology is deployed through the terms of licensing contracts, thus meeting both commercial and noncommercial goals.

- ✔ Well-crafted contracts based on best practices can be instrumental in achieving **global access**, provided the entire innovation process is given due consideration from the outset. This includes consideration of R&D capabilities, regulatory environment, manufacturing, IP management, access to markets, and trade-related concerns. Such an approach requires a lot of preparation and detailed knowledge of the processes related to developing a **mission-driven mindset** to establish optimum goals for the public sector.

- ✔ One of the central elements for public sector institutions is to have explicit IP policies and **demonstrated institutional capacity** to implement best practices in IP management. Any licensor, public or private, is more willing to license to institutions that proactively protect third-party property, which leads to confidence building and a higher degree of motivation to proceed with more licensing and technology transfer arrangements.

- ✔ Humanitarian licensing can benefit both the research and public service missions of a university or public sector research institution. Consider creating an institutional policy that standardizes the **reservation of humanitarian rights** on all technologies, including, as appropriate, the right to practice the invention for nonprofit goals. Potential licensees are less likely to resist if they know that the terms being requested are "standard" and part of the deal in doing business with an institution.

- ✔ Implementing the various best practices discussed and presented in this section is complex and requires experience. Public sector institutions need to plan and implement focused **capacity building in IP management**.

- ✔ **Networks** with individuals and organizations, such as foreign universities, corporations, product development partnerships (PDPs), and government agencies, should be seen as critical elements that enhance the innovative potential of any institution.

- ✔ Indeed, partnerships are an important way to fill in the capacities that are required to **make an institution innovative**. Few, if any, institutions have the entire range of capacities to bring ideas to market.

- ✔ Under many circumstances, patenting may be unnecessary and publication might offer the widest dissemination. The decision to **place inventions in the public domain** should be calculated and made on a case-by-case basis. **Open source** licensing might be another complementary component of an IP management strategy.

- ✔ Whenever possible, consider **nonexclusive licensing** as a strategy to maximize the utilization of research tools. On product patents, **exclusive licensing** may, in many circumstances, be more effective to reach broad dissemination, particularly if coupled with strong milestone clauses.

- ✔ Complementary strategies are the segmenting or apportioning of markets, whereby different licensees obtain exclusivity but only for one portion of the **field of use**. This strategy can also be used to implement tiered pricing.

Given that IP management is heavily context specific, these Key Implications and Best Practices are intended as starting points to be adapted to specific needs and circumstances.

FOR SCIENTISTS

- As the inventor, in most circumstances, you can **significantly influence how your technology is used**. For example, you can very reasonably request that your technology transfer office draft license terms that reserve for you the rights to continue research using your inventions or terms that reserve the rights for humanitarian uses of your technology.

- Notwithstanding the above, you must follow the IP policies of your institution. And there is no reason a priori that your interests in licensing practices should not be reflected in your institutional IP policies. Changing them, if necessary, **requires a dialogue** with senior management and technology transfer personnel and a good understanding of the purpose of intellectual property and how sound IP management can be put to work for the benefit of the public sector.

- Your interest and activities in licensing and partnership building can **raise the profile** of your research program and also of your institution. It may create goodwill, catalyzing additional scientific and development interest by partner organizations and individuals. And it can lead to earlier and more efficient translation of your research findings into useful products or services.

- In particular, **collaboration with private sector entities** can be a most valuable contribution to your institution's broader participation in innovative initiatives, particularly as it pertains to product development.

- The R&D work that you carry out in your program can often (perhaps serendipitously) lead to the invention of new **research tools**. But the patenting strategies for research tools may need to be different from those related to products if maximum dissemination and use are sought.

- One such avenue for research tools in particular may be open source licensing. This is a complex and evolving area in the biological sciences and requires further refinement to be effective and useful.

- Importantly, open source licensing is not the same as placing an invention into the **public domain. Open source entails contractual obligations.** An open source license may be extremely complex and may require your institution to agree to certain obligations. Several universities are unable to sign such open source licenses because they cannot, in good faith, agree to the conditions. Make sure you always consult your technology transfer officers before signing any agreement.

- Increasingly contracts will include **milestones**, which may affect your work, although quite often not directly. Research schedules and goals may be directly linked to specific milestones, and you need to know how such milestones might influence your program.

- Accessing other people's intellectual property can be facilitated through networks of committed professionals; your contributions in this area can be substantial, and **strong professional networks** will make you a more valued and essential member of the team.

Given that IP management is heavily context specific, these Key Implications and Best Practices are intended as starting points to be adapted to specific needs and circumstances.

KEY IMPLICATIONS AND BEST PRACTICES: SECTION 2

FOR TECHNOLOGY TRANSFER OFFICERS

- ✔ **Licensing and valuation practices between a public sector institution and product-development partnerships** show that one valuation formula is to ask for the licensees in developing countries to take over responsibility for future patent costs but to ask for no up-front fees, no milestone payments, and no running royalties. Any financial return to the university should be derived from opportunities in developed countries.

- ✔ **Both nonexclusive and exclusive licenses can be applicable to meeting socio-economic goals**. Within exclusive licensing, there are many feasible options, such as exclusivity limited to a certain field of use, or geography, or for limited periods of time.

- ✔ Certain **equitable access provisions** in licenses can be instrumental in enabling competition in low- and middle-income countries.

- ✔ The practice of **reserving rights for humanitarian use** may require additional work and will likely not generate licensing revenue; conversely, such provisions, if used in a strategic way, are unlikely to lead to loss of revenues.

- ✔ Potential licensors of intellectual property connected to critical agricultural and health care technologies will be **motivated by your institution's demonstrated IP capacity**, and will be more likely to enter into more licensing agreements.

- ✔ If you are a licensor, **put yourself in the position of the other party**. If the roles of licensor/licensee were reversed, would your position seem unreasonable? Inflexibility may be detrimental when the licensee has technologies you may wish to utilize.

- ✔ IP managers should **be cautious of simply imitating the open licensing procedures** of the software industry. Such licenses are not generic enough to cross fields of endeavor, and it is still unclear whether and to what extent biotechnology innovations in general will lend themselves to open source licensing.

- ✔ The **public sector must specify in writing exactly what it wants to accomplish with a commercial partner**, detailing when and how this will be achieved by specifying milestones—and related penalties should these milestones not be fulfilled.

- ✔ Avoid "best effort" clauses in agreements. Instead, make the extra effort to draft **comprehensive contracts with articulated milestones**. This up-front investment in time and effort will pay off if a problem arises. During the drafting and negotiation of agreements containing milestones, do not hesitate to involve people from other departments (including business schools), outside consultants, and experts in the relevant industries and markets.

- ✔ Developing meaningful **milestones that provide the appropriate balance of incentives, rewards, and penalties** requires detailed preparations, a sound understanding of the processes related to developing and marketing the product, realistic forecasting of product potential, persistence in quantitative forecasting and in putting together a master plan for the entire product rollout, and above all, a mission-driven mindset.

Given that IP management is heavily context specific, these Key Implications and Best Practices are intended as starting points to be adapted to specific needs and circumstances.

The Policy and Legal Environment for Innovation

Changes in both national and international legal frameworks have profoundly affected how companies manage their intellectual assets in furtherance of economic and strategic business objectives and how they pursue their R&D. Moreover, the changes have enabled a broader distribution of health-related technology to people in developing countries. Public sector institutions likewise have had to adapt to an increasingly globalized knowledge-based economy. One adaptation is the ever-increasing interaction between developed countries and developing economies, particularly more innovative developing countries (for example, Brazil, China, India, Korea, South Africa). A second adaptation is the increased complexity of interactions between public and private sector actors. A third involves an evolution in the judiciary: toward a clearer judicial structure with more reliable and predictable mechanisms of dispute resolution. Many innovative developing countries are undergoing far-reaching changes within the judiciary, and experience from the creation of the Court of Appeals for the Federal Circuit (CAFC) in the United States offers many useful insights.

In order to help revitalize flagging technological innovation in the faltering economy of the late 1970s in the United States, a fundamental **change in the judicial structure** took place in 1982. As a judge at the CAFC, Newman[1] describes the creation of the CAFC (the national appellate court that would hear all patent appeals) as the first profound change in over 100 years concerning IP-related dispute resolution. A single appellate court would better understand and **correct policy misperceptions**, largely created by judicial decisions that had negatively influenced investment incentives in relation to patenting. A uniform and predictable application of the law across the United States and a concomitant end to forum shopping would promote innovation. And indeed it has. The effects have been dramatic: industrial activity, based on strengthened patent incentives, has surpassed the most optimistic expectations.

In addition to the establishment of the CAFC, two other critical events at the beginning of the 1980s catalyzed the growth of the biotechnology industry. In 1980, the Supreme Court's landmark decision in *Diamond v. Chakrabarty*, despite dire predictions to the contrary concerning the patenting of life forms, opened the nation's economy to biotechnology as an industry, enabling investment and commercialization in this nascent field. Also in 1980, the Bayh-Dole Act catalyzed the revitalization of commercial products arising from government investments in academic research. The combination of the CAFC, the Chakrabarty decision, and the Bayh-Dole Act synergistically drove the biotechnology revolution in the United States.

Because of rapid scientific developments, new issues of law constantly arise. Advances in

Krattiger A, RT Mahoney, L Nelsen, JA Thomson, AB Bennett, K Satyanarayana, GD Graff, C Fernandez and SP Kowalski. 2007. 3: The Policy and Legal Environment for Innovation. In *Executive Guide to Intellectual Property Management in Health and Agricultural Innovation: A Handbook of Best Practices* (Krattiger A, RT Mahoney, L Nelsen et al.). MIHR (Oxford, UK), PIPRA (Davis, USA), Oswaldo Cruz Foundation (Fiocruz, Rio de Janeiro, Brazil), and *bio*Developments-International Institute (Ithaca, USA). Available online at www.ipHandbook.org.

health and agriculture raise legal questions for which there are no direct precedents and about which there is no consensus. Therefore, **the courts take an incremental approach** to such questions, building on indirect precedents that attempt to balance the competing visions of patent theory and that respond to the quick pace of scientific discovery. In other words, the present builds on the past to create a coherent and stable body of law.

When technology and biology are involved, the overview of jurisprudence (as well as decisions in individual cases) will affect the nation's economy and the public interest, and, additionally, have an even broader global impact. This *Handbook* arises from the premise that **the development of the products of science and technology profoundly benefits the public and that both scientific and industrial participation are required in order for their benefits to be realized.** This is a many-faceted concept; yet we exist in a time of such pervasive scientific and technological advance that the development of these benefits and their movement into commerce among nations warrant our most concerted efforts.

But are public research institutions really delivering public goods? This question might appear foolish. In the past 50 years, the intensity of research and the pace of discovery in the biomedical and health fields have accelerated dramatically, not only in the United States, but in many parts of the world, particularly in the more innovative developing countries. As a result, the number of safe and effective drugs, vaccines, and medical devices for a broad range of illnesses and conditions has skyrocketed, as have sales in developed countries. But in an increasingly global world—in which the risk of disease and the benefits of research can come from any corner—the **benefits from public sector health investment should be global**. Keusch and Nugent,[2] therefore, argue, on the basis of their experiences in the United States, that the public-benefit aspect of government-sponsored research investments should include (the poor) in countries outside the United States.

Because of the "public goods" aspect of health, governments should fund health research, and indeed they do. For similar reasons, they also fund agricultural research and extension services. Such publicly supported research fills knowledge gaps that private industry ignores, even though public sector inventions are usually brought to market by private sector product development. The choice of whether to develop new ideas into products is largely left up to the private sector. Thus, technology development from public research proceeds largely according to private sector priorities. So **what role do public agencies have in ensuring that the public benefits from its investments in health research?** The answer is not obvious. Under current arrangements, the public sector has limited capacity and experience in the *downstream* steps of developing and delivering biomedical products to patients. These steps typically require a significant investment of money. They are also not aligned with the public sector's comparative advantages.

The public sector, therefore, needs to be creative, and Keusch and Nugent outline several ways that decision makers can **strengthen and reorient the public sector's intellectual property (IP) strategies** to expand the ability of developing countries to access the benefits derived from public research investments. They discuss several strategies that public institutions can adopt to increase the resources and tools devoted to the public health needs of the developing world:

- At the *upstream* end, the public sector can direct funds toward research on diseases in developing countries and can partner with private and nonprofit entities wishing to do the same.
- At the *downstream* end, public sector institutions can directly provide products to users in poor countries, reduce barriers to the transfer of technology that benefit developing countries, or partner with industry and academia to expedite the development of products from research.

As much as the Bayh-Dole Act successfully created a large body of intellectual property from publicly funded research, it has reduced, in some regards, the availability of public goods for health and agriculture. Current practice undervalues the public-benefit aspect of the mandate, especially for the poor. **As developing countries increasingly consider implementing Bayh-Dole-related legislation, these countries should carefully study the conclusions from this chapter so they may improve upon the experiences in the United States.** It should be pointed out that the intent of Bayh-Dole was not to produce supplemental revenue streams to universities. Rather, it was to encourage private innovation and increase the use of technology for economic development.

Graff[3] surveys the opportunities available in 18 developing countries[4] for new technologies to flow to the private sector and the public policy issues needing to be addressed to facilitate this. Three key aspects of public policy are considered:

1. The availability of IP protection
2. The designation of IP ownership
3. The existence of the infrastructure needed to make IP protection and ownership a reality

The chapter reveals that strong IP protection capabilities are correlated with robust scientific research efforts, to the strength of the countries' IP laws through history, and support of, through membership in, international trade agreements, particularly the Agreement on Trade-Related Aspects of Intellectual Property (TRIPS) and the International Union for the Protection of New Varieties of Plants (UPOV). Regarding IP ownership issues, policies in most countries are moving toward granting the rights and responsibilities of ownership to research institutions. Finally, **sophisticated institutional IP management is correlated to research capacity and to government investment in public sector and university research and development.** Generally speaking, vigorous IP protection policies and the capacity to enforce and manage them are mutually strengthening. The biggest factor for this strengthening is the amount of research and development a country conducts, followed by **the ability of its economy to absorb new innovations into existing or new industries.** Those seeking to use intellectual property as part of an integrated strategy to grow the economy through public financing and commercialization of innovation will find these trends worth considering.

In the chapter by Finston,[5] the discussion moves away from broad national policies to specific technology transfer policies. Finston reveals, using many colorful examples, how such policies can have a broad-ranging, positive impact for a country, raising the standard of living, improving the economy, and opening many commercial opportunities. Primarily addressing government policymakers, this chapter defines the **policy conditions needed for a robust national technology transfer system:**

1. Government support of science education, research, and related infrastructure that together will create an enabling environment
2. Rule-of-law protections (predictable laws, fair enforcement, judicial remedies), including those relating to intellectual property
3. A reliance on market forces, which foster informed risk-taking and private sector investment, to determine which technologies and products should be developed

Finston argues that these three factors are mutually interdependent and should all be present to create a favorable environment for technology transfer. To support her claims, the experiences of five geographically and developmentally diverse countries with existing technology transfer policies are described: Brazil, Israel, Jordan, India, and Ireland. As a result of technology transfer reforms, these countries enjoyed growth in R&D, technology transfer, and economic activity. If one or more of the above three requirements were missing, a country would not have advanced as far technologically or economically.

The key lesson offered by her chapter is that the strength of government and of the private sector can be synergistically applied to improve the lives of all. Technology transfer works best when there is strong, consistent government support of basic research, including science education and technology-related infrastructure and robust IP protection.

Given the success of technology transfer in the United States, many countries' expectations of similar programs in their own countries are grossly overestimated (if not outright misdirected). Indeed, Heher[6] points out that many policymakers in developing countries proclaim that a technology transfer program should become a major source of income. Too often, such programs begin too optimistically, but within a few years, end up disillusioned.

Unless the central reasons for undertaking technology transfer (for long-term social and economic benefits) are understood, a boom-and-bust cycle, replete with unrealistic financial expectations, is likely to prevail at considerable cost to those involved. Indeed, countries have yet to develop answers to basic questions. For example, **what exactly is the nature of success in technology transfer**? And what precisely are the elements that make this success possible? This chapter uses international technology transfer benchmark data to benchmark and understand the implications of promoting technology transfer and the likely outcomes of a technology transfer initiative under realistic conditions.

Heher provides comprehensive data and concludes that income generation from technology transfer is an inadequate—if not inappropriate—reason for an institution to invest in technology transfer. Governments should not expect revenues from technology transfer to be able to fund research institutions. Indeed, the **financial benefits of technology transfer activities are captured primarily at the national economic level through business creation**, with national returns arising from indirect economic effects. The extended time period required for individual institutions to derive benefits together with the fact that the benefits are largely felt by the national economy suggest that appropriate national support measures are needed to encourage technology transfer.

Bringing this realistic approach to the institutional level is the topic of the chapter by Taubman and Ghafele.[7] First, the authors **strongly endorse the indigenous innovation potential of developing countries**. Second, they detail the **inadequacies of a top-down approach** to developing IP management policies and approaches, as such an approach would almost invariably ignore the unique strengths of a particular country or institution. Rather, to seize on such strengths, a thoughtful dialogue between policy-conscious practitioners and practically informed policymakers is advocated by the authors. This requires knowledge one to act flexibly.

Taubman and Ghafele insist that IP management for the public interest should go beyond licensing arrangements and consider the full range of two continuums: degrees of exclusivity and degrees of market engagement. To demonstrate, the efforts of Jordan and Indonesia to manage intellectual property for the public good are examined. Both countries have passed IP legislation and have developed their IP policies in relation to broader public policy goals. The authors conclude that success requires flexible use of market mechanisms and the strategic deployment of the full range of exclusive rights afforded by IP protection. This can lead to some creative solutions. **Public sector institutions should learn to use the rules at least as well as their private sector counterparts to achieve their public policy aims**. This has never been more urgent than with the coming into force of the TRIPS Agreement in most developing countries.

TRIPS mandates minimum IP protections for patented pharmaceutical products.[8] Within this requirement, countries have considerable freedom on many specific aspects of TRIPS, and it is wise (if not imperative) for developing countries to exercise these flexibilities to the maximum extent possible. Thus, TRIPS can have profound effects on innovation, on the scope and magnitude of R&D investments, and on product availability. Product price in low- and middle-income countries is vigorously debated. Predicting and measuring the impacts of TRIPS on innovation is an unwieldy task because of the numerous variables in play and also because TRIPS only came into force in 2005, in many of the innovative developing countries. This is a short time to measure specific impacts; a simple measure like the price of a product is but one of the factors that determine "access" to patented health and agricultural products. Another factor would be the types of drugs or vaccines that are becoming available. Indeed, based on a conference held in India, the conclusion was drawn by Eiss, Mahoney, and Satyanarayana[9] that much of **the impact of TRIPS will depend on how countries and institutions respond to the new IP regime.**

There is every indication that IP management skills appear to be one of the crucial elements for harnessing the positive potential of TRIPS, and mitigating the negative ones, and such skills will allow developing countries to gain access to emerging tools, technologies, and resources that can dramatically improve the health and welfare of their citizens. Effective IP management can allow public research institutions to use their own research products to benefit the poor and to enter into public-private partnerships that can direct the power of industry to the needs of the poor. Without knowledge of sophisticated IP management techniques, however, such efforts—and their benefits—will be impossible.

The World Trade Organization (WTO) offers many initiatives and instruments that seek to enhance IP capacities in the developing world. Specific initiatives and instruments are also aimed at mitigating the possible negative effects of TRIPS. These are discussed by Watal and Kampf[10] and include compulsory licensing, the Doha Declaration, elements of the Convention on Biological Diversity, traditional knowledge projects, technology transfer programs, and IP capacity building programs.

Of particular relevance here is the **Doha Declaration**, which sought to address the potential constriction brought about by TRIPS regarding access to patented medicines in developing countries. The Doha Declaration emphasizes that the TRIPS Agreement does not and should not prevent members from taking measures to protect public health, and it reaffirms the right of members to use, to the full extent, the provisions of the TRIPS Agreement that provide flexibility in terms of accessing medicines. The Declaration states that each member has the right to determine what constitutes a national emergency, or other circumstances of extreme urgency, and explains that public health crises, including those relating to HIV/AIDS, tuberculosis, malaria, and other epidemics, can represent such circumstances.

By virtue of WTO member states having signed and ratified the agreement, TRIPS constitutes international law. It must be translated or adapted into national law individually by each member state. Though not directly related to TRIPS, the chapter by Bremer[11] provides an overview of key legal provisions in the United States that have profoundly affected the evolution of IP rights and technology transfer. The fundamental basis underlying the transfer of technology as intellectual property is embodied in the country's Constitution, which embraces patents, copyrights, and trademarks within its scope. The terms and provisions governing these forms of intellectual property are codified in various statutes, but two pieces of legislation are especially important.

The first is the Stevenson-Wydler Act enacted to promote the utilization of technology owned by the government and generated with its help. It aids the transfer of that technology to the private sector and government at the state and local levels. The second law is the Patent and Trademark Amendment Act of 1980—known as the Bayh-Dole Act. The Bayh-Dole Act established a uniform federal patent policy and provided the first statutory authority for the U.S. government to take title to and hold patents through its agencies. The success of this act makes it of special interest to countries seeking to establish IP regulatory systems, and this chapter explains the structure and history of Bayh-Dole.

Bremer also discusses **the important interplay between patents and antitrust laws**, recognized as complementary tools that enhance competition. The laws are based on the important premise that patents *per se* do not convey market power. Only when coupled with other assets, or when acquired in order to build a monopolistic behavior, *can patents create* market power. Antitrust scrutiny is triggered when patents (and certain other market positions) are combined with apparent predatory practices that restrain trade and competition. The point of this chapter then becomes very clear: **a country strengthening its patent laws should, concurrently, strengthen its antitrust laws as well as its capacities to enforce them.**

One such enforcement is the capacity of a government to bring about a **compulsory licensing** action. A compulsory license is an authorization given by a national authority to a natural or legal person for the exploitation of the subject matter protected by a patent; the consent of the patent holder is not necessary. Compulsory licenses may be required to import or produce a given product or to use a patented technology for research. They are especially important when there are no close substitutes for a product or process and a research exception is not available or is too narrow. Compulsory licenses are granted in order to attain various public policy objectives, such as: to address emergencies and public health needs, to counteract anticompetitive business practices, or to permit the exploitation of patents that are not used.

Correa[12] discusses the usefulness of compulsory licensing and provides a **step-by-step guide to obtaining compulsory licenses** to ensure that the R&D of drugs needed by people in developing countries is kept free from unnecessary entanglements in the global IP system. His chapter provides many illustrations and a useful discussion of the patenting and licensing strategies of universities and other public sector research institutions. These institutions often hold patents on research tools, underscoring the importance of the public sector retaining research-use and humanitarian-use rights in all licenses.[13] While there are several ways to circumvent patented upstream technologies, and the compulsory license is especially powerful. But, perhaps due to its power, the strategy also has drawbacks. The flexibility of compulsory licensing should be considered in the context of all of the options available to TRIPS member countries. Importantly, applicants need to be certain that they have the capacity to exploit the licenses and the financial ability to remunerate the patent holder or holders. Nonprofit research institutions may often find this particularly difficult because, even with a compulsory license, commercial partners need to be willing to produce and distribute products developed under compulsory licenses. This is one reason for further investments in technological and IP capacity building and the establishment of strong institutional networks.

Institutional networks are most powerful when formed around geographic clusters. Innovation in health and modern agriculture relies on a sophisticated open system of knowledge sharing. Recent studies suggest that successful innovation indeed requires development of clusters of institutions, businesses, and personnel. "Location, location, location," the battle cry for property realtors everywhere, is heard increasingly with respect to innovation dynamics and knowledge-based growth.

A cluster is a group of similar things positioned or occurring closely together. Although companies and various not-for-profit entities in the same sector or product market have traditionally located themselves in close geographic proximity (rather than spreading out evenly across the geography or economy), the express search for ways to encourage clustering has only recently begun. One paradigm, as discussed extensively by Phillips and Ryan,[14] is that **local competition is the primary engine behind cluster development and sustainability.** Additionally, innovation now involves and generates significant externalities; innovators increasingly rely on an array of formal and informal collaborators, and the efficacy of those relationships will determine their ability to successfully launch an innovation into product development.

Offering an overview of recent research on clusters in Canada, this chapter observes that **one factor encouraging cluster formation is the development of a cost-effective, efficient IP management system**. Equally important is the use of ***social capital***, which can lead to less formal collaborations that can better disseminate and utilize discoveries. While the traditional strategy of protecting infant industries in order to develop them made some sense in the industrial world, its value in a knowledge-based world is unclear. Knowledge-based development is inherently different from traditional industrial development. Indeed, multiple types of knowledge are involved in such a system, and Phillips and Ryan address how clusters integrate four distinct types of knowledge: "know-why," "know-what," "know-how," and "know-who." A cluster's ability to use and share these types of knowledge is largely what empowers individual entities within the cluster to innovate. Basing their ideas on varied illustrations and deep analysis, the authors conclude that governments have an important role to play in the process of cluster formation and that ensuring a mix of "local buzz" and "global reach" is part of the recipe for success.

A specific experience of cluster development and the role of government are presented by Viljamaa,[15] who discusses the case of Turku, Finland. The city is home to a large concentration of biotechnology activities. This model can be described as a science-led strategy, led from above, with a range of important lessons for policymakers and institutional leaders alike. The experiences suggest that sharing facilities with companies and combining forces with other universities and R&D institutes are vital ways of building clusters and momentum in innovation. Active partnerships with larger entities are important, as is a global network of scientists. Viljamaa offers many ideas that are particularly pertinent to developing countries that wish to encourage the formation of clusters. One is that **building clusters from scratch is basically impossible; success comes from building upon existing strengths**. Many successful clusters have been based on older but related industries.

Probably the most famous example of a cluster that is grounded on entrepreneurship is the biotechnology cluster of the greater Boston area, which encompasses Massachusetts Institute of Technology (M.I.T.), Harvard University, Boston University, and others. M.I.T., some argue, has led the translation of university-generated research from the laboratory to the private sector through the cultivation of an entrepreneurial culture. Indeed, the entrepreneurial activities of M.I.T. have served as an incubator for generation after generation of entrepreneurial engineers and scientists who view risk as an opportunity. **Seeing risk and opportunity as two sides of the same coin**, students at M.I.T. don't utter, "Why do you want to do that?" but instead proclaim "Hey! Why not?" This positive attitude, this sense of self-confidence, typifies M.I.T.'s culture, from professors to students to its licensing professionals.

Nelsen[16] points out that M.I.T.'s licensing office also has served as a focal point for the formation of the greater Boston area's biotechnology cluster. By coordinating the management of M.I.T.'s intellectual property, the office contributes to the robust development of many companies that form the cluster. This promotes further development, economic progress, investment in innovation, creation of networks, and ultimately, success. Although Boston is quite unlike most developing-country cities, the fundamental principles that drive its economic development are universal.

With respect to working with developing countries, M.I.T. recognizes that there are often special circumstances requiring creative practices (for example, preferential pricing for developing country public sectors, strategic patent filing, and differential licensing practices). Hence, with M.I.T. licensing, **there are no rigid written policies guiding how technologies are handled** (the exception to this is clear and nonnegotiable conflict of interest policies and practices); instead, the choices are left open in order to creatively craft agreements to maximize access. This flexible management fuels the innovation engine, and this approach can be adapted by many other regions.

But what are the potentials for individual countries to develop thriving cluster complexes without misspending scarce funds? What variables are essential for cluster development? In another chapter, Phillips and Ryan[17] identify six factors: manufacturing capacity, domestic market, export market, R&D, an IP system, and a functioning drug regulatory system. The authors explore these factors across three development stages to measure a country's cluster capacity. The authors go further and provide a **five-stage process for realistic cluster building:**

1. Assessing capacities, resources, and opportunities
2. Choosing an anchor strategy (different cluster approaches will have different sets of requirements, leaders, and tactics, and different success rates.)
3. Identifying organizational and institutional leaders to take the lead in developing the cluster.
4. Adopting proactive tactics, spanning numerous areas, including having the necessary legal and social structures, efficient mechanisms to protect and adjudicate property, the lowest possible barriers for

entering or exiting key input and output markets, the ability to trade domestically and internationally, and effective tax, regulatory, and trade rules.

5. Sustaining the lifecycle of the cluster (Recognizing that the evolutionary dynamics of markets are unavoidable, clusters should re-invent themselves every now and then to prevent cluster decay.)

Importantly, clusters thrive when local strengths are nurtured rather than when companies are lured with subsidies. Building infrastructure does not fill the buildings with innovative enterprises, but rather, innovative enterprises make buildings happen. Hence different types and sources of capital flow are needed at different stages of cluster development. Government money sometimes gets in the way of private money and vice versa.

The real and most effective catalysts for change are key individuals who serve as ambassadors or entrepreneurs for geographic regions; they cross-fertilize public-private partnerships which, in turn, alert the public sector to market demands and provide companies with access to basic research, infrastructure, and people capacity. This is why many institutions look for "people policies" to nurture clusters. Indeed, people are at the center also of these intellectual assets. **Knowledge-based development is inherently different from traditional industrial development.** Today's innovation potential requires, above all, global, institutional, and personal links and networks. They are the necessary fertile ground that enables innovation to flourish. ■

All chapters refer to: *Intellectual Property Management in Health and Agricultural Innovation: A Handbook of Best Practices.* 2007. A Krattiger, RT Mahoney, L Nelsen, JA Thomson, AB Bennett, K Satyanarayana, GD Graff, C Fernandez, and SP Kowalski (eds.). MIHR: Oxford, U.K., and PIPRA: Davis, U.S.A. Available online at www.ipHandbook.org. The online version contains for each chapter a detailed Editor's Summary, Implications, and Best Practices.

1 Chapter 3.1 by P Newman titled The Courts and Innovation, p 147.

2 Chapter 3.2 by RA Nugent and GT Keusch titled Global Health: Lessons from Bayh-Dole, p. 153.

3 Chapter 3.3 by GD Graff titled Echoes of Bayh-Dole? A Survey of IP and Technology Transfer Policies in Emerging and Developing Economies, p. 169.

4 The countries are Argentina, Brazil, Chile, China, Ethiopia, India, Indonesia, Jordan, Kenya, Malaysia, Mexico, Philippines, Poland, Russia, South Africa, Tanzania, Uganda, and Vietnam.

5 Chapter 3.4 by SK Finston titled Technology Transfer Snapshots from Middle-Income Countries: Creating Socio-Economic Benefits through Innovation, p. 197.

6 Chapter 3.5 by AD Heher titled Benchmarking of Technology Transfer Offices and What It Means for Developing Countries, p. 207.

7 Chapter 3.6 by A Taubman and R Ghafele titled Public Sector IP Management in the Life Sciences: Reconciling Practice and Policy-Perspectives from WIPO, p. 229.

8 TRIPS requirements are broader than this. They also include the application of basic principles of the trading system and other international IP agreements; adequate enforcement of IP rights in member countries of the World Trade Organization (WTO); IP dispute settlement between Members of the WTO; and special transitional arrangements for the introductory period of the new system.

9 Chapter 3.7 by R Eiss, RT Mahoney, and K Satyanarayana titled Developing Countries and TRIPS: What Next? p. 247.

10 Chapter 3.8 by J Watal and R Kampf titled The TRIPS Agreement and Intellectual Property in Health and Agriculture, p. 253.

11 Chapter 3.9 by H Bremer titled U.S. Laws Affecting the Transfer of Intellectual Property, p. 265.

12 Chapter 3.10 by CM Correa titled Compulsory Licensing: How to Gain Access to Patented Technology, p. 273.

13 See also Section 2 of the Handbook discussed on pages 11-20 of this *Executive Guide*.

14 Chapter 3.11 PWB Phillips and CD Ryan titled The Role of Clusters in Driving Innovation, p. 281.

15 Chapter 3.12 by K Viljamaa titled What Does It Take to Build a Local Biotechnology Cluster in a Small Country? The Case of Turku, Finland, p. 295.

16 Chapter 3.13 by L Nelsen titled The Activities and Roles of M.I.T. in Forming Clusters and Strengthening Entrepreneurship, p. 309.

17 Chapter 3.14 by PBW Phillips and CD Ryan titled Building Research Clusters: Exploring Public Policy Options for Supporting Regional Innovation, p. 317.

FOR GOVERNMENT POLICYMAKERS

- ✔ Countries have considerable freedom to control the effects of TRIPS. Indeed, **the impact of TRIPS will depend on how countries and institutions respond to the new IP regime**. At a minimum, countries should take full advantage of the flexibilities offered by TRIPS, in line with the Doha Declaration. For example, a country strengthening its patent laws should concurrently strengthen its antitrust laws as well as capacities to enforce them.

- ✔ Technology transfer efforts can be powerful when combined with government's efforts to **reorient the public sector's IP strategies** to enable the poor to benefit from public investments in innovation. To be effective, this should acknowledge the **inadequacies of a top-down approach** to developing IP management policies and approaches. Each institution has its unique strengths. To seize on these strengths, thoughtful dialogue between policy-conscious practitioners and practically informed policymakers should be encouraged.

- ✔ Public institutions' IP policies should address the institution's obligation, whenever possible, to retain humanitarian-use rights to its inventions, and the **government's right to a license for technology developed with public funds**, in case the public benefit is not being served adequately. Under extreme, well-defined circumstances, this may include full "march-in rights". The potential for such government action will encourage companies to make products widely available in the market.

- ✔ **Public-private collaborations within publicly funded R&D programs** can be powerful arrangements for optimizing public research investment.

- ✔ Public-private partnerships aimed at product development are effective arrangements through which industry can invest and apply its expertise to address the needs of the poor. In many contexts such **product-development partnerships (PDPs) are now driving the drug-development pipeline in neglected-disease R&D**. National institutions in developing countries should be encouraged to participate in PDPs.

- ✔ The ability of the local and national economy to absorb new technologies into existing industry or business sector can be strengthened through the encouragement of **cluster formation**. They require a long-standing and durable commitment to science education, research and related infrastructure, a strategically situated anchor institution with a proactive technology transfer office, and reliance on market forces as the engine for technology transfer.

- ✔ Overall, **public funds should be directed at product development partnerships** that create collaborations, as opposed to buildings with bricks and mortar. Such strategies have proven most effective in strengthening and sustaining clusters.

- ✔ Governments should support local entrepreneurship with much attention given to endogenous development, specifically to **local, small- to medium-size enterprises and to spinouts**. An effective short-term strategy may be to attract foreign companies to the area. They will bring jobs and often knowledge and expertise.

Given that IP management is heavily context specific, these Key Implications and Best Practices are intended as starting points to be adapted to specific needs and circumstances.

KEY IMPLICATIONS AND BEST PRACTICES: SECTION 3

FOR SENIOR MANAGEMENT

(UNIVERSITY PRESIDENT, R&D MANAGER, ETC.)

- ✔ In an increasingly global world—in which the risk of disease and the effects of agricultural disasters span borders and the benefits of research can come from any corner—the society that **benefits from public sector health investment will be global**. The public-benefit aspect of government-sponsored research investments should include the poor in every society, including those of neighboring countries.

- ✔ There are many **strategies available to increase the resources and tools devoted to the public good** that do not run counter to economic development goals and private sector interests. At the upstream end, funds can be directed toward research in developing countries, and partnerships with private and nonprofit entities can be effective. At the downstream end, funds can directly provide products to users in developing countries, reduce barriers to the transfer of technology that benefits these countries, or partner with industry and academia to expedite the development of products from research.

- ✔ The main issue for universities is to ensure a high level of education, comprehensive partnerships with other universities, and **collaboration with the private sector**. This requires clear IP policies, transparent IP management practices, and sound management of conflicts of interest.

- ✔ **Public-Private Partnerships and Product-Development Partnerships** (PDPs) are novel, tightly focused organizations, dedicated to providing products to benefit the poor in developing countries. PDPs require that scientists put a priority on delivering global benefits and that universities fully embrace their larger role in society and the global community.

- ✔ A major policy objective is to **find a balance between public benefit and economic returns**. A university can include a public-benefit clause in its licenses to the private sector, invest part of its royalty stream in a foundation, establish an "ethical" investment fund, license technologies to nonprofits or others who would develop and manufacture products for developing countries, and bundle technologies to encourage development of medicines aimed at diseases of the poor.

- ✔ The ability of the local and national economy to absorb new technologies into existing industry or an entrepreneurial sector can be strengthened through the encouragement of cluster formation. But **robust innovation clusters** are not created from scratch. They require a long, durable commitment to science education, research, and related infrastructure; a strategically situated anchor institution with a proactive technology transfer office; and reliance on market forces as the engine for technology transfer.

Given that IP management is heavily context specific, these Key Implications and Best Practices are intended as starting points to be adapted to specific needs and circumstances.

FOR SCIENTISTS

- Global changes in IP regimes, especially changes that affect developing countries, have been tremendous. Within the evolving IP regime, your country has considerable freedom to control the effects of these changes. Indeed, much of **the impact of these changes will depend on how countries and institutions respond to the new IP regime**.
- An important response is the **creation of an effective technology transfer program**. Your role in this process is essential.
- As a scientist, you understand the **interrelatedness of science, R&D, technological advance, and commercial investment**. Share these insights with your institution's technology transfer office, as well as with its senior managers.
- Countries engaged in reforming their R&D and technology transfer efforts are today often including royalty-sharing provisions for scientists in publicly funded research institutions. This approach also comes with obligations to assign ownership rights to your institution and a duty to disclose inventions. All of these changes should be seen as **incentives to turn inventions into innovations that benefit society**.
- As your institution implements IP policies and patenting strategies, **your right to publish** is not jeopardized. IP protection and licensing are but one form of knowledge transfer that, if well undertaken, can very much be in the public interest.
- While access to foreign technology is integral to development, it is increasingly important to **focus directly on capturing the national (or indigenous) innovation potential** of developing countries. Through the activities of your research program, you may be positioned to facilitate such capture and development of the benefits arising from indigenous innovation and traditional knowledge. These efforts should be coupled with benefit-sharing provisions.
- Understand the obligations that are attached to different funding sources when funds are used within the same program. The **impact of joint public and private financial support** can be complex but will increase, particularly as your institution positions itself strongly within an innovation cluster and engages in product development.
- As a scientist, you play an **increasingly important role in knowledge-based innovation clusters**. Do not shy away from becoming an entrepreneur yourself.
- Collaboration is often based on establishing personal contacts, for example, **building close connections and networks** to other scientists and research groups in the same field via conferences and reciprocal visiting arrangements; these all foster the formation of collaborative research projects and are fundamental for effective sharing of know-how and show-how.

FOR TECHNOLOGY TRANSFER OFFICERS

- ✔ Traditionally, the mission of a technology transfer program was to bring university-generated intellectual property into use as rapidly as possible. But **public sector technology transfer has evolved to serve broader purposes**: to enhance the reputation of the institution. Successful technology transfer can help it achieve its missions of education, research, and community outreach; to ensure social impact; and to provide funds for further research.

- ✔ The laws relating to new technologies are evolving. Recent **court decisions may have an impact** on business and technological matters relevant to the operations of your technology transfer office (TTO).

- ✔ A TTO has much **responsibility in creating incentives to move discoveries into the product development arena**, motivating public sector researchers, not by the promise of revenue streams (which rarely appear), but by the satisfaction of seeing their work developed and applied to serve the public good.

- ✔ An understanding of not only the law, but also the public policy that underlies it. For example, with the Bayh-Dole Act in the United States, the **policy rationale is not directed toward revenue generation**, but rather toward moving publicly funded R&D into the marketplace to serve the public good.

- ✔ Financial **benefits from technology transfer can take many years to realize**—if they ever do materialize—so it is important to be realistic when making forecasts about expected income. International benchmark data indicate that a positive return can take eight to ten years to achieve. It is prudent not to justify the cost of technology transfer functions on the basis of financial returns.

- ✔ **The difficulties of managing and promoting technology transfer within a smaller research institution need to be recognized**, and the office should actively seek partnerships with other entities, such as local venture capital firms, incubators, and business development agencies. Alliances with other institutions, or a central TTO for several institutions, may also constitute viable alternative strategies.

- ✔ In a dynamic **innovation cluster**, authoritative IP management capacity, technology transfer, and licensing are all essential. Flexibility in licensing and partnership arrangements, and speedy action and decision making are equally important.

- ✔ TTOs are often ideally placed to **define and nurture an entrepreneurial culture** in the faculty. There can be large gains from such efforts.

- ✔ TTOs can, if appropriately structured, become a source of **creative networking and collaboration**, generating both academic and commercial success. Hence, this role in driving the success of clusters will be absolutely essential.

Given that IP management is heavily context specific, these Key Implications and Best Practices are intended as starting points to be adapted to specific needs and circumstances.

The IP Toolbox

Best practices in the management of intellectual property require a basic understanding of the various forms of IP protection that are available. The **forms of IP rights** in most countries include patents, trademarks, geographical indications, copyrights, and trade secrets. Special provisions for plants are also offered in many countries in the form of plant variety protection, or plant breeders' rights. A further emerging type of protection is that related to regulatory data, which can be protected from disclosure or acquisition for a certain period of time and offer data exclusivity. Although each of these statutory mechanisms of IP rights protect different forms of intellectual property, thus conferring different IP rights, when used alone or in combination, they provide a set of options for organizing and then making the most out of an organization's IP assets.

All of the above are reviewed in detail by Dodds and Krattiger[1] in Chapter 4.1, which includes short sections on institutional aspects including employee agreements, how to integrate the various rights, and how to identify infringement. Importantly, the form of protection chosen for a given invention should be guided by the mission of the institution (whether public or private), the purpose of the work it conducts, and the nature of the invention, or other intellectual property that will be subject to IP rights protections.

A **utility patent** is a type of statutory IP protection covering inventions, that is, a grant by the government to an inventor for any invention that is a new and useful process, machine, article, manufacture, or composition of matter or any new and useful improvement thereof. The invention, that is the intellectual property itself, is a product of the inventor's mind. The patent, then, confers certain rights to this property; it is the right to *exclude* others from making, using, selling, or importing the invention in the country where the patent is granted, normally for a period of 20 years from the date of the patent application.

In patents, many aspects of inventions are disclosed. Patents should thus not be seen as the exclusive domain of lawyers. Scientists in particular are well advised to be up-to-date on patents issued in their field of endeavor, and Nottenburg[2] in Chapter 4.2 provides a comprehensive guide to patents, using biotechnology patents as an example, that instructs scientists and others how to read utility patents.

Trademarks are a form of IP protection that serves to distinguish the products or services of one individual, company, or organization from the products or services of others. A trademark can be a word, phrase, symbol, design, or a combination thereof. Trademarks can even be sounds or colors, if they are in some way distinctive, that create an immediate association in the mind of the consumer between the trademark and the good. IP protection for a trademark confers an exclusive right to use the mark in commerce.

Krattiger A, RT Mahoney, L Nelsen, JA Thomson, AB Bennett, K Satyanarayana, GD Graff, C Fernandez and SP Kowalski. 2007. 4: The IP Toolbox. In *Executive Guide to Intellectual Property Management in Health and Agricultural Innovation: A Handbook of Best Practices* (Krattiger A, RT Mahoney, L Nelsen et al.). MIHR (Oxford, UK), PIPRA (Davis, USA), Oswaldo Cruz Foundation (Fiocruz, Rio de Janeiro, Brazil), and *bio*Developments-International Institute (Ithaca, USA). Available online at www.ipHandbook.org.

Trademarks are an often overlooked and undervalued form of intellectual property by the public sector and this is well argued by Needle[3] in Chapter 4.3 on the basis of many colorful examples. Trademarks should be a valuable component of any IP management strategy, complementing the protection afforded by other forms of statutory IP protection. Trademarks can complement other forms of IP protection, and, as in the case of many pharmaceuticals, serve to strengthen the period of proprietary rights to a product.

Geographical indications are signs used on goods that have specific geographic origin and possess qualities or a reputation that are derived from their place of origin. Geographical indications are another type of intellectual property, similar to trademarks in that they are source indicators. Most commonly, a geographic indication consists of the name of the place of origin of the particular goods (for example, Roquefort cheese or champagne).

Copyright is a type of statutory IP protection for the original works of authors, such as the chapters in the *Handbook*. Such works include literary, musical, dramatic, and architectural works. The copyright protects the work immediately after it is fixed in a tangible medium, for example, words on a page (what you are reading at this very moment) are copyrighted. The owner of the copyright, for example, we, the authors of this chapter, have certain rights to the work. Typically, these rights include moral rights (that is, having our names associated with the work) as well as the right to reproduce the work, to prepare adaptations of the work, and to distribute the work to the public. However, these rights can be either licensed or assigned to others. In the case of this chapter, we agreed to make it freely available to all through the Internet.

Trade secrets (in certain circumstances and jurisdictions called **know-how**) are an important form of intellectual property. Trade secrets protect know-how and any confidential information so designated. To be protected as a trade secret, the intellectual property must, of course, be kept secret, and must also confer some sort of commercial advantage to the holder. Enforcement of IP rights for trade secrets is possible when a competitor has misappropriated and/or stolen the trade secret.

A point often raised is when one should file for a patent or maintain the information as a trade secret. What is important to note is that patents and trade secrets are not in conflict with each other but are complementary IP assets. Depending on the nature of the know-how, or the invention, the organization may choose to either file a patent or to continue to hold as a trade secret. Dodds and Krattiger in Chapter 4.1 discuss trade secrets briefly, but they are fully discussed in the context of licensing in another section of the *Handbook*.[4]

The protection of intellectual property related to **plants, germplasm, and varieties** is covered in several chapters because there are many dimensions to the topic. Kesan[5] in Chapter 4.4 describes the various forms of intellectual property applicable to plants. These are utility patents (available in a few countries only), plant variety protection (or plant breeders' rights), plant patents, trade secrets, geographic indications, and trademarks. The strengths and weaknesses and pros and cons of each are discussed. It is worth noting that the use of one form of protection is not necessarily exclusive, in that a single plant may be simultaneously covered by several forms of IP protection.

Plant variety protection (PVP) is the most common tool for protecting varieties, and many countries have legislated and implemented a PVP system. Lesser[6] describes the PVP system in Chapter 4.5. PVP regimes are implemented in order to:

- provide breeders (both public and private sectors) with an opportunity to receive a reasonable return on past investments
- provide an incentive for continued or increased investment in future breeding research
- recognize the legal right of the innovator to be recognized as such
- acknowledge the economic right to remuneration for his or her efforts

In general, there are two exemptions to the protection provided: 1) a research exemption and 2) a farmer's exemption (this is not to be confused with farmers' rights). A research exemption allows for breeders to develop a new variety by using a protected variety; a farmer's exemption allows for the saving of seed for the sole use of replanting the farmer's land.

Given the advantages of a PVP system in attracting private investments and offering farmers a broader range of improved varieties, countries may gain substantially in internationally harmonizing their PVP regimes, as it lowers costs for users, simplifies the introduction of new varieties, and thus leads to the availability of more varieties and choices for farmers.

Lesser also points out that those countries that are members of the Agreement on Trade-Related Aspects of Intellectual Property Rights (**TRIPS**) agreed to implement special protection for plants (the so called *sui generis* protection). Many elect to follow the principles of the International Convention for the Protection of New Varieties of Plants (**UPOV**), an international treaty that provides an effective framework for PVP.

Plants can also be protected through trade secrets, trademarks, and geographical indications. Geographical indications might be used to communicate to consumers the association between a plant's special characteristics and the territory from which it originates. Trademarks can have particular value if the variety has market potential, and consumers come to specifically associate the trademark with desirable characteristics and qualities of the variety. Such value has important implications for developing countries that are exporters of agricultural commodities and products, as it can add significant additional value to their exports.

The management of intellectual property related to crops and germplasm is an essential function of a technology transfer office (TTO) and is the focus of Chapter 4.6 by Dodds and colleagues.[7] They also discuss specific issues related to IP management involving genebanks and practical aspects on the establishment of a PVP office. Blakeney[8] in Chapter 4.7 reviews international aspects, including the international exchange of germplasm. A working knowledge of the relevant **international treaties related to genetic resources** is important for anyone dealing with genetic resources as they increasingly affect the international exchange and use of germplasm.

One such treaty is UPOV. Significantly, the latest revisions of 1991 expanded the scope of protections that could be granted to include *essentially derived* varieties. This has important implications for genetically modified organisms. These revisions also allow countries to limit farmers' rights, allowing them only to save seeds for use on their own land.

Another agreement is the **Convention on Biological Diversity**, particularly the provisions concerned with informed consent to use of biological materials and equitable benefit sharing following access. Some people argue that these requirements may be in conflict with this requirement of TRIPS. In practice, however, whereas UPOV and related *sui generis* systems focus on plant varieties, the Convention on Biological Diversity essentially deals with wild genetic resources. Exceptions include the reach of the Convention into genetically modified crops through the Cartagena Protocol on Biosafety, but this is not related to intellectual property.

The International Treaty on Plant Genetic Resources for Food and Agriculture (the Treaty) is a recent addition to international agreements. The Treaty establishes a multilateral system that embodies a sort of genetic commons within which the exchange of germplasm in major crop varieties between member states is facilitated. Conditions limit the rights of recipients to seek IP rights in material obtained and support the rights of donors to share in some form of benefit. The Treaty further recognizes the contribution of farmers and indigenous peoples' **traditional knowledge** to agricultural biodiversity. This is accomplished through the development and conservation of landraces, in primitive varieties developed to deal with local climate and diseases and to appeal to local tastes, by interbreeding locally occurring undomesticated plants with cultivated plants, as well as by exchanging different genotypes among farmers and farms. Again, some argue that certain terms of the Treaty may not be compatible with UPOV standards but overall this assessment seems unlikely.

A distinct but closely related topic is that of **information resources.** These include **computer software** and systems, **databases, geographic information systems** (GIS), **remote sensing** (RS) information, and **library** resources. The integration of these is increasingly prevalent in advanced agricultural systems such as the forecasting of disease and harvests. Dodds and colleagues[9] in Chapter 4.8 discuss the various IP elements related to information resources and how they can be managed effectively. The chapter also addresses licensing elements.

A very different topic is that of **data protection** and **data exclusivity.** These systems of protection are especially important in pharmaceuticals and agricultural chemicals. Two chapters review these complex topics, Chapter 4.9 by Clift[10] and Chapter 4.10 by Cook.[11] In short, regulatory data are the data that the researcher or manufacturer of a product must provide to the appropriate regulatory agency in order to prove that the product is safe and efficacious. Regulatory data are protected from disclosure or acquisition for a certain period of time, usually five to ten years from the product's first authorization to market, during which time no other applicants are allowed to use it to obtain marketing authorization for the same product.

Regulatory data protections are substantively different from other sorts of intellectual protection, including confidential information protection and patents. The provisions in Article 39.3 of TRIPS, concerning the protection of regulatory data, are broad and subject to interpretation. Both the United States and the European Union have interpreted and implemented the obligations in different ways (as explained in Chapter 4.10).

The chapters by Clift and Cook also examine data exclusivity from the perspective of specific TRIPS requirements (Article 39) which essentially include three obligations on governments:

- protect data on new chemical entities, the collection of which involved considerable effort, against unfair commercial use
- protect such data against disclosure, except where necessary to protect the public
- protect such data against disclosure, unless steps are taken to ensure that the data are protected against unfair commercial use

It is important to note that these requirements do not create new IP rights (other than defining the reach of trade secrets). Article 39.3 only articulates widely accepted trade secret and unfair competition law and is not an invitation to create new IP rights per se for test data. ■

All chapters refer to: *Intellectual Property Management in Health and Agricultural Innovation: A Handbook of Best Practices.* 2007. A Krattiger, RT Mahoney, L Nelsen, JA Thomson, AB Bennett, K Satyanarayana, GD Graff, C Fernandez and SP Kowalski (eds.). MIHR: Oxford, U.K., and PIPRA: Davis, U.S.A. Available online at www.ipHandbook.org. The online version contains for each chapter a detailed Editor's Summary, Implications, and Best Practices.

1 Chapter 4.1 by J Dodds and A Krattiger titled The Statutory Toolbox: An Introduction, p. 337.

2 Chapter 4.2 by C Nottenburg titled How to Read a Biotech Patent, p. 351.

3 Chapter 4.3 by W Needle titled Trademark Primer, p. 361. Trademarks and trademark licensing are further discussed in detail in Chapter 11.6 by WT Tucker and GS Ross titled Use of Trademarks in a Plant-Licensing Program, p. 1059.

4 Chapter 11.5 by KF Jorda titled Trade Secrets and Trade-Secret Licensing, p. 1043.

5 Chapter 4.4 by JP Kesan titled The Statutory Toolbox: Plants, p. 371.

6 Chapter 4.5 by WH Lesser titled Plant Breeders' Rights: An Introduction, p. 381.

7 Chapter 4.6 by J Dodds, A Krattiger, and SP Kowalski titled Plants, Germplasm, Genebanks, and Intellectual Property: Principles, Options, and Management, p. 389.

8 Chapter 4.7 by M Blakeney titled Plant Variety Protection, International Agricultural Research, and Exchange of Germplasm Legal Aspects of *Sui Generis* and Patent Regimes, p. 401.

9 Chapter 4.8 by J Dodds, S Somersalo, SP Kowalski, and A Krattiger titled IP and Information Management: Libraries, Databases, Geographic Information Systems, and Software, p. 419.

10 Chapter 4.9 by C Clift titled Data Protection and Data Exclusivity in Pharmaceuticals and Agrochemicals, p. 431.

11 Chapter 4.10 by T Cook titled Regulatory Data Protection in Pharmaceuticals and Other Sectors, p. 437.

FOR GOVERNMENT POLICYMAKERS

- ✓ The statutory **tools of IP**, such as patents, copyright, trademarks, trade secrets, geographic indications, and plant variety protection, are tools that can be used to achieve a goal. The tool in itself is neutral; what matters is how the tool is used.

- ✓ When setting up a **patent office**, notwithstanding considerable latitude provided under TRIPS, there are advantages in implementing practices that are consistent and compatible with the practices of other countries. Doing so will facilitate greater opportunities for international collaboration in R&D and technology transfer. Particularly important is making patent applications and issued patents available online. This furthers innovation and licensing.

- ✓ **Copyright** is also an important form of IP that can be used to encourage innovation. The recent trend, at least in the United States, to provide for ever-increasing duration of protection (now exceeding four generations) should be avoided as this approach prevents the availability of important commercially but insignificant works.

- ✓ The use of **trademarks** is important for building integrity and stability in commerce and for offering new opportunities for national innovations. Trademarks can also be highly valuable for public sector entities.

- ✓ Judicious **plant variety protection** of new varieties will encourage investments in the development of crops that are essential for food security, a better environment, and economic development. As with patents, domestic innovation, the transfer of foreign varieties for increased production and productivity, and spurring national investments in crop breeding can be enhanced significantly through membership in international bodies, such as UPOV. This can lead to the earlier availability of improved varieties.

- ✓ Notwithstanding the above, countries can exercise significant latitude in regulating access to certain categories of plant genetic resources they consider strategically important. Plant breeding, however, and the enhancement of crops, is based on the stepwise improvement of existing varieties, and this requires **broad access to genetic material**. Related to this are geographic information systems and corresponding **data protections** that can add substantial value to biodiversity resources and **traditional knowledge**.

- ✓ Introducing stringent confidentiality of data and exclusivity laws can prevent **early introduction of generics and promote competition critical for improving access to life saving drugs.** There is a need to balance the various competing interests.

- ✓ IP protection mechanisms, however, depend upon effective and **equitable enforcement** by national governments. This requires effective, transparent, and enforceable **contract law** that can be implemented to protect natural, cultural, and economic resources, all by furthering useful interactions with the global community. This balance is critical.

Given that IP management is heavily context specific, these Key Implications and Best Practices are intended as starting points to be adapted to specific needs and circumstances.

KEY IMPLICATIONS AND BEST PRACTICES: SECTION 4

FOR SENIOR MANAGEMENT

(UNIVERSITY PRESIDENT, R&D MANAGER, ETC.)

- ✔ The implementation of a **broad institutional IP policy**, consistent with the institution's mission, can foster the integration of the various forms of IP protection in furtherance of an institution's mission and goals.

- ✔ For public sector institutions, **trademarks** can be a valuable element in an institutional strategy that aims at fostering a positive image (or brand) and generating value. Because of the broad value of trademarks, they assist institutions in maintaining a good image and brand, thus serving as a tool for senior management in maintaining and enhancing the institution's reputation, standing, and value.

- ✔ Scientists can gain a lot from regularly reviewing newly issued **patents** from around the world. Patents often disclose much more than scientific publications but are generally overlooked as valuable sources of scientific and technical know-how. Such an information-gathering approach requires appropriate staff training and the availability of good Internet connections, which make it possible for patents to be downloaded. All patent office Web sites provide patents free of charge as does the Worldwide Intellectual Property Organization (WIPO).

- ✔ A sound **patenting strategy** is an extremely useful tool to bring inventions to fruition that make an impact on economic development and meeting public sector goals.

- ✔ Although many public sector institutions have for years provided their improved germplasm free of charge or at nominal costs to breeders and farmers, the **protection of improved varieties** can be a critical tool in furthering broad access and simultaneously meeting commercial and humanitarian objectives through appropriate "market segmentation."

- ✔ In many countries, but not in the United States, patent law includes a broad **research exemption**. This should not be confused with possible restrictions on materials obtained through material transfer agreements. Although extremely useful, material transfer agreements should be used judiciously, particularly when intellectual property is also embedded in material. This aspect also requires well-trained licensing/technology-transfer personnel and good management systems.

- ✔ The delivery of innovation for the greater **public good** by public sector institutions is not necessarily inconsistent with appropriate patent and other forms of IP protection. **Trade secret protection** in particular may be a valuable—and cost effective—means of achieving greater accessibility by disadvantaged members or groups of society. Whereas academic institutions in particular may regard such protection as inappropriate, it should be remembered that their mission is gradually shifting and increasingly include the delivery of products. This requires adjustments in the way information and know-how are managed. In turn, the changes require much internal discussion and sometimes culture change.

- ✔ Senior management's backing of the **technology transfer office** is important as is its support in the implementation of rigorous IP-related policies and procedures (such as those related to confidentiality).

Given that IP management is heavily context specific, these Key Implications and Best Practices are intended as starting points to be adapted to specific needs and circumstances.

FOR SCIENTISTS

- ✔ Research endeavors can go much further, in certain circumstances, if **appropriate IP protection** is sought. If appropriately managed, this is not in conflict with the broad dissemination of research results but encourages that your inventions serve humanity.

- ✔ **Patents** often disclose much more technical and scientific information than do academic publications. Make it a habit of regularly reading up on newly published patent applications or issued patents in your field. You can access this information for free on the Internet (such as on www.uspto.gov).

- ✔ Your institution's good reputation and standing can be used as a valuable **trademark or brand**. Maintaining the high reputation requires strict adherence to your institution's policy and best practices.

- ✔ Good **data management**, especially accurate record keeping through comprehensive notebooks, is the foundation for building a portfolio of IP assets. Essentially, best practices in scientific record keeping should be precisely the same as best practices in record keeping for purposes of IP management.

- ✔ Conversely, you should always know the origin and possible restrictions of **data and information** you use, no matter how insignificant they might seem. Make sure you document the source of important data and information in your laboratory notebook. If you have questions, never hesitate to contact your technology transfer office for help or clarifications.

- ✔ Particularly if your research is related to product development, the **confidentiality of your data** may be critical in ensuring global access. Data is a valuable form of intellectual property that can be used to obtain certain price or access terms in licensing negotiations. Whereas, as a researcher in an academic environment you may regard such protection as inappropriate, remember that it is the goals of your research that should drive the IP tools applied to your inventions. If you are engaged in the delivery of products, adjustments in the way your information and know-how are managed may be necessary to speed-up the translation of your research findings into innovative products or services.

- ✔ If a given invention cannot be patented in your own country (for example, a **biological invention**, including gene sequences), the invention may still be patentable in another country. The United States and Canada tend to have the broadest interpretations with respect to the patenting of organisms and biological materials. In pursuing patenting elsewhere, under certain circumstances, your research endeavors may leverage additional investments required to bring the fruits of your research to benefit your country and society at large and may also lead to additional research grants.

Given that IP management is heavily context specific, these Key Implications and Best Practices are intended as starting points to be adapted to specific needs and circumstances.

FOR TECHNOLOGY TRANSFER OFFICERS

- ✔ Intellectual property is often perceived as constraining research, particularly in public sector institutions. **Your role in communicating** the importance of judiciously using patents, trademarks, trade secrets, and so forth, and the benefits of good IP management, is critical. Such communication should be tailor-made to senior management, and even to your institution's board, as well as to scientists. Each responds to a different language. (And different colleagues will require different degrees of understanding. For example, your discussions on patents will necessarily differ with scientists and patent counsel. Choose your words and the level of detail you provide judiciously.)

- ✔ In many institutional settings, making better use of **patents** and other forms of intellectual property requires a culture change to a greater or lesser degree. This may include establishing an expectation for scientists in your institution to regularly review patents. Encouraging scientists to share a **broader IP awareness and culture** will be potentially powerful and valuable.

- ✔ **Trademarks** are a critical, and often overlooked, option for IP protection. They can be used as stand-alone IP protection, or they can be integrated into an overall strategy for integrated IP protection, for example, a strong trademark for a patented product or process.

- ✔ Your job requires a judicious balance of work that relates directly to your benchmarks and targets, and of **contributing to the overall IP culture** of an organization. The latter is often not spelled out in your job description but it is important nonetheless. The greater the general level of awareness related to intellectual property, the more likely it is that the value of IP assets can be captured and utilized. And your job also becomes easier when you gain a broader understanding of intellectual property.

- ✔ Genebank management and that of **genetic resources**, in general, is increasingly becoming a sensitive issue. An organized, stepwise approach is vital for effectively managing a genebank and for avoiding difficulties. Ownership of genetic resources can be tricky, so rigorous documentation and clear procedures on incoming and outgoing genetic resources may be critical.

- ✔ The above point applies equally to **data**, both incoming and outgoing. Particularly if your institution conducts research related to product development (especially clinical trials), the **confidentiality of data** may be critical in ensuring global access. Specific data are a valuable form of intellectual property that can be used to obtain a certain price or access terms in licensing negotiations.

Given that IP management is heavily context specific, these Key Implications and Best Practices are intended as starting points to be adapted to specific needs and circumstances.

Institutional Policies and Strategies

The boundaries of any property must be clear. As a farmer needs to know where his or her field begins and ends, an innovator must know exactly the definition of his or her invention. But it is more difficult by far to delineate where rights to something intangible begin and end. In addition, intangible assets can be difficult to keep track of, to share, and to use. Yet in a research-and-development environment, intangible assets are often the most valuable and important ones. How then, can they be leveraged to reinforce the mission of an institution? How can the specific objectives be achieved more effectively and efficiently through the incorporation of best practices in IP (intellectual property) management? What are these principles and practices? And how does an IP policy relate to an operational IP strategy?

This section offers insights into achieving and maintaining **clarity about the ownership of intellectual property in public sector institutions** and stresses the value of IP strategies and IP policies. These are important for achieving success and have been increasingly encouraged—or required—by certain donors as a strategy to **ensure global access.** Global access, especially by the poor, and ownership of intellectual property go hand in hand. Indeed, the most important aspect of IP protection is that it bestows *control* over intellectual assets. If an organization—especially a public institution—fails to obtain IP rights for its inventions, it risks losing control over them. Failure to maintain rights may result in private entities appropriating elements of the value without major regard to the mission of public institutions, or it could lead to the intellectual assets becoming useless due to lack of further investment and development. This is the most important reason why the public sector should take IP management more seriously than it traditionally has. IP management is a fundamental element in the public sector's strategy of putting intellectual property to work for the public good.

Appropriately, the first chapter in this section of the *Handbook* is a comprehensive discussion on IP strategy by Pitkethly.[1] His definition of strategy relevant to IP management is:

- the formulation and adoption of courses of action enabling the reaching of long-term goals and objectives of an institution
- the allocation of resources (financial and human) necessary for carrying out these actions

By extension, **IP strategy is an integral part of an overall business strategy that uses IP rights to manage technology.**

Pitkethly begins by mapping out how IP rights systems fulfill four purposes:

- providing incentives for innovation
- allowing for the packaging of intellectual assets into innovative processes
- encouraging the diffusion of technical information

Krattiger A, RT Mahoney, L Nelsen, JA Thomson, AB Bennett, K Satyanarayana, GD Graff, C Fernandez and SP Kowalski. 2007. 5: Institutional Policies and Strategies. In *Executive Guide to Intellectual Property Management in Health and Agricultural Innovation: A Handbook of Best Practices* (Krattiger A, RT Mahoney, L Nelsen et al.). MIHR (Oxford, UK), PIPRA (Davis, USA), Oswaldo Cruz Foundation (Fiocruz, Rio de Janeiro, Brazil), and *bio*Developments-International Institute (Ithaca, USA). Available online at www.ipHandbook.org.

- enabling the capturing of added value (economic and/or humanitarian) through the control of intellectual assets

Viewed in this way, IP rights systems can be instrumental in enabling the diffusion of technological information. For example, patent specifications provide detailed embodiments of inventions, which are available for all to see. Using the Internet, these records can be accessed for free anywhere in the world. This availability of and access to information greatly facilitates innovation, since others will be able to work to improve or invent around the disclosed patented invention. Patents can also be useful sources of information for scientists, since the patent application may be the first and only publication about a competing innovation.

Organizations should have both **external and internal IP strategies**. Broadly speaking, this is referred to as litigation, licensing, and learning.

An external IP strategy involves exploiting inventions (by developing them in-house, selling them, or licensing them). Litigation denies IP rights to others; licensing allows rights to others. Learning can be a part of technology out-licensing, since it not only gives others access to these technologies but also provides learning opportunities for the organization. This strategy is especially effective if an institution's aim is to diffuse technology as widely as possible and ensure global access.

As a result, **donors are increasingly requiring grantees to take IP management seriously.** Ballantyne and Nelki[2] of the Wellcome Trust (the Trust) in the United Kingdom, a major charitable funder of biomedical research, is one such pioneering funding entity. The Trust requires institutions that have produced intellectual property using the Trust grants to determine whether the public will benefit from the protection of that property and whether all participants in the process receive proceeds proportional to the amount of money, equipment, knowledge, or labor they contributed. But above all, the Trust insists that intellectual property arising from its grant awards be adequately exploited; the Trust does not shy away from taking over activities if grantees fail to adequately exploit the intellectual property. The chapter ends with a series of case studies that illustrate the practical aspects of IP strategies.[3]

Grantees more and more frequently will need to present IP management strategies as part of their proposals. In the case of the Bill and Melinda Gates Foundation, certain grantees are required **global access strategies**[4] that outline how recipients will manage new and existing intellectual property. Certain minimum standards may also be required, such as:

- keeping the research field open by prohibiting any licensee from enforcing intellectual property against universities and research institutions that carry out noncommercial activities
- retaining licensable research rights to any invention developed with donor funding
- obtaining freedom to operate for all background intellectual property owned by collaborating institutions
- ensuring good IP management, which includes exploitation of intellectual property, with the goal of ensuring its use in developing countries

Some granting agencies also require prospective grantees to explicitly state their **IP policies.** Kowalski[5] reviews and discusses several institutions' policies and concludes that, at a minimum, an IP policy should define IP ownership, outline the patenting policy, describe the manner in which an institution will handle confidential information, set out the principles of its IP licensing and marketing approaches, explain how income arising from intellectual property will be distributed, and delineate the rights and obligations of inventors and the institution, as well as any rights the institution will retain (such as for research and for humanitarian uses).

Any new or revised IP policy (and IP strategy) will have to be "sold" to people both inside and outside an institution. It is important to explain *what* the policy contains and *why* the policy is designed the way it is. And perhaps staff at multiple levels should be involved in developing and revising, as needed, the IP policy. This group will be able to have extensive discussions about the role and function of intellectual property in the organization. These discussions will be an effective mechanism for building capacity and staff support of the policy. Some of the most controversial issues can be resolved before they become an obstacle, such as: Who owns what? Who benefits and how?

These sometimes-troublesome questions are discussed by Weidemier,[6] who reviews how universities in the United States are handling these aspects. Her chapter examines eight possible cases that illustrate and clarify the somewhat abstract principles of **ownership of university inventions** and are followed by a series of hypothetical scenarios. Weidemier, among others, concludes that universities should require all employees and visitors to sign **invention assignment agreements** on their date of arrival. Neither an employee handbook that discusses patent assignment nor a published university patent policy may be enough to ensure that the university is assigned ownership.

With the assignment of ownership rights to an employer comes the duty to disclose that an invention has been made. Indeed, an inventor is responsible for and has much to gain by making timely disclosures of his or her invention to the technology transfer office (TTO), the first step in enabling technology transfer. Di Sante[7] points out that successful commercialization is built on a foundation of good relations between inventors and technology transfer professionals. Such relationships should be established long before the transfer services of the technology transfer office are required, since this will enable technology managers to negotiate both faculty and business concerns about licensing agreements whenever the opportunity arises. **The role of the inventor in the entire IP protection and IP licensing/transfer process cannot be overstated** and should continue throughout the life of the technology.[8] For example, years after a patent has been licensed, the inventor may be the best-placed person to alert a TTO that a certain product being sold may infringe the patent.

But dealing with inventors is not always easy.[9] Inventors are prone to fall in love with their own creations, and, perhaps unreasonably, anticipate that theirs is the next great thing. It is the technology transfer professional's responsibility to tactfully ensure that the inventors' expectations are kept in line with reality.

Establishing good relationships with inventors is an important way to identify the intellectual property being generated in the research institution. But from an institutional point of view, a more comprehensive perspective on intellectual property is often warranted. This applies particularly to times when an organization develops a strategic plan or IP management strategy. In this context, IP audits can be essential and often form the basis for an internal review and a revision of IP strategy. Blakeney[10] provides a comprehensive overview of IP audits. Indeed, the **importance of IP audits** is becoming more and more apparent, in the private sector as well as in the public sector, as public entities increasingly deal with other parties' intellectual property.

An IP audit seeks to accomplish three broad objectives for an institution. First, it seeks to identify the intellectual property generated by its researchers. This intellectual property is an asset, with value that an institution ought to identify, assess, and manage. Second, an audit seeks to identify and review the management of third-party intellectual property as a way of avoiding liability for misuse. The IP audit is thus a systematic, methodical identification of the intellectual property within the institute.[11] As the chapter shows, the audit follows a procedure, from start to finish, so that at the audit's conclusion senior managers are able to frame and implement good IP management practices. This is the third broad objective of an IP audit: to contribute to the formulation and execution of the IP policy and IP strategy.

From a practical point of view, an IP audit reviews a number of existing practices and establishes the context in which intellectual property is being handled. For example, a research institute's ownership and control over any intellectual property will depend on its legal status as an entity. An IP auditor will review the incorporation documents to identify what powers the institute has to own and to deal with intellectual property. For universities and government institutions, such a review will also include the prevailing government policies.

The IP auditors will also scrutinize the IP policy of the institute, if indeed there is a policy. It should ask questions, such as: Where is it posted? What does it say? Are new employees required to read it? IP audits may also uncover potential conflicts of interest. Bennett[12] offers a primer on issues related to the management of conflict of interest and conflict of commitment. **Conflict of interest** occurs when the financial interests of an institution's researchers are incompatible with the institution's mission, policy, or goals. **Conflict of commitment** may arise when the time a researcher spends in external activities related to, for example, downstream technology development, interferes with his or her attention to duties to the institution (for example, teaching or extension responsibilities).

While **conflicts of interest should not be seen in a negative light**, making exceptions to the rules is both dangerous and potentially harmful. Someone with a potential conflict of interest is not guilty of anything; rather, he or she may actually be a more valuable "asset" *because* of the potential conflict. This applies most strikingly when a professor has an interest in forming a spinout company based on university research. Potentially, larger issues will arise due to undisclosed conflicts of interest. What a university needs is to define a clear chain of command and in rare circumstances to establish oversight committees. Committees tend to slow the process with significant delays in time, which typically makes the process unmanageable and useless.

Most conflicts arise when *potential* conflicts are not disclosed. Conversely, a major tactic in managing conflict of interest is to disclose *potential* conflicts. And most conflicts of interest can be managed fairly easily provided the policy is clear and precise. M.I.T., for example, manages an unusually large number of spinouts and, therefore, has a very strict conflict of interest policy (this is discussed by Nelsen[13]). A technology transfer officer's role is to creatively craft arrangements within the rules, not to use these rules as deterrents. Put differently, Nelsen describes M.I.T.'s operating motto: "A firm wall between university and industry—but a wall with many doors… In sum, **technology transfer inevitably brings conflicts of interest. The challenge is to manage them.**"[14] ■

All chapters refer to: Intellectual Property Management in Health and Agricultural Innovation: A Handbook of Best Practices. 2007. A Krattiger, RT Mahoney, L Nelsen, JA Thomson, AB Bennett, K Satyanarayana, GD Graff, C Fernandez, and SP Kowalski (eds.). MIHR: Oxford, U.K., and PIPRA: Davis, U.S.A. Available online at www.ipHandbook.org. The online version contains for each chapter a detailed Editor's Summary, Implications, and Best Practices.

1 Chapter 5.1 by R Pitkethly titled IP Strategy, p. 459.

2 Chapter 5.2 by Z Ballantyne and D Nelki titled Management Policy: A Donor's Perspective, p. 475.

3 The following are reviewed in detail: the development of a typhoid vaccine, malaria drugs, the single nucleotide polymorphisms (SNP) consortium, and the International HapMap Project.

4 For an example of how global access strategies can be incorporated into product development partnerships and vaccine development, see Mahoney RT, A Krattiger, JD Clemens and R Curtiss III. 2007. The Introduction of New Vaccines into Developing Countries IV: Global Access Strategies. *Vaccine* 25(2007): 4003–4011.

5 Chapter 5.3 by SP Kowalski titled Making the Most of Intellectual Property: Developing an Institutional IP Policy, p. 485.

6 Chapter 5.4 by BJ Weidemier titled Ownership of University Inventions: Practical Considerations, p. 495.

7 Chapter 5.5 by AC Di Sante titled The Role of the Inventor in the Technology Transfer Process, p. 507.

8 An important exception to this is licensing. A TTO officer typically would not have the inventor participate during negotiations with potential licensees, although rare exceptions may apply.

9 See also Section 8 dealing more exhaustively with inventions and inventors.

10 Chapter 5.6 by M Blakeney titled Conducting IP Audits, p. 515.

11 This may include: patentable biological assets, such as germplasm resources, DNA libraries and enabling technologies (marker genes, probes); technological know-how; confidential information; patents, utility models and industrial design rights in equipment; copyrighted information (database rights, computer programs, and databases); publications; CD-ROMs; video materials; online materials; trademarks; and more.

12 Chapter 5.7 by AB Bennett titled Conflict of Interest and Conflict of Commitment Management in Technology Transfer, p. 527.

13 See Box 1 for M.I.T.'s policy (Chapter 3.13 by L Nelsen titled The Activities and Roles of M.I.T. in Forming Clusters and Strengthening Entrepreneurship, p. 309).

14 Chapter 6.1 by L Nelsen titled Ten Things Heads of Institutions Should Know about Setting Up a Technology Transfer Office, p. 537.

FOR GOVERNMENT POLICYMAKERS

- ✔ There is an ongoing debate over how IP systems can achieve the optimal **balance between private rights and public benefits**. However, experience suggests that IP rights systems, if soundly applied and used by the public sector as well as by the private sector, are better than any of the proposed alternatives in achieving public-goods objectives. Government policies can be instrumental in helping public sector institutions find the right balance.

- ✔ As a specific requirement, public sector institutions should be required to develop and publish their **institute-specific IP policies** that adapt broader principles to the specific context of the institution's mission and strategy. Such policies should include clear conflict-of-interest policies.

- ✔ Incorporating the goals of dissemination of public-sector-generated R&D to benefit primarily the poorer segments of the population are goals that **do not run counter to benefiting economically** from inventions. Much will depend on the specific context and how these seemingly contradictory goals are managed.

- ✔ Public sector institutions can achieve little with their intellectual property in the absence of an enforceable **system for protecting and promoting local innovation** that includes clear assignment rules regarding ownership of inventions.

- ✔ A government may wish to analyze the interface between its **laws governing charitable organizations** and how the laws may impact the freedom of nonprofit institutions in owning and licensing both their own and third-party intellectual property.

- ✔ Policymakers should consider the promotion of legislation that clarifies under what circumstances employees in the public sector, including those at universities, shall **assign patent rights to their employers**. This aspect has ramifications for statutory laws in many countries with respect to "hired to invent," "shop-right," and other matters.

- ✔ Technology transfer and IP management are complex, requiring the creative input and participation of different professionals from varying fields of expertise. Therefore, it is important to recognize that **investments in public sector education and training programs** need to consider many aspects, including scientific, business and entrepreneurial, legal, judicial, and policy.

Given that IP management is heavily context specific, these Key Implications and Best Practices are intended as starting points to be adapted to specific needs and circumstances.

FOR SENIOR MANAGEMENT

(UNIVERSITY PRESIDENT, R&D MANAGER, ETC.)

- ✔ A sound **IP policy** should address, among others issues, clear ownership of intellectual property generated, conflicts of interest and conflicts of commitment, the manner in which an institution will handle confidential information, the principles of the institution's IP licensing approaches, how income arising from intellectual property will be distributed, and any rights the institution will retain (such as for research and for humanitarian uses).

- ✔ An **IP strategy**, on the other hand, describes the courses of action enabling the reaching of long-term goals of the institutions and the allocation of resources necessary to carry out these actions. Public sector institutions may wish to specifically address in their IP strategies how their research endeavors, in general, and IP management strategies, in particular, will achieve global access of their products and how the endeavors will benefit humanitarian objectives.

- ✔ Components of such an IP strategy may include how the institution deals with **incoming** third-party intellectual property, how it deals with **internally generated** intellectual property (patenting and other protection strategies that should include how the institution balances the public sector component of its mission with economic imperatives), and how it will **out-license** its intellectual property to third parties.

- ✔ Particular emphasis should be placed on **global access strategies**, not only because philanthropic funding agencies increasingly require grantees to address them, but because this approach is especially effective if an institution's aim is to diffuse technology as widely as possible.

- ✔ The **process by which an IP policy and IP strategy are developed** may be valuable in bringing about internal culture change and create strong support from staff.

- ✔ Successful IP commercialization is built on a foundation of good **relations between inventors and technology transfer professionals**. Such relationships should be established long before the transfer services of the technology transfer office.

- ✔ The **importance of IP audits** is becoming more and more apparent, in the private and in the public sector, as public entities increasingly deal with third-party intellectual property. IP audits can be useful mechanisms that form the basis for an internal review and revision of an institution's IP strategy and IP policy.

- ✔ Technology transfer invariably brings **conflicts of interest**. The challenge is to manage them in a transparent and consistent manner. Importantly, potential conflicts of interest should not be viewed in a negative light. Most real conflicts arise when **potential** conflicts are not disclosed.

FOR SCIENTISTS

- ✔ As the creator of inventions and technologies, **your role in technology transfer** is critical. So please read on!

- ✔ Your role can best be carried out if you have good relations with the technology transfer office and officers. But fulfilling your role also requires a good knowledge of and understanding of your institution's **IP policy**. The policy will likely articulate ownership of intellectual property, conflict of interest, the handling of confidential information, and more. Become familiar with the content and the meaning of the various provisions and how they may affect you.

- ✔ The purpose of such a policy, and more importantly of your institution's **IP strategy**, is not just to protect your inventions, but also to control technologies and IP assets in such a way as to allow you and the TTO to determine how your inventions can—and should—be used to spur economic growth and contribute to the greater public good.

- ✔ Remember, few inventions will lead to blockbuster products, make millions of dollars, or save billions of people. Have realistic expectations, especially regarding what it will take for your invention to make a difference. It is not bad to love your own creations as long as you have **realistic expectations**.

- ✔ More and more philanthropic donors expect to find IP management components in **grant applications** and to understand how intellectual property will be used to achieve global access and humanitarian benefits. This is just one reason why a close relationship with your TTO is important, and becoming even more so.

- ✔ When your institution conducts or commissions an **IP audit**, view this as an opportunity to better identify the intellectual property generated in your research program, to improve and streamline the management of third-party intellectual property (allowing you to concentrate more on research), and to contribute to the formulation and execution of an IP strategy that benefits your program and its (global) impact.

- ✔ One of the most important responsibilities you have is to disclose any **potential conflict of interest**. You are not guilty of anything if you have a potential, perceived, or even real conflict of interest. Most problems arise when conflicts are not disclosed. Clear conflict of interest policies that are followed and implemented in a transparent manner is all that is required to manage them.

FOR TECHNOLOGY TRANSFER OFFICERS

- ✔ An **IP policy** should address, at a minimum, ownership of intellectual property, conflicts of interest and conflicts of commitment, the handling of confidential information, the principles of IP licensing approaches, the sharing of income derived from intellectual property, and any rights the institution will retain (such as for research and for humanitarian uses).

- ✔ Public sector institutions will increasingly be expected to define an institutional **IP strategy** that specifically addresses how IP management will be used to achieve global access/humanitarian benefits of the inventions and products developed at your institution. It should include how the institution deals with **incoming** third-party intellectual property, how it deals with internally generated intellectual property, and how it will **out-license** its intellectual property to third parties.

- ✔ The **process by which an IP policy and IP strategy are developed** may be valuable in bringing about internal culture change and create strong support from staff.

- ✔ Successful IP commercialization is built on a foundation of good **relations between inventors and technology transfer professionals**. Such relationships should be established long before the establishment of transfer services of the technology transfer office.

- ✔ The **importance of IP audits** is becoming more and more apparent, in the private sector and even in the public sector, as public entities increasingly deal with third-party intellectual property. IP audits can be useful mechanisms that form the basis for an internal review and revision of an institution's IP strategy and IP policy.

- ✔ Technology transfer invariably brings **conflicts of interest**. The challenge is to manage them in a transparent and consistent manner without granting any exceptions, irrespective of the prestige of the scientist or the amount of funding they attract. Importantly, potential conflicts of interest should not be viewed in a negative light, provided they are disclosed (and managed). Most problems arise when **potential** conflicts are not disclosed. Few conflicts of interest are well managed by committees.

- ✔ All employees (and visitors in some cases) should be required to sign an **invention assignment agreement** on their date of arrival. Neither an employee handbook that discusses patent assignment nor a published university patent policy may be enough to ensure that the university is assigned ownership.

Given that IP management is heavily context specific, these Key Implications and Best Practices are intended as starting points to be adapted to specific needs and circumstances.

Establishing and Operating Technology Transfer Offices

Technology transfer is the process of converting scientific findings into useful products or services for society. A complex endeavor, it takes place in the broader context of innovation. Any country or institution can undertake health or agricultural innovation to varying degrees, but some developing countries that are more scientifically advanced are starting to reap the benefits of decades of investments in education, health research infrastructure, manufacturing and production capacity, and regulatory institutions. Increasingly referred to as **innovative developing countries**,[1] these countries are characterized by, among other things, sustained government support for research, the availability of venture capital, functioning regulatory systems, and an ability to partner with local and foreign public and private research organizations. All of this requires sound IP (intellectual property) management, which makes such partnerships more effective and allows technologies to be transferred not just in one direction but in more complex and valuable ways, to benefit more people.

Technology transfer is thus a rewarding process for research-based institutions and the people who make it happen. It leads to new products, services, and jobs. But it is also a multifaceted process with important policy, economic, and managerial ramifications. Discussing these aspects in detail, Nelsen,[2] who leads the Technology Licensing Office of the Massachusetts Institute of Technology (M.I.T.) offers practical, timely advice about some of the most important policy and strategy imperatives for an institution starting up a technology transfer office (TTO) or intending to strengthen its current endeavors in technology transfer.

Viable strategies to set up and operate **a TTO must be firmly grounded in realistic economic expectations**. Technology transfer will not really make your university or research institution rich because building a robust technology transfer program will take sustained financial investment. It takes time (eight to ten years) to build an IP portfolio, establish contacts, and develop skills in technology transfer. And it may take up to two decades or more before a university technology transfer program (including entrepreneurial spinouts) substantially affects the local economy. The ultimate impact, however, may be very large—both economically and culturally—for the university, its graduates, and the wider community.

Successfully implementing these plans will require visible and sustained support—fiscal and otherwise—from senior administration to set the program's mission, policies, and priorities. Clear mandates will help technology transfer professionals choose among competing priorities and the ever-present trade-offs between business and academic values. IP ownership policies, the roles of researchers in interactions with industry,

Krattiger A, RT Mahoney, L Nelsen, JA Thomson, AB Bennett, K Satyanarayana, GD Graff, C Fernandez and SP Kowalski. 2007. 6: Establishing and Operating Technology Transfer Offices. In *Executive Guide to Intellectual Property Management in Health and Agricultural Innovation: A Handbook of Best Practices* (Krattiger A, RT Mahoney, L Nelsen et al.). MIHR (Oxford, UK), PIPRA (Davis, USA), Oswaldo Cruz Foundation (Fiocruz, Rio de Janeiro, Brazil), and *bio*Developments-International Institute (Ithaca, USA). Available online at www.ipHandbook.org..

and other ground rules should be set up before the program begins because conflicts of interest, both real and perceived, are inevitable. For the same reason, clear policies and a well-understood review-and-appeal process need to be put in place early. Finally, **technology transfer is a talent-based business**. It is difficult to find people who can speak the two languages of academia and industry and who also have the creativity to craft agreements that meet the needs of both sides.

The chapter's conclusion discusses some technology transfer pitfalls caused by unrealistic expectations. It emphasizes the role of senior management in the evolution of the culture (which must begin with top-level administration), the need for transparent conflict-of-interest policies, and the importance of sufficient autonomy and infrastructure support for technology transfer officers. A TTO can create many benefits for the university, industry, and the surrounding community, but it requires carefully planned and consistent long-term financial and administrative support. And above all, **it requires TTO officers that are able—and willing—to take risks and university presidents to support them.**

Moving into the particulars of how to establish a TTO, Young[3] stresses the importance of a strong mission statement, attention to staffing needs, and to the unique operating contexts of each institution. Based on a lifelong experience in establishing and running TTOs, with a chapter that provides many examples of TTO launches from around the globe (Australia, India, China, Japan, England, South Africa, Russia, and the United States), Young concludes that **efficient and effective TTOs possess the following key characteristics:**

- An articulated mission
- Transparent policies and procedures
- Entrepreneurial staffing and an entrepreneurial environment
- Customer-friendly relations with both internal and external constituents
- A highly supportive university administration and community
- Strong links to potential industry partners
- Access to risk, or venture, capital

Even so, there is no "right" way to set up an office, but success does require considering some specific issues, as discussed by Campbell.[4] One of these issues is **establishing business processes at the outset**. Adequate attention should be paid to information management and realistically setting budgets. Offices tend to be either a department within the institution or a subsidiary company. Either way, accountability lines will need to be transparent. Like the preceding authors, Campbell also stresses that the **core element for successful technology transfer is people**. The TTO should be led by an individual who understands the details of running a business. It is also useful to have staff with experience working in the relevant business sectors. To be able to recognize new opportunities, the technology transfer manager needs to win the confidence of academics, which is why **it is helpful for the TTO to be embedded within the institution**. Likewise, staff should be exposed to both academics and business people.

Campbell discusses several examples of TTO structures and policies based on her experience in the United Kingdom, in particular King's College London (KCL). She also shares useful lessons from Switzerland's experiences with Unitechtra, a subsidiary nonprofit technology transfer company **jointly owned by the two universities of Bern and Zürich**. This model from a small, but highly innovative country (when the number of patents per capita is used as one measurement), is particularly relevant for developing countries with limited resources, where several institutions may consider establishing a joint TTO to ensure economies of scale and critical mass.

Fernandez,[5] from **Chile**, develops a specific, yet potentially powerful model for establishing or improving technology transfer operations for universities and research institutes within developing countries. The model takes into account several key insights about technology transfer. Namely, there is a critical mass of R&D activity necessary to justify the costs of a fully functioning TTO. Some estimates would put this figure within the range of US$100 to $500 million in research expenditures annually. While it is uncommon for a single university within a developing country to attain a financial critical mass, a group of universities together can attain it relatively easily. In addition, some of the typical functions of technology transfer are more easily scalable than others, and are thus more easily shared by a group of universities.

The model developed in this chapter essentially requires sharing the costs of technology transfer services among a consortium of universities, with an additional startup subsidy provided by government. **A "hub-and-spokes" configuration is proposed, which allows essential policy decisions and scalable functions to be moved to the center and keeps essential context-specific and unscalable functions on campuses at universities.** This also allows for a more efficient distribution of scarce resources and of key personnel who have the necessary skills, allowing a few experienced professionals to selectively, yet effectively, manage and mentor technology transfer staff across a range of institutions.

Fernandez openly points out the challenges of implementing such a model, emphasizing the need for articulated policies to be shared by all consortium members in at least three areas. These areas each represent potential points of real conflict within such multi-institutional systems. The first area is a clear **policy of ownership** that ensures that everyone involved in the process knows who bears ultimate responsibility for a given technology. The second is a clear policy on the **distribution of income** from commercialized technology. Such a policy should provide incentives that will elicit the cooperation and support needed from multiple players in the technology transfer process. This is also important so that realistic financial expectations are established, helping to avoid disputes that could threaten the viability of the system as a whole. The third area regards the **prevention and resolution of conflicts of interest.** Such a policy is important to maintaining the integrity of the university's main educational and research functions amid increased commercial opportunities, so that any disputes that do arise can be resolved in a systematic and fair manner, likewise protecting the viability of the system.

An important feature of this model is that it does not lock in its participating member institutions; it is **open to competition.** As the economy grows and member universities' R&D activities increase, their local TTOs can take on more and more of the functions that had been delegated to the central office. A university would thus be able to "graduate" from the system as its own TTO becomes self-sustaining. Even prior to that, the central TTO may be relied on only if it provides effective management of technology transfer projects that member institutions cannot replicate themselves or obtain elsewhere cost effectively. If the centralized system fails to be competitive, the member institutions can simply elect to manage the commercialization of technologies in their own offices or through other more competitive channels.

Moving into the **organizational aspects** of establishing a TTO, Dodds and Somersalo[6] review the specific requirements in terms of human infrastructure, physical infrastructure, and operational plans for the office. The latter is essential for defining office protocol for various topics: patents, trademarks, copyrights, and trade secrets; plant variety protection; contracts, agreements, and licenses; policy development; technology evaluation; invention marketing; conflict analysis; negotiation support; and strategy inputs. Importantly, any TTO must **emphasize the importance of confidentiality** in all its operations. Other wide-ranging organizational matters that need to be addressed early include; a coordinated *staffing plan* detailing authority, responsibilities, and work plans; a *staff employment handbook* that explains the ethical standards that employees must follow; a plan for addressing *governmental and state filing requirements*; a *tax plan* (including accounting standards and auditing); considerations to establish an *advisory panel*, and criteria for drawing on *external expertise* (such as consulting contracts, patent attorneys, general legal counsel, licensing specialists, marketing specialists, and database specialists).

The specific **organizational and administrative aspects** of a TTO are discussed by Hines,[7] based on her experience at Stanford University's Office of Technology Licensing. This chapter offers similar offices in developing countries a detailed outline of how an office can be structured in terms of personnel and human resources. The structure consists of a director, seven licensing associates, eight licensing liaisons, one copyright licensing specialist, and the equivalent of eight and one-half administrative staff, in addition to other administrative staff and industrial contracts officers. Much action in such an office revolves around the important technology licensing associates who work with the inventors (professors, graduate students, and research staff) and with prospective licensees. In addition to providing a list of key personnel for a TTO's operation, the chapter also defines, and gives examples of complex cases, job descriptions, and a comprehensive list of standard operating procedures (for licensing agreements and for invention disclosures). Like the other authors in this section, Hines stresses the importance of a well-trained staff.

Continuing and practical training should be an integral part of any TTO's yearly work plan. For this, Pefile and Krattiger[8] present a few focused case studies that can be incorporated into short courses in IP management. They stress the importance of **hands-on training programs** whereby participants play specific roles within difficult case studies tailor-made to serve the particular needs of the TTO or its staff. Such approaches allow participants to see how a specific professional role can affect the complex process of crafting beneficial and creative partnerships that lead to mutually beneficial solutions. Even for those not involved in deal making, this approach has great utility because it enables participants to view their respective tasks in a broader context, and thereby gain a perspective as to the challenges presented at various stages in the overall process.

Indeed, the **importance of building networks in the technology transfer and licensing community cannot be overstated.** To illustrate this, Hersey[9] uses the experience of the Association of University Technology Managers (AUTM) as an extended case study of the building of dynamic, productive, and sustainable networks. Her chapter contains numerous lessons applicable anywhere. Networking among peers in any profession is generally understood to be beneficial, as it cements relationships between individual practitioners and helps build and strengthen the profession itself. By working through networks, practitioners exchange ideas and experiences, forming best practices that become performance standards for both individuals and their institutions. Networks thereby contribute to IP management capacity building at both the individual and institutional levels, and this then feeds back to further support and expands the network.

As groups of like-minded, mission-driven professionals, networks can be formed at different geographical levels in order to serve various functions. This multilevel approach allows organizations to address different aspects of their respective missions. *Local networking* creates opportunities to work with colleagues in the immediate vicinity. *National networking* can be a mechanism for working with colleagues to encourage national legislation addressing intellectual property and technology transfer, as well as for designing and implementing systems for appropriate IP management, training, and education. *Regional networking* provides opportunities to work with neighboring countries in coordinated R&D endeavors and related IP management and technology transfer initiatives. Importantly, *international networking* will become increasingly important as globalization advances. Building networks with colleagues around the world will provide opportunities for many forms of technology transfer and IP management capacity building.

This is particularly relevant in the context of TTOs working with external patent counsel as developing countries increasingly wish to file patents in other jurisdictions. From a licensee point of view, having patent coverage in the more lucrative markets of the United States, Europe, and Japan may be a prerequisite for licensing a technology from a developing country. Similarly, if a patent is drafted poorly or does not provide adequate coverage for the technology, licensing opportunities may either be lost or significantly devalued. The costs associated with inadequately drafted patents can be significant, so it is important for a TTO to carefully select a patent attorney whose work will enhance the institution's prospects for obtaining optimal licensing arrangements.

Goldman[10] reviews the **process of selecting, hiring, and interfacing with patent counsel.** Of course, central to this relationship is ensuring that patent counsel can prepare and prosecute patent applications in a manner that achieves positive results in a cost-effective fashion. The chapter presents the steps in this complex process, and the responsibilities that both counsel and the TTO should assume. Patent attorneys can also provide general counseling, resolve inventorship issues, provide licensing and agreement support, and resolve disputes. By selecting a qualified patent attorney and developing a good working relationship, a TTO can develop a resource that will ease the workload and facilitate its missions. This choice of patent counsel is therefore essential for operating a viable TTO and should be approached thoughtfully. There are several factors that should be carefully considered and weighed:

- size of the attorney's firm[11]
- scope of the attorney's legal experience and capabilities
- the attorney's experience with academic institutions
- the attorney's technological background
- the firm's location

After a patent attorney is selected, determining how work will be allocated is important. Generally, the less work that is sent to the attorney, the lower the TTO's legal fees. Still, the more work the TTO retains for itself, the less time its staff will have for other matters. Another critical aspect of the relationship between the patent attorney and the TTO is payment for services. The chapter presents several possible methods and schedules for payment, and also cautions against certain related practices. For example, a letter of retainer can, among other things, specify billing procedures, such as fixed fees, hourly billing rates, and equity combinations. Another feature of the retainer letter will be a specification of the bill content: an acceptable bill will include an indication of which attorney or attorneys worked on a particular project, the amount of time spent daily on that project, and what that work involved. This will make clear the services for which the TTO is being charged.

Dodds[12] considers a broader picture of **how to hire an IP lawyer without going bankrupt**. While the process can be complex and costly, Dodds outlines various strategies for how TTOs can make the best use of attorneys. He points out the value of retaining an IP attorney, especially when a TTO is just getting established. A critical initial role of the lawyer should be to work closely with the TTO to develop an IP strategy that most effectively delivers benefits to the office. If your TTO strategy is ill conceived, all the remaining activities will be irrelevant.

Any emerging TTO will have a wide range of legal matters to be addressed. These include the types of IP protection to be provided, when to apply trademarks and copyright, how much to rely on trade secrets, the development of contracts and agreements, license reviews, and negotiations support. The lawyer can also be used to think in innovative ways about how to capture value from an IP portfolio.

Importantly, the legal relationship between a lawyer and a client is protected under a special set of legal rules that make up the concepts of **client confidentiality and legal privilege**. This umbrella of confidentiality and legal protection from disclosure is a very important part of the relationship. The importance of confidentiality and trust cannot be underestimated. Such confidentiality requires excellent record keeping.

Indeed, a TTO needs to have **a systematic way of managing agreements and many other forms of data** as the amount of data will increase significantly and year by year. Unfortunately, too many TTOs still try to accomplish this task with a paper filing system, which is cumbersome, slow, and inflexible. Above all, this type of system severely limits the ability to analyze data creatively. Using electronic systems, a manager can rapidly formulate questions that in a physical file environment would be unthinkable, due to the time required to locate, assemble, and analyze the information sets.

Electronic filing systems also provide shared communication links and can utilize advanced spreadsheet applications. The chapter by Sloman[13] considers **the relative merits of spreadsheets, flat-file databases, and relational databases as TTO data management tools/systems.** It emphasizes the benefits of the latter, highlighting both their ability to transfer entire projects from one manager to another with the click of a button and their unprecedented power to allow managers to look at data and business models in creative ways. Such a system requires less data entry and can be easier to maintain and audit. For all of these reasons, the relational database is frequently the preferred system.

The sooner **a functional contract management system** is implemented, the easier it will be to keep track of contracts and make the most of them, both for the organization and for its collaborators. Two chapters, one by Hamzaoui[14] and one by Potter and Rygnestad[15] discuss the importance, design, and implementation of contract management systems from different, but equally pragmatic perspectives. Hamzaoui bases her chapter on the practices of the Whitehead Institute for Biomedical Research. Both chapters review the specific approaches for actually implementing a contract management system, including:

- accessibility (e.g., hard-copy filing, electronic filing, database systems)
- security (e.g., loss prevention, unauthorized access)
- resources for implementation
- personnel time, training, and management

The value of the proper management of contracts and agreements is usually only seen in its absence—lost deals, a poor reputation and, in the worst case, lawsuits. Early investments to prevent these sorts of problems are like any prophylactic measure and the savings will certainly be substantial.

Graciously, **the Whitehead Institute agreed to make available, for free, their proprietary agreement management system, called *WIIPS*™, through the online version of the *Handbook*.**[16] Users can download a fully functional version that they can also modify and adjust to their particular institutional needs. *WIIPS*™ is a relational database designed to automate essential IP management and technology transfer functions. It simplifies record keeping and generates useful reports for technology disclosures, patent applications, joint invention agreements, licenses, and material transfer agreements. In addition, the system stores essential information on every inventor, owner, and licensee who has interacted with a given TTO.

Finally, the chapter by Pefile[17] takes a broad view and considers the **mission of a TTO in the context of knowledge transfer**. Indeed, making money will always be a consideration when setting objectives, but technology transfer adds value in other important ways; as a resource to facilitate innovation for the public good and as a way to broker the exchange of knowledge between the business and public sectors for society's benefit. Transferring knowledge across such disciplines as the humanities, law, and social sciences is as important as transferring knowledge and technology across the applied sciences, and TTOs should be set up to have the flexibility to accomplish this broader knowledge-transfer objective.

An effective evaluation system should strengthen an institution's ability to maintain leadership across the frontiers of scientific knowledge. The evaluation system also will stimulate partnerships that promote investments in fundamental science and engineering, as well as the overall more-effective use of physical, human, and financial resources for social and economic benefit. Without a measurement process, institutions cannot justify their efforts in R&D, IP management, commercialization, and technology transfer in relation to their economic and social goals. Finally, Pefile calls upon all TTO managers to take the time to reflect upon their operations and ways in which they can be made more effective and beneficial for all. ■

All chapters refer to: *Intellectual Property Management in Health and Agricultural Innovation: A Handbook of Best Practices*. 2007. A Krattiger, RT Mahoney, L Nelsen, JA Thomson, AB Bennett, K Satyanarayana, GD Graff, C Fernandez, and SP Kowalski (eds.). MIHR: Oxford, U.K., and PIPRA: Davis, U.S.A. Available online at www.ipHandbook.org. The online version contains for each chapter a detailed Editor's Summary, Implications, and Best Practices.

1 Morel CM, T Acharya, D Broun, A Dangi, C Elias, NK Ganguly, CA Gardner, RK Gupta, J Haycock, AD Heher, PT Hotez, HE Kettler, GT Keusch, AF Krattiger, FT Kreutz, S Lall, K Lee, R Mahoney, A Martinez-Palomo, RA Mashelkar, SA Matlin, M Mzimba, J Oehler, FG Ridley, P Senanayake, P Singera and M Yun. 2005. Health Innovation Networks to Help Developing Countries

Address Neglected Diseases. *Science* 309:401-404.

2 Chapter 6.1 by L Nelsen titled Ten Things Heads of Institutions Should Know about Setting Up a Technology Transfer Office, p. 537.

3 Chapter 6.2 by TA Young titled Establishing a Technology Transfer Office, p. 545.

4 Chapter 6.3 by AF Campbell titled How to Set Up a Technology Transfer Office: Experiences from Europe, p. 559.

5 Chapter 6.4 by C Fernandez titled How to Set Up a Technology Transfer System in a Developing Country, p. 567.

6 Chapter 6.5 by J Dodds and S Somersalo titled Practical Considerations for the Establishment of a Technology Transfer Office, p. 575.

7 Chapter For 6.6 by S Hines titled Administration of a Large Technology Transfer Office, p. 581.

8 Chapter 6.7 by S Pefile and A Krattiger titled Training Staff in IP Management, p. 597.

9 Chapter 6.8 by K Hersey titled Building Networks: The National and International Experiences of AUTM, p. 617.

10 Chapter 6.9 by ML Goldman titled How to Select and Work with Patent Counsel, p. 625.

11 Due to its resources and personnel, a large law firm is generally able to handle most legal problems that confront a technology transfer office. On the other hand, smaller firms might have the advantage of lower cost, while still having attorneys with the skills needed to serve the technology transfer office.

12 Chapter 6.10 by J Dodds titled How to Hire an IP Attorney and Not Go Bankrupt, p. 635.

13 Chapter 6.11 RG Sloman titled Technology Transfer Data Management, p. 641.

14 Chapter 6.12 by A Hamzaoui titled WIIPS™: Whitehead Institute Intellectual Property System (A Relational Database for IP Management and Technology Transfer) p. 649.

15 Chapter 6.13 by R Potter and H Rygnestad titled Organizing and Managing Agreements and Contracts, p. 651.

16 www.ipHandbook.org. In order to run the database, users must have access to Microsoft® Access® and a Windows XP (or higher) operating system. Users are authorized to copy and/or modify both the software and its user guide, as long as they acknowledge that the resulting product is based on *WIIPS*™.

17 Chapter 6.14 by S Pefile titled Monitoring, Evaluating, and Assessing Impact, p. 659.

FOR GOVERNMENT POLICYMAKERS

- ✔ Technology transfer is the process of converting scientific findings into useful products or services for society. Hence **encouraging public institutions and private sector enterprises** to work together is an important element in any national strategy aimed at strengthening innovation.

- ✔ In the increasingly interlinked and globalized worlds of science, technology, and commerce, such collaborations should **extend beyond national borders**, as success will increasingly be measured by the ability to form dynamic, integrated, and mutually beneficial networks that span countries and institutions.

- ✔ Government policies and laws regarding technology transfer ought to be **flexible** so that each institution can shape its approach according to its own culture, mission, and context.

- ✔ Laws regarding **IP ownership** are essential for successful technology transfer.

- ✔ National institutions often require **governmental encouragement and sustained funding** to develop technology transfer offices (TTOs), as well as IP policies (conflict of interest management, allocation of revenues, and so forth).

- ✔ Viable strategies to set up and operate **a TTO must be firmly grounded in realistic economic expectations**. Technology transfer will not make any institution rich because building a robust technology transfer program will take sustained financial investment. It takes time (ten years or more) to build an IP portfolio, establish contacts, and develop skills in technology transfer. And it may take up to two decades or more before a university technology transfer program (including entrepreneurial spinouts) substantially affects the local economy.

- ✔ A certain **critical mass of R&D activity is necessary to justify the costs of a fully functioning TTO**. Some estimates would put this figure within the range of US$100 to $500 million in research expenditures annually.

- ✔ Several alternative models to an institutional TTO can be successful. **Costs can be shared among a consortium of universities or research institutions**. Such hub-and-spokes configurations allow essential policy decisions and scalable functions to be moved to the center, while keeping essential context-specific and unscalable functions embedded within individual institutions. This allows for a more efficient distribution of scarce resources and of key personnel who have the necessary skills, allowing a few experienced professionals to selectively, yet effectively, manage and mentor technology transfer staff across multiple institutions.

- ✔ Recognizing that **technology transfer is a talent-based business**, the importance of building networks in the technology transfer and licensing community cannot be overstated. Governments should encourage the creation and operation of national technology transfer associations that concurrently build international linkages.

Given that IP management is heavily context specific, these Key Implications and Best Practices are intended as starting points to be adapted to specific needs and circumstances.

FOR SENIOR MANAGEMENT

(UNIVERSITY PRESIDENT, R&D MANAGER, ETC.)

- ✔ Successfully establishing and operating a technology transfer office (TTO) will require **visible and sustained support—financial and otherwise—from senior administration**, which can set the program's mission, policies, and priorities. Clear mandates will help technology transfer professionals choose among competing priorities.

- ✔ A TTO can create many benefits for the university, industry, and the surrounding community, but it requires carefully planned and consistent long-term financial and administrative support. And above all, **it requires TTO officers able—and willing—to take risks and senior management to back them**.

- ✔ Efficient and effective TTOs must have an articulated TTO mission, transparent TTO policies and procedures, entrepreneurial staffing and an entrepreneurial environment, customer-friendly relations between TTO staff and internal and external constituents, a **highly supportive administration**, strong TTO links to potential industry partners, and TTO access to risk, or venture, capital.

- ✔ The **core element for successful technology transfer is people**. The TTO should be led by an individual who understands the details of running a business. Additionally, staff members with experience working in the relevant business sector are required.

- ✔ An important factor for a successful TTO is **the institution's entrepreneurial culture**.

- ✔ Strategies to set up and operate **a TTO must be firmly grounded in realistic economic expectations**. Technology transfer will not make any institution rich because building a robust program will take sustained financial investment. It takes time (ten+ years) to build an IP portfolio, establish contacts, and develop skills in technology transfer.

- ✔ **A critical mass of R&D activity is necessary to justify the costs of a fully functioning TTO**. Some estimates would put this figure within the range of US$100 to $500 million in research expenditures annually.

- ✔ Several alternative models to an institutional TTO can be successful. **Costs can be shared among a consortium of universities or research institutions**. Such hub-and-spokes configurations allow essential policy decisions and scalable functions to be moved to the center, while keeping essential context-specific and unscalable functions embedded within individual institutions.

- ✔ Implementing a consortium model of a TTO across institutions presents many challenges. These can be managed with clearly articulated policies of **ownership**, the **distribution of income** from commercialized technology, and mechanisms for the **prevention and resolution of conflicts of interest**.

- ✔ An important feature of this model is to **allow for a certain level of competition**, a locally embedded TTO officer, and an evolution of the model. As the member institution's R&D activities increase, local TTOs can take on more and more of the functions that had been delegated to the central office.

Given that IP management is heavily context specific, these Key Implications and Best Practices are intended as starting points to be adapted to specific needs and circumstances.

FOR SCIENTISTS

- ✔ Your power to **shape institutional policy** should not be underestimated, especially in the ways that the fruits of your research can be made to increase economic development and benefit humanity.
- ✔ Know your institutional **conflict of interest policy**. Most conflict of interest issues arise when procedures are not properly followed.
- ✔ Work with your TTO to ensure that your institution's **disclosure of information** form is simple and easy for you to use.
- ✔ Understand why you might benefit from **engaging in technology transfer** and what you want to get out of the relationship with the TTO.
- ✔ A national or regional consortium of universities to develop a technology transfer system could be beneficial to you and your colleagues because it would be more cost effective and would have **greater latitude and leverage** in exploiting commercialization opportunities than would a single campus office.
- ✔ Establishing **networks** among colleagues will increase your awareness of opportunities and also help you understand the broader implications of your research.
- ✔ Keep your TTO informed about your networking activities, particularly if there is a possibility of **shared research endeavors**. These collaborative research projects often form the foundation of networks for technology transfer and licensing opportunities.

Given that IP management is heavily context specific, these Key Implications and Best Practices are intended as starting points to be adapted to specific needs and circumstances.

FOR TECHNOLOGY TRANSFER OFFICERS

- ✔ You have a duty to ensure that senior management understands that a successful technology transfer office (TTO) requires **visible and sustained support, financial and otherwise**. Work with senior management on the definition of clear mandates that will help you choose among competing priorities and the ever-present trade-offs between business and academic values.
- ✔ Above all, ensure that senior management knows that **it requires TTO officers who are able and willing to take risk and senior management to support you**.
- ✔ Efficient and effective TTOs must have an articulated TTO mission, transparent TTO policies and procedures, entrepreneurial staffing and an entrepreneurial environment, customer-friendly relations between TTO staff and internal and external constituents, a **highly supportive administration**, strong TTO links to potential industry partners, and TTO access to risk, or venture, capital.
- ✔ One of the most important factors for a successful TTO is **the institution's entrepreneurial culture**. This is determined most often by the attitude and degree of support from senior management.
- ✔ A TTO must **emphasize the importance of confidentiality** in all its operations.
- ✔ Any TTO needs to have, from the outset, **a systematic way of managing agreements and many other forms of data** as the amount of data will increase significantly and year by year. The sooner a functional contract-management system is implemented, the easier it will be to keep track of contracts and make the most of them, both for the organization and for its collaborators.
- ✔ The Whitehead Institute's **proprietary agreement management system** (called *WIIPS*™) may constitute a viable software option for emerging and established TTOs. *WIIPS*™ **can be downloaded for free** from the online version of the *Handbook*.
- ✔ The importance of continued **hands-on training programs** of TTO staff cannot be overstated.
- ✔ Similarly, the **importance of building networks in the technology transfer and licensing community is critical**. By working through networks, practitioners exchange ideas and experiences, forming best practices that become performance standards for both individuals and their institutions.
- ✔ When **recruiting personnel** to staff your office, consider key qualifications. The importance of having the best professionals working for you cannot be overstated. Staffing can have a significant impact on the success of your office.
- ✔ Any TTO will have a wide range of legal matters to be addressed, and procedures for **working with external patent counsel and general counsel** should be well established. Make sure you are in, and stay in, the driver's seat.

Given that IP management is heavily context specific, these Key Implications and Best Practices are intended as starting points to be adapted to specific needs and circumstances.

Contracts and Agreements to Support Partnerships

A contract follows the deal that has been made between two or more parties. Put differently, form (contract) follows function (deal). And any deal has a purpose: it is a means of transferring *value* between parties. Ideally, the trust between partners permeates their relationship and interactions, from the negotiations of an agreement, through project implementation, and into future agreements, collaborations, and projects. However, in practice, circumstances are rarely perfect. So **negotiating agreements should be seen as an initial step toward longer-term, productive, and mutually beneficial relationships.**

There are many types of agreements, as noted and discussed by Mahoney and Krattiger,[1] and those **related to collaborations** fall into one of four broad categories:

- *Confidentiality agreements* protect confidential information from disclosure to third parties.
- *Materials transfer agreements* (MTAs) protect samples (tangible property) from misuse by or unauthorized distribution to third parties.
- *Co-development agreements* and *collaboration agreements* outline the specific contributions of different parties who work toward a mutual goal.
- *Patent licenses* and *technology licensing agreements* allow one party to further develop, use, make, or sell the patented (and/or trade-secret-protected) technology of another party. Patent licenses may be specific to one or several patents. Technology licenses usually include the transfer of know-how (which may or may not be a trade secret) and sometimes materials. Such licenses may provide for the further development of a technology or limit production/manufacture of a good or provision of a service. Commercialization licenses are discussed in part 12 of the *Executive Guide*: Dealmaking and Marketing Technology to Product-Development Partners.

In addition, parties may engage in research agreements and distributorship agreements that contain elements of the four types of agreements listed above. Many agreements, such as MTAs contain confidentiality provisions; patent/technology licenses often contain confidentiality and material transfer provisions as well. But certain standard elements are integral to most agreements:

- *recitals, preamble, and "whereas clauses"* that lay out the broad motivations and goals of the agreement
- *a list of the parties* entering into the agreement
- *definitions* of terms used in the agreement
- *confidentiality clauses*
- *territory and exclusivity clauses* that define the geographic regions in which the licensee is permitted to make, use, and/or sell the technology in question

Krattiger A, RT Mahoney, L Nelsen, JA Thomson, AB Bennett, K Satyanarayana, GD Graff, C Fernandez and SP Kowalski. 2007. 7: Contracts and Agreements to Support Partnerships. In *Executive Guide to Intellectual Property Management in Health and Agricultural Innovation: A Handbook of Best Practices* (Krattiger A, RT Mahoney, L Nelsen et al.). MIHR (Oxford, UK), PIPRA (Davis, USA), Oswaldo Cruz Foundation (Fiocruz, Rio de Janeiro, Brazil), and *bio*Developments-International Institute (Ithaca, USA). Available online at www.ipHandbook.org.

- *liability clauses* that establish who will accept liability for a product and that set out terms of liability
- *payment clauses* that define the forms of payment, if any, of up-front fees and/or royalties (Terms will ideally balance the licensor's need for short-term income and the licensee's capacity to make further investments for longer-term development.)
- *an arbitration clause* that establishes how disputes will be handled
- *term and termination*, establishing how long the agreement will last and under what conditions the agreement may be terminated
- *jurisdiction, warranties, and notices*, which specify (1) where disputes are to be resolved, (2) that the licensor does, in fact, own the intellectual property to be licensed, and (3) where official communications are to be directed
- *illegal/unenforceable provisions* that specify which terms can be discontinued due to invalidity without rendering the agreement void
- *subject law*, which specifies where the parties wish to have the agreement interpreted and adjudicated
- *signatories* (These are representatives, called agents, who have the authority to bind their respective organizations, called principals, to the terms and provisions of the agreement.)

Public sector research institutions can use a variety of agreements to protect and manage intellectual property. These agreements are powerful tools to foster competition in the private sector and reduce prices for consumers in developing countries. The authors emphasize **the importance of establishing and maintaining trust when negotiating and implementing agreements.**

Although no agreement will ever be perfect, there are good and not-so-good agreements (and poorly written and highly ineffectual agreements). **The best agreements are generally those that do not use technical or legal jargon**, that use short, clear sentences free of vague adjectives and written in active voice. Business people (who have extensive technical knowledge but generally limited legal knowledge) and judges (who generally have limited technical knowledge but extensive legal knowledge) should find such documents accessible.

Agreements that serve the parties best may take longer to negotiate, but each time two parties have successfully developed an agreement, the development of subsequent agreements should be easier. Taking time to think through and discuss the terms of an agreement fosters communication between the partners, although often template agreements[2] can be used as starting points in building trust.

During collaborations, and in some types of licensing agreements, materials and confidential information are passed from one party to another. **Confidentiality, or nondisclosure, agreements are contracts that govern the disclosure of confidential information by one party to another party and can be useful for building trust.** Kowalski and Krattiger[3] explain that disclosures may be unilateral, bilateral, or multilateral, and confidential information is valuable precisely because it is not known to business competitors or to the public. Key provisions in a confidentiality agreement include:

- *definition* of the information
- important *exceptions*, which describe circumstances under which the obligation of confidentiality is inapplicable (for example, the information was already in the public domain or is commonly known)
- *conditions* on the use of the confidential information, which is a detailed description of the ways in which the receiving party may and may not use the information
- *requirements for documentation*, which describe the requirements for written records (may include keeping track of the disclosed information, whether disclosed in writing or orally)

An organization that enters into a confidentiality agreement must ensure that all who have access to the confidential information understand that they must keep sensitive information confidential.

Traditionally, scientists have freely shared information as well as research materials. However, with fundamental research and commercial development merging ever closer, in both health and agricultural research, materials that once would have been used exclusively for fundamental research increasingly have direct commercial value. Therefore, universities, for-profit corporations, and nonprofit research institutions now realize that they must obtain proprietary protection for their research materials. Unrestricted transfers of research materials between scientists are becoming less and less common, particularly transfers between scientists in industry and those in academia. Biological materials are transferred still, though not as freely as before, but now with conditions attached as part of MTAs. Increasingly important in the life sciences, **MTAs delineate the terms under which tangible materials are transferred between two or more parties.** Technically, MTAs are bailments because they involve the transfer of possession but not of title. In other words, the party that transfers the materials retains full ownership, and the party that receives the materials holds them "in trust"; an analogy for such a transfer might be the act of leaving a watch at a watch-repair shop or a suit with a dry cleaner.

Bennett, Streitz, and Gacel[4] explain that an MTA specifies the term of a transfer, delineates how materials may and may not be used, and provides for other related issues such as confidentiality. An MTA may also contain licensing provisions for the transfer of embedded IP rights (such as patent rights). Thus, an MTA can be a hybrid instrument, covering the transfer of both tangible property (via bailment and contract) and intangible intellectual property (via licensing of patent rights). Thus, MTAs can be quite complex. Besides the usual clauses included in MTAs, the following are the perhaps the most critical to consider, especially for public sector institutions:

- **reach through clauses**, which describe the extent to which the supplying party may "reach"' into the research and into new intellectual property or material generated from work with the supplied materials
- **derivatives**, clauses that explain who will own modifications if the receiving institution makes modifications to the material

Particular consideration should be given to these all-important clauses, and receiving institutions must understand the implications of such clauses if indeed they are included. As a result, material transfers between private- and public-sector institutions are typically much more complex than MTAs between two universities. It is less problematic for universities to transfer materials than it is for materials to be transferred between industry and academia. If a problem does occur under a transfer arrangement, it is usually because IP rights attached to the materials transferred have been exclusively licensed and the terms of that agreement impose constraints on the institution providing the material.

Collaborative research agreements and sponsorship agreements are generally more complex than MTAs. A collaborative research agreement, for example, often involves multiple partners (who are increasingly a mixture of private- and public-sector actors) working together on a research project. The partners each contribute an amount of money, talent, and technology into a central pool that all draw from. Chapters by Bair Steinbock[5] and Gold and Bubela[6] explain how to write collaborative research agreements and covering myriad issues, including licensing provisions. The authors point out that the following elements are the most critical in these agreements (besides the usual must-haves for any agreement):

- *statement of objectives,* which explains what the parties want to accomplish together and why their collaboration is important
- *statement of work,* which explains the research plan, outlining approaches and methodologies, specifying who will be responsible for work product, and delineating time frames, benchmarks, and delivery dates
- *work plan* that specifies what each party will be expected to contribute, how necessary changes to the work plan will be made, and how communication between the parties is to take place
- *dispute resolution plan,* which explains procedures and mechanisms that would be used in turn, should a dispute arise

Focusing on patent licensing, Krattiger[7] points out that not every patent needs to be licensed per se. A special form of agreement—which can also take the form of a public statement—is the nonassertion covenant (nonassert). Such an agreement certifies that a party or parties in possession of intellectual property will not assert and defend IP rights (typically patents). Such nonasserts can be used in a broad range of IP management scenarios. For example, **nonassertion covenants are particularly useful for granting developing countries access to essential innovations in health and agriculture**, since such agreements offer simple and effective ways of dealing with three major constraints common in agri-biotechnology transfer and licensing:

1. Nonassertion covenants can be used to circumvent liability associated with licensing.
2. Nonassertion covenants can make research tools available.
3. Transaction costs can be reduced because such costs associated with nonassertion covenants are lower than those associated with bilateral and multilateral licensing agreements.

From a legal perspective, nonasserts are preemptive patent-infringement settlement agreements that are designed and drafted with the purpose of resolving future infringement disputes. There are no compelling reasons why nonasserts could not become more widely used to foster important advances and innovation that address needs in developing countries.

A patentee's public declaration of nonenforcement of a patent via a nonassert can have wide-ranging implications in terms of enhancing public sector R&D. This would be the case especially with patent rights covering research tools, and particularly in the United States, due to limitations on research exemptions. These are critical for accelerating the development of essential biotechnological applications in both the health and agricultural industry sectors. Carefully drafted, targeted nonasserts permitting the use of these tools—anywhere in the world—to address humanitarian needs (including in a commercial setting) could have broad-ranging and significant positive impacts. The approach reduces transaction costs, encourages innovation to help the poor, and accomplishes this without much cost, time, or loss of commercial opportunity. ■

All chapters refer to: *Intellectual Property Management in Health and Agricultural Innovation: A Handbook of Best Practices.* 2007. A Krattiger, RT Mahoney, L Nelsen, JA Thomson, AB Bennett, K Satyanarayana, GD Graff, C Fernandez, and SP Kowalski (eds.). MIHR: Oxford, U.K., and PIPRA: Davis, U.S.A. Available online at www.ipHandbook.org. The online version contains for each chapter a detailed Editor's Summary, Implications, and Best Practices.

1 Chapter 7.1 by RT Mahoney and A Krattiger titled Agreements: A Review of Essential Tools of IP Management, p. 675.

2 The online version of the *Handbook* offers 16 confidentiality agreements, 22 MTAs, 25 collaboration agreements of various types, 39 licenses of various types, 17 other types of agreements ranging from nonasserts to consultancy agreements, and several IP-related clauses in employment agreements. These are real agreements from various institutions around the world that can be used as starting points for generating template agreements. Note, however, that whereas template agreements can be useful, they should be used cautiously. They are seeds for further discussion and negotiation. No generic template will be appropriate in every cultural and legal climate.

3 Chapter 7.2 by SP Kowalski and A Krattiger titled Confidentiality Agreements: A Basis for Partnerships, p. 689.

4 Chapter 7.3 by AB Bennett, WD Streitz and RA Gacel titled Specific Issues with Material Transfer Agreements, p. 697.

5 Chapter 7.4 by M Bair Steinbock titled How to Draft a Collaborative Research Agreement, p. 717.

6 Chapter 7.5 by ER Gold and T Bubela titled Drafting Effective Collaborative Research Agreements and Related Contracts, p. 725.

7 Chapter 7.6 by A Krattiger titled The Use of Nonassertion Covenants: A Tool to Facilitate Humanitarian Licensing, Manage Liability, and Foster Global Access, p. 739.

FOR GOVERNMENT POLICYMAKERS

- ✔ A public sector institution can use a variety of agreements to both manage and protect intellectual property, regardless of whether that intellectual property is owned by the public sector institution or by licensing partners in the private sector. The key issue is to allow for **maximum flexibility whereby institutions can set, or negotiate, the terms that best fit the mission and goals of the institution** and the purpose of the partnership.

- ✔ It is important to encourage partnerships that accelerate the development and use of new technologies, whether they are domestic or foreign, and to **provide support and encouragement during negotiation in the form of tangible commitments to capacity-building**, as well as to broader IP management training in patenting, licensing, and technology transfer, for example.

- ✔ **Confidentiality agreements** are meant to protect sensitive data that one party transfers to another. They do not run counter to public sector missions or to publishing important research findings. Many organizations, including public sector institutions, often have information that is legitimately kept confidential. Such information can include business plans, research proposals, and databases containing business contacts.

- ✔ **Confidentiality agreements rely on a culture of trust, not a culture of secrecy.**

- ✔ Predictably enforced and fairly construed contract laws will greatly facilitate the formation and enforcement of contracts. A functioning court system is essential to encouraging partnerships. Indeed, suppliers of biological materials (and of confidential information and intellectual property) will be encouraged to enter into agreements if the suppliers are confident that their property rights will be protected and that agreements will be enforced. Such **confidence fosters collaborative research, and drives international collaboration**.

- ✔ It is important that the courts adjudicate contract disputes efficiently and fairly because the quality of the judicial system will influence the **quality and quantity of a country's international partnerships and agreements** and also will influence the complexity and sophistication of technologies transferred to and from a given country.

Given that IP management is heavily context specific, these Key Implications and Best Practices are intended as starting points to be adapted to specific needs and circumstances.

FOR SENIOR MANAGEMENT

(UNIVERSITY PRESIDENT, R&D MANAGER, ETC.)

- ✔ A public sector institution can use a variety of agreements to both manage and protect intellectual property, regardless of whether that intellectual property is owned by the public sector institution or by licensing partners in the private sector. The key issue is to allow for **maximum flexibility whereby institutions can set, or negotiate, the terms that best fit the mission and goals of the institution** and the purpose of the partnership.

- ✔ It is important to encourage partnerships that accelerate the development and use of new technologies, whether they are domestic or foreign, and to **provide support and encouragement during negotiation in the form of tangible commitments to capacity-building**, as well as to broader IP management training in patenting, licensing, and technology transfer, for example.

- ✔ **Confidentiality agreements** are meant to protect sensitive data that one party transfers to another. They do not run counter to public sector missions or to publishing important research findings. Many organizations, including public sector institutions, often have information that is legitimately kept confidential. Such information can include business plans, research proposals, and databases containing business contacts.

- ✔ **Confidentiality agreements rely on a culture of trust, not a culture of secrecy.**

- ✔ No agreement will ever be perfect. Technology transfer officers who negotiate agreements that are in keeping with an institution's policy, ought to be given **full support by senior management**, especially when deals are criticized from outside of the organization.

- ✔ Senior management can be instrumental by signing off on certain **template agreements** that can be used as a basis for negotiating deals. But a template agreement should be used *only* as a starting point for discussion.

- ✔ Any institution should have clear guidelines stating **who is authorized to sign agreements.**

Given that IP management is heavily context specific, these Key Implications and Best Practices are intended as starting points to be adapted to specific needs and circumstances.

FOR SCIENTISTS

- ✔ A public sector institution can use a variety of agreements to both manage and protect intellectual property, regardless of whether that intellectual property is owned by the public sector institution or by licensing partners in the private sector. The key issue is to allow for **maximum flexibility whereby institutions can set, or negotiate, the terms that best fit the mission and goals of the institution** and the purpose of the partnership.

- ✔ **Confidentiality agreements** are meant to protect sensitive data that one party transfers to another. They do not run counter to public sector missions or to publishing important research findings. Many organizations, including public sector institutions, often have information that is legitimately kept confidential. Such information can include business plans, research proposals, and databases containing business contacts.

- ✔ **Confidentiality agreements rely on a culture of trust, not a culture of secrecy.**

- ✔ **No agreement will ever be perfect**. Technology transfer officers who negotiate agreements that are in keeping with an institution's policy are making their best effort at getting deals that respect and strengthen the institution's mission. Your role, however, is to share with those who negotiate agreements all of the relevant information and your insights. In some cases, especially with collaborative research agreements, you may be an integral member of a team that will address issues such as research plans and purpose.

- ✔ In most cases, you will not be authorized to sign certain types of agreements without review by counsel or by your technology transfer office. Make sure you know **whether or not you are authorized to sign certain agreements**.

- ✔ Everyone in your group or laboratory should know—and understand—the **obligations entered into through certain agreements** that affect information, data, and materials used in your laboratory and research program. This is especially important for material transfer agreements and confidentiality agreements.

- ✔ You will need to keep track of data and information related to **confidentiality agreements**. Understand what can and cannot be disclosed and to whom information can be disclosed. If you have questions, do not hesitate to contact your technology transfer office for guidance.

- ✔ An MTA should not be viewed as a barrier to materials access. In fact, **MTAs are tools for gaining greater access to materials from a wider range of sources** (scientists from the public and private sectors, both in your own country and abroad). However, not all clauses in an MTA may be appropriate. Which clauses are appropriate will depend on the circumstances, the purposes of the transfer, and the institution from which the material is being received.

Given that IP management is heavily context specific, these Key Implications and Best Practices are intended as starting points to be adapted to specific needs and circumstances.

KEY IMPLICATIONS AND BEST PRACTICES: SECTION 7

FOR TECHNOLOGY TRANSFER OFFICERS

- ✔ A public sector institution can use a variety of agreements to both manage and protect intellectual property, regardless of whether that intellectual property is owned by the public sector institution or by licensing partners in the private sector. The key issue is to allow for **maximum flexibility whereby institutions can set, or negotiate, the terms that best fit the mission and goals of the institution** and the purpose of the partnership.
- ✔ Recognizing that no agreement will ever be perfect, you will need to **work with senior management to obtain their full support and backing**, especially when deals are likely to be criticized from the outside.
- ✔ Certain terms should be **"negotiated" internally prior to negotiating with third parties**. Senior management can be instrumental by signing off on certain template clauses that can be used as the basis for negotiating deals.
- ✔ **A template agreement** should be used *only* as a starting point for discussions.
- ✔ Contracts should be tailored to fit local customs and business practices. Be **sensitive to cultural and linguistic differences among parties** to a contract.
- ✔ Your office ought to be the **official repository** of all agreements dealing with incoming and outgoing biological materials.
- ✔ Legal jargon in agreements should be avoided. Instead, **use short, clear sentences that are free of vague adjectives and are written in the active voice**. The vocabulary should be accessible both to business people (who have extensive technical knowledge but limited legal knowledge) and judges (who have limited technical knowledge but extensive legal knowledge).
- ✔ Confidentiality agreements rely on a culture of trust, *not* a culture of secrecy. Make sure that **confidentiality agreements contain the necessary exceptions** appropriate for the mandate of your institution. A tricky question is how broadly the term *confidential information* should be defined. Too narrow a definition may leave out important information; too broad of a definition may prevent the parties from getting on with their work.
- ✔ MTAs call for extra caution with respect to clauses that deal with **reach through and the ownership of derivatives**. These clauses need not be negative. In fact, you may wish to impose certain reach through clauses yourself. These decisions will depend on the circumstances.
- ✔ When negotiating **collaborative research agreements**, you should involve the scientists to the maximum extent possible. Also, pay particular attention to a clear and detailed work plan, how communication is to happen among the parties, how modifications to the work plan are to be agreed upon, and how disputes are to be resolved.

Given that IP management is heavily context specific, these Key Implications and Best Practices are intended as starting points to be adapted to specific needs and circumstances.

Inventors and Inventions

Universities, inventors, and inventions—and by extension intellectual property—are all inextricably intertwined. Scientists are the central force behind the research, teaching, and extension missions of universities. And inventors can be considered the central force behind intellectual property, since they generate patentable inventions. Dealing with scientists, and inventors, at least from an IP management perspective, is not always straightforward or easy, but the technology transfer process and licensing are made easier if **scientists know some basics of IP management** and of patenting, and are somewhat familiar with best practices. This applies equally to high-flying attention-seekers and low-key geniuses.

A university research program must frequently make decisions about whether its researchers' discoveries should be protected. The process leading to this decision can place a tremendous strain on the relationship between the technology transfer office (TTO) and scientists. And it is one of the principal reasons why scientists should be given an opportunity, from the day they join a research institution, to learn the very basic concepts of IP management to better understand the process and the challenges faced by TTOs and TTO officers. How are decisions made as to when to patent, what to patent, and how to protect an invention? When it comes to new inventions, unclear and non-transparent procedures lead to tensions that can be costly in terms of money, time, missed opportunities, and relationships between scientists, heads of departments, and TTOs. So it is important that university leaders and administrators—and scientists—work in concert on policies and mechanisms that establish the procedures used for recording inventions, for invention disclosure, and for sharpening the interface between scientists and technology transfer and licensing offices.

Three conspicuous but usually neglected issues deserve special consideration up front. First, it is often difficult to know when an invention has actually been made. Training of scientists with respect to IP management and an institutional atmosphere that encourages inventions and invention disclosures are essential for making the most of scientific endeavors. Second, inventors need to have a clear understanding of their rights—and responsibilities—when it comes to their inventions. Unrealistic expectations that are either too high or too low will get in the way of optimally productive research and can be a source of conflicts with TTOs. Third, **inventions, per se, are not necessarily innovations, though they may become innovations.**[1]

The section's opening chapter by Mutschler and Graff[2] provides essential information that university scientists and inventors need to know in order to manage new and existing intellectual property and to deal with TTOs. University

Krattiger A, RT Mahoney, L Nelsen, JA Thomson, AB Bennett, K Satyanarayana, GD Graff, C Fernandez and SP Kowalski. 2007. 8: Inventors and Inventions. In *Executive Guide to Intellectual Property Management in Health and Agricultural Innovation: A Handbook of Best Practices* (Krattiger A, RT Mahoney, L Nelsen et al.). *bio*Developments-International Institute (Ithaca, USA), MIHR (Oxford, UK), PIPRA (Davis, USA), and Oswaldo Cruz Foundation (Fiocruz, Rio de Janeiro, Brazil). Available online at www.ipHandbook.org.

faculty and scientists anywhere need a working knowledge of what intellectual property is and what can be done with it so that they are able to make decisions about their laboratories' IP issues. These issues range from how to start and run a research program (from an IP management perspective), how to handle new inventions produced by that research, and how to manage the property of their collaborators. Graduate students and postdoctoral scientists should also acquire a basic understanding of intellectual property, since this knowledge will be valuable to them no matter what their future careers hold, whether in government, academia, non-governmental organizations, or industry.

Universities typically have—or should have—**institutional IP policies** that must be complied with by all personnel. In the United States, these IP policies must conform to the guidelines outlined by the Bayh-Dole Act of 1980. This act was intended, among other purposes, to promote private sector investment in federally funded research to facilitate the transfer of federally funded research to industry. For these reasons, university employees in the United States must sign agreements that state that all intellectual property generated under the university's aegis belongs to that university (though an inventor is typically given a share of the revenues that come from the sale of any intellectual property that he or she generates). One major exception to the policy of assigning IP rights to the university, however, is copyrighted material (books, papers, drawings, paintings, sculptures, and so forth).

A university's **IP office or TTO** is typically responsible for protecting and developing commercial applications (including out-licensing patents) for inventions developed at the university. Its functions typically include:

- determining the most effective way to protect inventions
- evaluating the commercial potential of inventions
- obtaining the appropriate protection for inventions
- locating suitable commercial-development partners and marketing inventions to them
- negotiating and managing IP licenses
- encouraging and assisting the formation of new companies around university-generated intellectual property (start-ups)

In addition to briefly reviewing the above, Mutschler and Graff familiarize scientists with issues they face on a daily basis:

- how to deal with confidential information
- how to deal with materials from third parties
- what constitutes a public disclosure of scientific finding
- how the patenting process works
- understanding the basics of intellectual property

If scientists read only one chapter in the entire *Handbook*, Chapter 8.1 is the one they should read. But on a daily basis, a scientist's responsibility goes further than having a basic understanding of IP management. For authoritative IP management, university faculty, staff, and students need, at a minimum, to appropriately document their research findings, use of intellectual property not owned by the university, dealings with collaborators outside the university, and places and times of public disclosures of research results. **Good record keeping** is not only important for preparing publications, reports, and grant proposals, it is also essential for preparing IP-protection documents and supporting IP rights. Universities must be very careful when they use materials and methods that do not belong to them, in order to avoid infringing on the property rights of others. In a worst case scenario, third-party materials may be used in research, which would mean the new intellectual property generated through use of that material would belong to that third party. Ownership would depend on how the material was obtained. Issues related to ownerships (material transfer agreement) are discussed elsewhere.[3] But irrespective of the terms of access, good record keeping in general and laboratory notebooks in particular are essential to possibly later disentangle ownership issues.

Scientists should be familiar with the **ins-and-outs of keeping a laboratory notebook to document research.** As a matter of institutional policy, the contents of laboratory notebooks should be treated as confidential and valuable. Notebooks should be stored in a safe place and any loss or theft should be reported immediately. A laboratory notebook is owned by the institution where work is conducted (essentially by the employer of the scientist). Therefore, when a scientist leaves an institution permanently, he or she should be required to turn notebooks over to supervisors (though copies can generally be kept by departing scientists).

Thomson,[4] a scientist herself, shares the practical aspects of laboratory-notebook keeping and her chapter offers a sample policy. Crowell[5] examines the entire range of invention documentation. It cannot be overemphasized how important is good documentation of research. It is a critical component of best practices in IP management for the following reasons:

- Well-kept laboratory notebooks are one of the most important *sources of documentation*. A laboratory notebook should contain detailed records of every experiment that has been planned or executed (including the date it was performed), the reasons for performing it, the methodology used in performing it, the results of the experiment, and the significance of the results.
- Laboratory notebooks are important instruments of *institutional memory*. Laboratories invariably have high personnel turnovers: scientists move on, post docs move up, students graduate, and technicians are promoted.
- Consistent documentation is important to *determine patentability*, and may even be essential for determining inventorship, for drafting and prosecuting patent applications, and (if necessary) for protecting patents from third-party challenges such as prior-art challenges and (in the United States) patent-interference proceedings.

Once scientists think they may have patentable inventions, they should file an **invention disclosure** to their TTO. McGee[6] examines the entire invention-disclosure process from a practical and logistical point of view and stresses the importance of involving inventors throughout the protection and commercialization process. He discusses how IP professionals can best work with inventors to develop high-quality invention disclosures. An invention disclosure is a description of something novel and nonobvious that would allow anyone of ordinary skill in the corresponding art to reproduce the invention. It may be simple in scope and include most details in an attached draft of a scientific paper (McGee also notes, quite appropriately, that carefully kept laboratory notebooks can be used in place of an invention disclosure). Importantly, an invention disclosure irrefutably establishes the date and scope of an invention, as well as the identity of the inventor(s). Disclosures are essential for managing intellectual property, preserving IP assets, "harvesting" inventions and securing IP protection for those assets, and eventually translating the inventions into innovative products or services.

Thus, an invention disclosure is the beginning of what is sometimes a long but often rewarding process that can benefit the institutions where the disclosures are made, the society at large, and the inventors in particular. ■

All chapters refer to: *Intellectual Property Management in Health and Agricultural Innovation: A Handbook of Best Practices*. 2007. A Krattiger, RT Mahoney, L Nelsen, JA Thomson, AB Bennett, K Satyanarayana, GD Graff, C Fernandez, and SP Kowalski (eds.). MIHR: Oxford, U.K., and PIPRA: Davis, U.S.A. Available online at www.ipHandbook.org. The online version contains for each chapter a detailed Editor's Summary, Implications, and Best Practices.

1 See, also in this *Executive Guide*, part 9: Evaluation and Valuation of Technologies.

2 Chapter 8.1 by M Mutschler and GD Graff titled Introduction to IP Issues in the University Setting: A Primer for Scientists, p. 747.

3 See, also in this *Executive Guide*, part 7: Contracts and Agreements to Support Partnerships.

4 Chapter 8.2 by JA Thomson titled How to Start–and Keep–a Laboratory Notebook: Policy and Practical Guidelines, p. 763.

5 Chapter 8.3 by WM Crowell titled Documentation of Inventions, p. 773.

6 Chapter 8.4 by DR McGee titled Invention Disclosures and the Role of Inventors, p. 779.

PROCESS FOR GOVERNMENT POLICYMAKERS

- ✔ People and institutions typically look after their possessions in a much more serious manner than if they have no stake in them. This is applicable to physical property and to intellectual property. For this reason, governments should consider enacting legislation or, as appropriate, implementing **policies that clearly spell out how public sector institutions can protect, own, and license inventions** made in their institutions. This equally applies to government research centers and to universities.

- ✔ Arguably, the minds of scientists operate differently from those of lawyers, politicians, bankers, and policymakers. Similarly, those engaged in managing the intellectual property in public sector institutions face different challenges than the scientist-inventors. These differences can be the source of much tension, but such tension can often be preempted if **scientists are given an opportunity to learn the basics of IP management**, including best practices, in terms of data and information management related to inventions. Public sector institutions should have the resources to offer limited, but essential, training to every scientist when they join an institution.

- ✔ Such training programs can be given as a series of short seminars or even half-day orientation courses. These are most effective if the institutions have clear IP policies that include matters related to ownership of inventions, the **duty to disclose inventions**, and **laboratory notebook keeping**. The latter is common practice in any private sector R&D center. Comprehensive research records are fundamental to best practices in science, IP management, and in the regulatory process.

FOR SENIOR MANAGEMENT

(UNIVERSITY PRESIDENT, R&D MANAGER, ETC.)

- ✓ People and institutions typically look after their possessions in a much more serious manner than if they have no stake in them. This is applicable to physical property and to intellectual property. For this reason, **work with your governments to implement policies** (or enact legislation, as appropriate) that clearly spell out how public sector institutions, including government research centers and universities, can protect, own, and license inventions developed at your institution.

- ✓ Arguably, the minds of scientists operate differently from those of lawyers, politicians, and university presidents (although many a president is a former scientist). Similarly, those engaged in managing intellectual property in public sector institutions face different challenges than do scientist-inventors. The differences can be a source of much tension, but such tension can be preempted if **scientists are given an opportunity to learn the basics of IP management**, including best practices, in terms of data and information management related to inventions. Public sector institutions and companies alike should offer and require limited, but essential, training to every scientist, student researcher, and technician when he or she joins a research program.

- ✓ Such training programs can be provided as a series of short seminars or even half-day orientation courses. And they are most effective if the institutions have clear IP policies that include matters related to ownership of inventions, the **duty to disclose inventions**, and **laboratory notebook keeping**. The latter is common practice in any private sector R&D center. Comprehensive research records are fundamental to best practices in science, IP management, and in the regulatory process.

- ✓ University faculty, staff, and students do not have to become IP experts. The IP management training programs is best offered by the technology transfer personnel that will be interacting with scientists rather than by lawyers and outside consultants can be useful facilitators. Part of the aim of such training is **team building** that encourages communication between the scientists, technology transfer personnel, and senior management. It is part of creating a culture of IP awareness.

- ✓ Many scientists at public institutions often do not (initially, at least) appreciate the importance of laboratory notebooks and documentation protocols. For private sector R&D centers, this is done as a matter of routine. Some argue that good laboratory notebook practices lead to better science. **Laboratory notebooks** surely lead to better invention disclosures, prevent fraud, clarify inventorship, facilitate patent applications, and ultimately, pay off for individuals and institutions in the long term.

- ✓ If an invention is protected, then much can be gained if inventors are actively involved in all phases of the protection and marketing of their inventions. Inventors not only have intimate knowledge of their inventions; they may also have useful leads and contacts in companies or have ideas about how an invention could be incorporated into existing products or services. The practice of occasional seminars by technology transfer personnel for scientists is a practice that will **strengthen the interest and involvement of scientists** in this process.

FOR SCIENTISTS

- ✔ **IP management is an important element in facilitating the translation of research** into useful products or services that benefit your community and country.

- ✔ Encourage your technology transfer office (TTO) to organize **occasional seminars on the basics of IP management**. Ideally, your institution should provide an IP management primer when you join the institution that will help you understand the basic elements of IP protection and smooth the interface with your TTO. Even if you have taken such primers or seminars before, attend those offered by your new employer and encourage those in your group to do so as well. This will facilitate communication with your TTO staff and answer your questions about IP management.

- ✔ One potentially controversial issue faced by many TTOs involves keeping **laboratory notebooks**. For private sector R&D centers, this is done as a matter of routine. Make it a habit to use laboratory notebooks, as doing so can lead to better science and easier invention disclosures and can facilitate patent applications.

- ✔ **Good practices in laboratory notebook keeping** should include the signing of each page by a supervising scientist, occasional spot checks, and the setting aside of time for recording experiments and results. This applies to research assistants, students, post docs, and everyone else working in a laboratory.

- ✔ **Good record keeping** is important. It includes linking research proposals with material transfer agreements, publications, invention disclosures, and so forth. It promotes both scientific goals (it facilitates the writing of publications and grant proposals) and legal goals (good records make it easier to obtain and defend patents).

- ✔ Good record keeping goes beyond publications and IP management. Especially in institutions dealing with the **development of products** and clinical trials in health, or biosafety research in agriculture, record keeping may be essential for providing regulators the necessary evidence that good laboratory practices have been followed and may underpin regulatory filings. In many cases, experiments conducted years before regulatory filings can become valuable for those filings and, unless laboratory detailed notebooks were kept, experiments may have to be repeated at great cost and may also delay filings.

- ✔ Invention disclosures are the first step in protecting intellectual property. **Disclose early and often**. Rather than wait until your scientific paper is accepted, make it a habit every few months to think what might be disclosed and what should be disclosed, and then disclose it. But expect only a small portion of your invention disclosures to lead to patent applications.

- ✔ **Recognize when you actually have an invention**. Often, it is much earlier than you think. By filing an invention disclosure with your TTO, you are initiating a dialogue. Even if the TTO does not immediately file a patent based on your first invention disclosure, it is a process that has started, and follow-up invention disclosures will be much easier.

- ✔ Ideally, you should **invite your TTO liaison to visit your laboratory occasionally** and discuss with you and your research team what you have been doing. Discussions with technology transfer experts, especially patent attorneys, can help you to identify inventions.

FOR TECHNOLOGY TRANSFER OFFICERS

- ✔ Arguably, the minds of scientists operate differently from those of bankers, politicians, and licensing executives. Similarly, those engaged in managing intellectual property in public sector institutions face different challenges than do scientist-inventors. The differences can be a source of much tension, but such tension can be preempted if **scientists are given an opportunity to learn the basics of IP management**, including best practices, in terms of data and information management related to inventions. Public sector institutions and companies alike should offer and require limited, but essential, training to every scientist, student researcher, and technician when he or she joins a research program.

- ✔ Such training programs can be provided as a series of short seminars or even half-day orientation courses. And they are most effective if the institutions have clear IP policies that include matters related to ownership of inventions, the **duty to disclose inventions**, and **laboratory notebook keeping**. The latter is common practice in any private sector R&D center. Comprehensive research records are fundamental to good research practices in science, IP management, and regulatory areas.

- ✔ University faculty, staff, and students do not have to become patenting experts. Keep any such training programs simple and practice oriented. Generally, the intricacies of patenting legislation is not what motivates a scientist; rather, it is a vision of how his or her invention can eventually make a difference in people's lives. The IP management training programs should thus be practical and offered by technology transfer personnel that will be interacting with scientists rather than by lawyers. Contractors can be useful as facilitators. Part of the aim of such training is **team building** that encourages communication between your office and the scientists in your institution. It is part of creating a culture of IP awareness.

- ✔ It is good practice to **include senior management as participants** in the training sessions. This is especially useful when the training program includes case studies.

- ✔ Prepare simple brochures and Web sites that encourage scientists to contact you with their questions and inventions. Similarly, make an effort to attend seminars given by the researchers in your organization. It is a great way to show your interest in their activities and to build a good understanding of what the researchers actually do. Overall it helps to **get scientists involved in all phases of protecting and marketing their inventions.**

Evaluation and Valuation of Technologies

Oftentimes, the first question that arises with an invention is whether or not it should be patented. When devising a patent strategy, three questions should be considered:

- Should a patent application be filed or should the invention simply be published?
- Should the invention be marketed to existing companies or be used to develop a spinout company?
- What is the potential value of the invention?

Should a patent application be filed or should the invention simply be published? The answer depends on a number of factors: the needs (and dynamism) of the market, the uniqueness and usefulness of the invention, the likelihood that patent protection can be obtained, the specific mission of an institution, and the "attitude" of the inventor, that is, whether he or she is inclined to assist the technology transfer office (TTO).

This latter point merits discussion. For practical purposes, an invention can be defined according to patent statute: it must be novel, contain an inventive step (be nonobvious), and be useful or have industrial applicability. But often an invention is many years away from "working" in the real world. In other words, an invention is not an innovation until the new knowledge and invention are introduced into and utilized in an economic or social environment. **Determining how to translate an invention into an innovation that makes a difference in people's lives (economically or socially or both) is one of the principal reasons for which technology transfer offices exist.**

Should the invention be marketed to existing companies or be used to develop a spinout company? Each approach has advantages and disadvantages. An existing company usually has an established infrastructure, as well as access to financial instruments and distribution networks. Its financial health can be readily assessed. However, connecting with existing companies might be challenging, partly because they already have research agendas, networks, and priorities. The biggest risk with an established company is that it will lose interest in the technology before anything develops. Spinout companies, on the other hand, are focused on their own inventions, but because the companies are nascent, they are also fragile.

According to Nelsen,[1] who has led the M.I.T.'s Licensing Office for the past 20 years, a number of other factors should also be considered when deciding whether or not patenting a new invention is in the public interest. For example:

- Is the technology self-evidently useful as-is? Will it be widely used even if it is not patented but instead released into the public domain?
- Can the patent-holding institution devise a nonexclusive licensing strategy that will bring in revenue without restricting the availability of the technology?

Krattiger A, RT Mahoney, L Nelsen, JA Thomson, AB Bennett, K Satyanarayana, GD Graff, C Fernandez and SP Kowalski. 2007. 9: Evaluation and Valuation of Technologies. In *Executive Guide to Intellectual Property Management in Health and Agricultural Innovation: A Handbook of Best Practices* (Krattiger A, RT Mahoney, L Nelsen et al.). MIHR (Oxford, UK), PIPRA (Davis, USA), Oswaldo Cruz Foundation (Fiocruz, Rio de Janeiro, Brazil), and *bio*Developments-International Institute (Ithaca, USA). Available online at www.ipHandbook.org.

- If the technology will not be useful without substantial high-risk investment, then it will have to be patented and exclusively licensed in a developed country in order to bring in the adequate revenue. However, if this path is taken, then an additional question must be asked: should patents be pursued in developing countries in order to encourage companies to produce competing (lower-cost) generics in those countries?
- Can the patent holder require the licensee to sublicense in order to promote low-cost manufacture and distribution of the technology?
- If the drug (or vaccine) is expected to be used only in developing countries, will the patenting and distribution of a limited number of licenses attract sufficient investment and create a market? Put differently, will such a market be sufficiently profitable that it will encourage further development and testing of the drug or vaccine?
- Should the patent holder reserve for itself the unrestricted use of a patented research tool?

Answers to the above questions will act as guidance for patenting decisions. Above all, **institutions should determine whether or not patenting is the most effective way to ensure global access to their technologies**. Broad licensing strategies should also be considered at this stage, and through licensing only comes in later. In this context, it should be remembered that it can be challenging to negotiate a licensing agreement that is fair to everyone—licensor, licensees, and the public sector. However, Nelsen asserts that it is far better to make an imperfect deal than no deal at all. People do not benefit until technology is developed and brought to market.

What is the potential value of the invention? While many inventors believe that their product is of supreme importance, many fail to see the potential value of the product. Putting a "price tag" on an invention is difficult. In other words, determining the value of an invention and the resulting technology, product, or service will be complicated and at times nearly impossible. Fortunately, **the "full worth" of an invention need not be determined at the time the invention is made, nor even when the invention is transferred or licensed** to a third party through a patent license. Value can be realized through the use of royalties—payments to the inventor based on the specific (negotiated) contribution of the invention to a new product or service that are made when the product or service is sold. Royalties involve a trade-off and add new complications involving decisions about what to base the royalty rate on, how they should be calculated, and so forth. Royalties mean revenues are coming years later, but many institutions prefer payments now.

With respect to specific valuation techniques, Potter[2] reviews five major approaches, provides illustrations based on agricultural technologies, and discusses a hypothetical negotiation between a university and a company.[3] Potter outlines five approaches to valuing technology.

Costs approach. The pricing of a product is based on the cost of developing the product. This approach is rarely used to assign a value to a technology because the *cost* of research is not usually correlated with the *value* of the intellectual property that was the basis for the technology.

Income approach. The value of a technology is determined by a pure income approach, whereby future anticipated revenues (cash flows) are discounted to present value. The big drawback to this approach is that, for a new technology, there are generally no sales, markets, or cost data that can be used to predict future revenues.

Market approach. The value of a technology is determined based on the value of a similar or comparable technology. The inherent weakness of this method is that it is difficult to find a comparable technology if the technology in question is truly novel.

Hybrid approach. The value of a technology is determined by a combination of the income and market approaches. This method will deliver both the benefits and the drawbacks of both methods.

Royalties approach. The value of a technology is calculated based on royalty rates that have been applied to similar technologies. With this method, the inventor would typically receive a return on sales of the final product, with risk being shared between the inventor and the developer.

Case Studies

Table of Contents

Agriculture, Forestry, Bioprospecting and the Environment

Vaccine Development

Health Products and Diagnostics

Somatic Embryogenesis of Grapes: Fundación Chile

Fundación Chile is a private nonprofit organization. Its mission is to add economic value to Chile's products and services by promoting innovation and technology transfer for Chile's natural resource, agricultural, and manufacturing sectors. Fundación Chile's primary strategy is to develop new technology-based companies in Chile that can have a significant economic and social impact. These new companies are generally joint ventures with strategic partners, although other models, such as licensing, are used. The main activities are focused in the area of agribusiness, marine resources, forestry and forest products, environment, information technology, education and human resources, and tourism.

Fundación Chile is unusual as a nonprofit institution that participates in the creation of innovative private companies. In fact the foundation is involved in a wide range of activities relevant to different stages of development of new businesses, including technology services, R&D, incubation, scale-up, seed capital, and financial innovation. Fundación Chile's activities are focused on Chilean production of goods that can be exported or that can replace imports, but possibilities for production in additional territories that can increase the volume and value derived from Chilean production are also considered.

Since 1997, Fundación Chile has been active in developing applications of biotechnology that can improve productivity, add value to existing products, and promote introduction of new products.[1] Biotechnology activities are mainly focused in forestry, horticulture, and aquaculture, with increasing emphasis on quality enhancement. Biotechnologies used include recombinant proteins, tissue culture, molecular genetics, functional genomics, and genetic engineering. Strategic alliances in biotechnology in the private sector include a licensing agreement for a salmon vaccine with Syngenta, a strategic alliance in forestry biotechnology with CellFor Inc. (Vancouver, BC, Canada), a collaboration in stone fruit biotechnology with Okanagan Biotechnology Inc. (Summerland, BC, Canada), and a joint venture in grape biotechnology with Interlink Associates LLC (Princeton, NJ, U.S.A.). Fundación Chile seeks to establish strong IP positions through the licensing of key existing IP and the development of new intellectual property in areas of specific strategic importance in Chile.

Fundación Chile's biotechnology activities involve an extensive network of Chilean and foreign research centers and universities, as well as participation in key international consortia. Collaborators within Chile include Fundación Ciencias para la Vida, the Chilean National Institute for Agricultural Research, the University of Chile, the University of Concepción, the University of Santiago, the University of Talca, University Federico Santa Maria, Andres Bello University, and Austral University. Alliances with foreign research centers and universities include the University of California, Cornell University, the University of Florida, the U.S. Department of Agriculture (USDA), New Zealand HortResearch, and New Zealand Forest Research. Fundación Chile is a member of PIPRA (the Public Intellectual Property Resource for Agriculture) and the California Institute of Food and Agricultural Research and is a participant in the ALCUE-Food Specific Support Action funded by the 6th European Framework.

As a result of this networking, Fundación Chile has been able to participate in the development of products within a relatively short time frame. A recombinant protein vaccine for salmon, developed in a collaboration of Fundación Chile and Fundación Ciencias para la Vida, has been licensed to Syngenta and is being

Fernandez C. 2007. Somatic Embryogenesis of Grapes: Fundación Chile. In *Executive Guide to Intellectual Property Management in Health and Agricultural Innovation: A Handbook of Best Practices* (Krattiger A, RT Mahoney, L Nelsen et al.). MIHR (Oxford, UK), PIPRA (Davis, USA), Oswaldo Cruz Foundation (Fiocruz, Rio de Janeiro, Brazil), and *bio*Developments-International Institute (Ithaca, USA). Available online at www.ipHandbook.org.

introduced into the market. Elite clones of radiata pine developed through somatic embryogenesis in collaboration with CellFor are in advanced stages of testing and are being scaled up for market introduction by the Fundación Chile company GenFor. Other biotechnology programs of Fundación Chile, including genetic engineering of varieties of pine trees, peaches, and grapes, are in earlier stages of development.

THE TECHNOLOGY

Importance of institutional support for a long-term R&D program

Agricultural biotechnology R&D programs are long term, expensive and controversial; an institution undertaking such a program must be committed to the process for the long term. In the late 1990s Fundación Chile made a strategic decision to invest in development of biotechnology applications for strategic sectors of the Chilean economy, particularly forestry, agriculture, and aquaculture. Genetic engineering was clearly a key technology with large potential impact, as demonstrated by the rapid adoption of genetically engineered varieties of maize, soybeans, and cotton in some parts of the world. However, these major crops play a relatively minor role in Chile. Little effort was being expended to make improvements in perennial crop species, such table grapes, in which Chile is a major player.

Building a foundation for the program

Typically, three different types of technological components are needed for development of a genetically engineered plant product:

- germplasm that provides a competitive genetic background
- specific genes that confer new traits of interest
- enabling tools, such as genetic markers, promoters, tissue culture and regeneration systems, and transformation methods

In addition, human resources, laboratory infrastructure, and financing are needed to carry out the R&D required to adapt and combine these components to produce a product. Laboratory infrastructure existed in Chile, but improvements were needed. There were capable researchers in Chile, but they were limited in number. Research efforts were spread across many different objectives, and sustained support for any one specific program was rare.

In the case of grapes, the foundation technologies were not available in the local R&D institutions at the start of the program, except, to a limited degree, germplasm. A global search led to the identification of sources of technologies and expertise. The availability and priority of different components were assessed, and efforts were initiated to access, license, and transfer the key components.

IP and freedom to operate

The IP and freedom-to-operate issues confronted were complex, largely due to the need to address the situation in Chile and the situations in Chile's major export markets, the long and uncertain time frames for development and commercialization of genetically engineered perennial fruit crops, and the concentration of rights to core technologies in the hands of companies with little or no interest in the development of minor crops. A complete solution was not possible in the short term with the resources available. However, it was possible to establish a position in key technologies that maximized the likelihood of being competitive within a specific niche.

A critical aspect was the active involvement of personnel with professional experience in commercial R&D programs and major agri-biotech research centers in other countries, as well as experience in the licensing of agricultural biotechnologies. Practices vary from country to country and from institution to institution within a country. At the initiation of the program there was little experience in Chile with patenting and licensing technologies developed in public research institutions. The involvement of personnel with international experience, providing appropriate examples drawn from a number of sources, played an important part in bridging gaps in experience and expectations.

Establishment of a grape biotechnology platform

At the time the program was initiated there were only a few published reports of transformation of *Vitis vinifera*. In order to be able to obtain R&D funding from public and private sources, and to be considered seriously as a potential licensee by technology providers, it was considered critical to demonstrate the ability to reproducibly transform the target species. For many transformation systems, an important factor is the availability of a robust tissue culture system that makes it possible to regenerate plants efficiently. In our experience, tissue culture systems involve considerable art and are often difficult to reproduce in other laboratories. Thus, establishment of a strong position in grape tissue culture was given the highest initial priority. The process and progress in this area are discussed below. The second priority was access to specific gene candidates for engineering a trait of commercial interest in the Chilean market. This was carried out in parallel in order to ensure that the tissue culture and transformation platform developed could be applied to the production of prototypes with traits of interest with a minimum lag.

Identification of suitable laboratories

The search used different and complementary channels, including reviews of research publications, project databases, conference proceedings, patent documents, news items, and personal contacts. All of them are relevant, and each provides unique and useful kinds of information.

Access to many of these sources has been facilitated by the rapid improvement of the Internet, both in terms of content and ease of access. Even for people without good Internet access, the availability of high-quality documents in electronic form has greatly reduced the cost of access.

Open sites such as PubMed (www.ncbi.nlm.nih.gov) and HighWire (highwire.stanford.edu) provide convenient access, not only to bibliographic information, but to many full papers. More and more, full papers are available at no charge, some can be downloaded for a fee from sites of journal publishers or specialized clearinghouses. Even for people without good Internet access, the availability of high-quality documents in electronic form has greatly reduced the cost of access.

Online databases such as those at the World Intellectual Property Office (www.wipo.int/ipdl), the European Patent Office (www.espacenet.com), the U.S. Patent and Trademark Office (www.uspto.gov), and many other national patent offices provide increasingly convenient access to issued patents and published applications.

Less widely appreciated, but valuable due to their more specialized content, are online databases of research projects. These often include information that is otherwise difficult or impossible to find. Examples include the European Union Community Research & Development Information Service (cordis.europa.eu), the Current Research Information System of the USDA (cris.csrees.usda.gov), the FAO-BioDeC database of biotechnology projects in developing countries (www.fao.org/biotech/inventory admin/dep/default.asp), and a database of biotechnology activities, by country, of the Red de Cooperación Técnica en Biotecnología Vegetal para America Latina y el Caribe (www.redbio.org). In Chile, the Web sites of the major funding agencies for R&D, CONICYT (www.conicyt.cl), CORFO (www.corfo.cl), and FIA (www.fia.cl), include databases of projects. Many research institutions provide databases of internal research activities and funded projects, which may be useful once specific institutions of interest have been identified.

Negotiation of a research and option agreement

Once the identification of the laboratory or institution has been made, documents are typically exchanged via e-mail. Most large private companies and universities have standard forms that are adapted to the specific needs of a project. Typically, research agreements will include the following information:

- date
- parties
- definitions of terms such as project, project proposal, sponsor, and joint and recipient intellectual property
- reports and conferences for proper follow up of activities
- costs, payments, and other support
- publications
- intellectual property
- grant of rights
- confidentiality and publicity
- term and termination
- insurance and indemnification
- governing law
- assignment
- agreement modification
- notices
- counterparts and headings

It is important to emphasize that this standard form was designed for use in the United States. Intellectual property laws vary among countries, so, it is important that the content of any agreement is reviewed by a local lawyer knowledgeable in IP matters.

Most universities in the United States, and many other public research institutions, will require that the public institution be able to continue to use the technology for research and education purposes even if exclusive rights for commercial use are granted.

Our general approach has been to negotiate agreements that provide rights to use technologies for R&D, along with an option for a future commercial license. We want to avoid situations where resources are invested in research if the results cannot be commercialized. Due to the high degree of uncertainty in the development and commercialization of agri-biotechnology products, we also want to avoid paying at the outset for full commercial rights, if in the end they will not be used. In technology access agreements we have generally tried to structure compensation in ways that reduce the up-front costs in favor of sharing any benefits eventually realized after commercialization. This is important for making effective use of the resources currently available, but, more importantly, it helps to align the interests of the technology provider with our interests. The agreements typically contain modest up-front payments, milestone payments based on successful transfer of the technology, additional milestone payments if a commercial license is entered into and a product is introduced to market, and royalties based on revenue derived from commercialization of products produced using the technology.

In the case of grape tissue culture technology sought by Fundación Chile, the university at which the technology had been developed already had agreements in place with a private company. Thus, initially we had to negotiate a sublicense agreement with that company. Later, changes in the scope of that company's activities led to a return of the IP rights to the university. We then entered into additional negotiations with the university. Similar events affected other agreements related to the project. It is important to recognize that management of such agreements is a dynamic process.

Material transfer agreements (MTAs)

In addition to intellectual property, the transfer of agricultural biotechnologies often requires, or is at least facilitated by, the transfer of actual biological materials such as plant tissue cultures, plasmids, vectors, or reagents. The physical transfer and use of the materials are generally covered by an MTA.

In countries with limited international innovation programs, lawyers have not been exposed to or do not have enough experience on matters related to MTAs. In Fundación Chile's case, the most practical approach was to use, as a reference, MTA forms prepared by the technology transfer offices of universities in the United States and other countries with experience in these matters. Some of these offices have sample forms posted on their Web sites.[2]

An MTA should be carefully reviewed. In the past, investigators have sometimes carelessly accepted terms that could have critical affects on the value of the R&D being conducted, terms such as reporting requirements and rights given to the provider of the material to use information generated by the recipient. It is also critical to consider whether the material provided incorporates materials or technologies already owned by third parties. If so, it is advisable to request clarification of any restrictions that my be "inherited" with those materials.

Importation of materials

Each country has its own regulations regarding the importation of biological materials. In Chile, there are forms and procedures that must be followed. Samples of grape tissue culture were imported following these procedures without major obstacles, although significant time and resources were required.

Exchange of professionals between laboratories

Good communication between parties is essential for a successful outcome. For transfer of some technologies, the exchange of written information and materials supplemented by phone calls and e-mails may be sufficient. However, in many cases, successful transfer is greatly facilitated by the active participation of investigators from the provider and recipient laboratories in activities in both laboratories.

In the case of the grape tissue culture system, a Chilean investigator first spent time in the laboratory of the inventor, to get hands-on experience with the procedures, and then returned to set up the system locally. Several months later, the inventor spent a full week working side by side with local investigators, reinforcing the training and providing an opportunity to resolve issues that had arisen during initial implementation. Some time later, the project leader visited the inventor's laboratory to observe the procedures there, with experience accumulated in Chile providing a foundation for increased "receptivity." At the end of each exchange, written reports were prepared, disseminated, and discussed.

CONCLUSIONS

Currently the lab in Chile has been able to master grape embryogenic tissue culture and regeneration techniques and apply them to genetic engineering. The genetic transformation of grape tissue cultures has allowed the production of thousands of transformed grape lines, from which several promising lines have been advanced to the field for additional testing. ■

For further information, please contact:
CARLOS FERNANDEZ, *Director, Strategic Studies, Foundation for Agriculture Innovation (FIA), Loreley 1582, La Reina, Santiago, Chile. carlos.fernandez@fia.cl*

---

1 Fernandez C and MR Moynihan. 2007. A Model for the Collaborative Development of Agricultural Biotechnology Products in Chile. In *Intellectual Property Management in Health and Agricultural Innovation: A Handbook of Best Practices* (eds. A Krattiger, RT Mahoney, L Nelsen, et al.). MIHR: Oxford, U.K., and PIPRA: Davis, U.S.A. Available online at www.ipHandbook.org.

2 The online version of *Intellectual Property Management in Health and Agricultural Innovation: A Handbook of Best Practices* provides many sample forms from a host of different organizations around the world (see www.ipHandbook.org).

The Groundnut Story: A Public-Private Initiative Focused on India

Groundnut, or peanut (*Arachis hypogaea*) is a staple oilseed crop grown for food and for forage in India. It is cultivated on 7.5 million hectares with annual production of about eight million tons. More than five million small and marginal farms depend on this crop for their viability.

During the monsoon season of 2000, a new groundnut disease emerged in India. The spread of the disease grew to epidemic proportions causing crop loss corresponding to more than US$65 million. The causal agent of this devastating disease was found to be tobacco streak virus (TSV), which causes stem necrosis in the groundnut plant resulting in complete destruction of the crop. In addition, TSV infects several other economically important crop plants, such as sunflower and marigold, and lives in many weed hosts. Parthenium, a prevalent weed, is a symptomless carrier of TSV and plays a major role in the perpetuation and spread of the disease. The constant threat of TSV outbreak has caused food shortages and financial insecurity for groundnut farmers.

By nature, groundnut plants show little resistance to TSV. Moreover, all currently grown cultivars are susceptible to TSV infection. Therefore, a nonconventional method of incorporating disease resistance in the cultivars was needed to control the disease. Transgenic crop plants that express the coat protein (CP) gene of the target virus pathogen have been shown to provide a high degree of resistance to many plant viruses. The Agricultural Biotechnology Support Project II (ABSPII), which focuses on safe and effective development and commercialization of bioengineered crops in order to benefit resource-poor farmers in developing countries, decided to fund the bioengineering of groundnut genotypes to incorporate the CP gene for conferring TSV resistance.[1]

LICENSING ARRANGEMENTS

Sathguru Management Consultants, the regional coordinator of the ABSPII project in South Asia, approached the Donald Danforth Plant Science Center (the Danforth Center) for the development of a vector construct containing the TSV-resistance gene for conferring viral resistance to groundnut plants.

The CP technology for conferring resistance to viral infection is owned by Monsanto Company. A patent nonassertion agreement[2] from Monsanto for the CP technology to be used for nonprofit public good was obtained by the Danforth Center. This non-assert was facilitated by the ABSPII project. The Danforth Center further developed the technology for TSV-CP-mediated-resistance in groundnut to be deployed in South Asia and Southeast Asia.

A consortium of public institutions was formed by ABSPII with International Crops Research Institute for the Semi-Arid Tropics (ICRISAT) and Acharya N. G. Ranga Agricultural University (ANGRAU) in the state of Andhra Pradesh. These institutions were the primary licensees of the technology developed by the Danforth Center for TSV-resistant groundnut cultivars.

With Sathguru Management Consultants as facilitator of the technology transfer, a nonexclusive licensing agreement was penned for nonexclusive licensing of the CP technology, free of royalties and upfront payments, to public institutions planning to develop the varietal groundnut. A tripartite agreement was arranged, with the Danforth Center as the technology licensor and Sathguru Management Consultants and ICRISAT as licensees. Development efforts of TSV-resistant groundnut by the public research institutions are underway and slated for commercialization in 2009.

Medakker A and V Vijayaraghavan. 2007. The Groundnut Story: A Public-Private Initiative Focused on India. In *Executive Guide to Intellectual Property Management in Health and Agricultural Innovation: A Handbook of Best Practices* (Krattiger A, RT Mahoney, L Nelsen et al.). MIHR (Oxford, UK), PIPRA (Davis, USA), Oswaldo Cruz Foundation (Fiocruz, Rio de Janeiro, Brazil), and *bio*Developments-International Institute (Ithaca, USA). Available online at www.ipHandbook.org.

Similar nonexclusive licensing arrangements have been made with private organizations for the development of hybrid groundnut cultivars. These licenses include upfront and royalty payments and an understanding with regard to benefit sharing.

POLICY COMPONENTS

Because groundnut is a so-called orphan crop, there was little interest in producing and selling open-pollinated varieties owing to their susceptibility to viral infection. Moreover, private industry lacked the motivation to commercialize hybrid varieties. Key policy makers for the ABSPII project secured financial support for developing and distributing the TSV-resistant groundnut and for facilitating the project through planning and implementation.

KEY LESSONS LEARNED

Technology can be a major force in alleviating poverty and increasing food security in developing countries. Moreover, investment gains can be multiplied by adopting technologies in different regions through the creation of synergic partnerships for product development, implementation, and commercialization. ■

For further information, please contact:
AKSHAT MEDAKKER, *Associate Consultant-Technology Management, Sathguru Management Consultants Pvt. Ltd., 15 Hindi Nagar, Punjagutta, Hyderabad 500034, India. akshatm@sathguru.com*

VIJAYVIJAYARAGHAVAN, *Founder and Director, Sathguru Management Consultants Pvt. Ltd., 15 Hindi Nagar, Punjagutta, Hyderabad 500034, India. vijay@sathguru.com*

1 www.absp2.cornell.edu.

2 See also, in the *Handbook*, Chapter 7.6 by Anatole Krattiger titled, The Use of Nonassertion Covenants: A Tool to Facilitate Humanitarian Licensing.

Golden Rice: A Product-Development Partnership in Agricultural Biotechnology and Humanitarian Licensing

IP (intellectual property) constraints are often perceived as barriers to market entry, especially when it comes to developing countries. This case study examines the IP management component in the development of Golden Rice[1] (or beta-carotene-containing rice) and the transfer and introduction of Golden Rice to developing countries.

Rice, one of the most widely grown food crops, contains neither vitamin A nor beta-carotene, yet it is a staple food crop for billions of people, especially in Asia. Here, and in other developing countries, vitamin A deficiency (VAD) is a major problem affecting primarily children under age five and pregnant and lactating women. Thousands of impoverished people lose their eyesight because of VAD. Severe VAD (xerophthalmia, or night blindness) leads to permanent blindness: 500,000 people, 250,000 of them children, lose their sight every year due to VAD.[2] The deficiency also leads to a depressed immune system that increases the incidence and severity of infectious diseases and infant mortality rates.

There are several avenues for mitigating VAD, including programs to fortify food with vitamin A and beta-carotene and to distribute vitamin A supplements to affected populations. For the supplement distribution, more than US$100 million are spent every year. An alternative, and complementary, approach is to insert relevant genes in rice. This allows farmers to grow beta-carotene-rich rice. By enhancing those varieties primarily grown or consumed by poor people, beta-carotene can be delivered at essentially no cost once the Golden Rice has been developed and bred into local varieties.

Interestingly, rice plants synthesize beta-carotene in foliage and other parts of the plant, but not in the grain, and all but two steps of the biosynthetic pathway are present in the grain. By the addition of only two genes, phytoene synthase (*psy*) and phytoene desaturase (*crt* I), the pathway is reconstituted and beta-carotene accumulates in the endosperm (the endosperm being the edible part of the grain).[3]

INTELLECTUAL PROPERTY FEATURES OF THE CASE

The development of Golden Rice led to a significant change in the relationship between the public sector and intellectual property. A freedom to operate (FTO) review of pro-Vitamin A-containing Golden Rice was commissioned by the International Rice Research Institute, a center of the Consultative Group on International Agricultural Research (CGIAR), with funding from the Rockefeller Foundation (led by one of us [AK]). The review showed that about 70 patents and patent applications were applicable to the improved rice when *all* patents issued in or applied for in *all* countries, including patents on commercially accessed research tools, were considered.[4] The published analysis also showed, in accordance with analysis by Zeneca (which later merged with Novartis to form Syngenta) that, in practice, only a few, if any, patents pertaining to Golden Rice were applicable in developing countries, together with a few material transfer agreements.

Obtaining Freedom to Operate

Fortunately, these potential—and arguably perceived—constraints were resolved in a few months in the year 2000 by a straightforward IP management strategy comprising four goals:

- identification of major IP components (the above-mentioned FTO review)
- interpretation, with Zeneca, of the relevance of the FTO review to the proposed humanitarian use in developing countries

Krattiger A and I Potrykus. 2007. Golden Rice: A Product-Development Partnership in Agricultural Biotechnology and Humanitarian Licensing. In *Executive Guide to Intellectual Property Management in Health and Agricultural Innovation: A Handbook of Best Practices* (Krattiger A, RT Mahoney, L Nelsen et al.). MIHR (Oxford, UK), PIPRA (Davis, USA), Oswaldo Cruz Foundation (Fiocruz, Rio de Janeiro, Brazil), and *bio*Developments-International Institute (Ithaca, USA). Available online at www.ipHandbook.org.

- in licensing for humanitarian use, led by Zeneca, of IP components it did not already own
- licensing by Zeneca, as Syngenta, via the inventors of the assembled (or bundled) intellectual property to public sector institutions in developing countries that could use the rights for the benefit of resource-poor farmers, and others, deficient in vitamin A

The patented key technologies for Golden Rice production include core patents related to the specific biosynthetic pathway. These patents were filed by the inventors, Potrykus and Beyer. Their work built on myriad other technologies that were published in issued patent documents and scientific literature. These core patents were licensed to Zeneca, which already owned its own plant-biotechnology-related patents. Zeneca then negotiated access to all possibly necessary patents, including intellectual property from Bayer AG, MonsantoCompany, Novartis AG, Orynova BV, and Zeneca Mogen BV.

All of these companies, including Zeneca (which, coincidentally, almost immediately merged with Novartis Agribusiness to form Syngenta), provided access to their technologies, free of charge, for defined humanitarian research and use of Golden Rice in developing countries. It is important to note that, contrary to what many commentators state, the licensing process was relatively uncomplicated, with the involvement of commercially experienced people.

Licensing

Within a short time, 16 further licenses, including licenses with the right to further sublicense (for example, the license issued to IRRI), were issued to public sector licensees. Thus national programs in Bangladesh, China, India, Indonesia, the Philippines, South Africa, and Vietnam obtained licenses for use of the technology in local rice varieties important in VAD areas.

Terms of the humanitarian license agreement

The Golden Rice Humanitarian Board, although not a legal entity, provides a forum for discussion of strategic and tactical issues relating to the humanitarian project. Both Potrykus and Beyer have the right to issue licenses. Two licensees also have that right, as does Syngenta, which has not exercised its right. All the licenses are in the same form, as proposed by Syngenta and agreed to by the inventors.

The essential elements of the licenses include the following points:

- Syngenta retains commercial rights, although it has no current plans to commercialize Golden Rice. Humanitarian use, and research leading to it, is allowed.
- Humanitarian use is defined as use in developing countries by resource-poor farmers (earning less than US$10,000 per year from farming).
- The technology must be introduced into public seed varieties, as a way to optimize public sector benefit and use.
- No technology fee (or surcharge) may be charged for Golden Rice, as a way to optimize public sector benefits.
- Sale of Golden Rice is authorized by farmers, as a way to reach urban poor.
- Farmers are allowed to reuse harvested seeds.
- Golden Rice may not be released in a country that lacks biosafety regulations and where official government review has not been made to ensure health and environmental safety.
- Export of Golden Rice is not permitted, except to other licensees for humanitarian research and subsequent use. (Export of crops is a commercial activity. The purpose of the humanitarian project is to assist resource-poor people in overcoming VAD).
- With regard to improvements to the Golden Rice technology:
    - Humanitarian use of any improvements to Golden Rice is guaranteed under the same terms of the original agreement (and thus any improvements to the technology will serve the humanitarian purpose). Syngenta has acted on this—donating to the humanitarian project new transformations, including the intellectual property and results reported in Paine and colleagues.[5]
    - Commercial rights to improvements of the technology are granted back to Syngenta.
- No warranties are given by the licensor or licensors (as is common for licenses), and each party is responsible for what it controls.

KEY LESSONS LEARNED

The rapid resolution of the IP constraints surrounding Golden Rice demonstrated, first of all, how effective IP management, coupled with strong collaborations between the public and private sectors, can help achieve global access to new technologies and products for humanitarian goals. The IP constraints identified by Kryder and colleagues[6] did not delay the development of the product, and their clarification and resolution required only managerial and influencing skills and the resulting goodwill of IP owners.

More specifically, three specific lessons have been learned:

1. Intellectual property and patents did not delay the development and introduction of Golden Rice by a single day. Notwithstanding this, the resolution of the potential IP constraints could not be ignored.
2. Other constraints are much more critical to the introduction of Golden Rice, in particular, and to potentially life-saving food biotechnology

applications, in general. These constraints are, in decreasing order of importance:

- the necessity of governments to establish a sustained and positive policy priority for the adoption of all relevant, including novel, technologies in agriculture
- the importance of the establishment of affordable, workable, and science based regulatory systems designed to comply with international obligations *and* to address local needs and concerns (The unnecessarily burdensome, overly politicized regulatory requirements for genetically modified organisms [GMOs] and the absence of consideration of benefit has led to years of delay in the introduction of Golden Rice technology. Yet there is no evidence to justify such a burdensome regulatory system.)
- the need for the capacity and funding of national agricultural rice research institutions to keep segregated different versions of genetically modified crops, including conducting field trials with them
- the anticipated need to develop effective seed distribution systems for reaching farmers in remote areas, including the presence of private sector entities willing to invest in seed distribution systems (However, a major aim is also to have farmers pass the seed on to neighboring farmers to reach "infrastructure remote" areas often associated with VAD.)

3. Recognizing that universities are not set up to develop products, Syngenta was instrumental in converting the proof-of-concept results generated at ETH Zurich and University of Freiburg into deliverable products. Although Syngenta retained commercial exclusivity for the technology, the company decided not to develop a commercial product of Golden Rice for markets in developed countries. Syngenta's continued support of the project with advice and scientific know-how has proven absolutely essential for the success of the product-development partnership.

From a broader perspective, the FTO review of Golden Rice, in particular before "commercial analysis," served as a wake-up call to the public sector to pay more attention to IP management as a powerful tool for meeting public sector goals. Concern about potential constraints on public sector research and innovation in agriculture spurred the public sector's interest in intellectual property. One important response was work that led to the formation of the Public Intellectual Property Resource for Agriculture (PIPRA). Supported by, among others, the Rockefeller and McKnight foundations, PIPRA is a public sector initiative that recognizes that continuing and enhancing relationships with the private sector, and between the public sector institutions, are critical components of the utilization of intellectual property to meet public sector goals. As part of its initial work, PIPRA began a study of the structure of IP ownership in agricultural biotechnology. In the words of the study's authors, Richard C. Atkinson and colleagues: *This study found that roughly one-fourth of the patented inventions were made by public-sector researchers, which is substantially larger than the IP portfolio held by any single agricultural biotechnology company. It is, however, highly fragmented across institutions and across technology categories. And much of this IP has been licensed, often under terms that are confidential but which have likely resulted in greatly restricted access to the underlying technologies.*[7] *This study suggested that, apart from a few important exceptions, public-sector scientists have invented many of the types of technologies that are necessary to conduct basic biological research and develop new transgenic plant varieties. For instance, they have developed technologies to transfer genes into plant cells; have characterized specific DNA elements that drive unique patterns of gene expression; and have identified many genes that confer important plant traits. Such discoveries underscore the fact that public-sector research institutions have been significant sources of technological innovation.*[8]

We believe that this study involving Golden Rice shows how public and private sector innovations can be put to work directly to help the poor with more focused public sector IP management. Indeed, IP management is merely one of the components needed to bring innovation to the poor.[9] Other factors, such as regulatory requirements, can be much more costly and do constitute tremendous barriers to the poor benefiting from innovations that are becoming commonplace in much of the world. ■

For further information, please contact:

ANATOLE KRATTIGER, *PO Box 26, Interlaken, NY 14847, U.S.A.* *afk3@cornell.edu*

INGO POTRYKUS, *Im Stigler 54, 4312 Magden, Switzerland.* *ingo@potrykus.ch*

JORGE E. MAYER, *Golden Rice Project Manager, Campus Technologies Freiburg, Stefan-Meier-Str 8, 79104 Freiburg, Germany.* *jorge.mayer@goldenrice.org*

---

1 Golden Rice was invented by Ingo Potrykus, then at ETH in Zurich, Switzerland, and Peter Beyer of the University of Freiburg, Germany. See also www.goldenrice.org.

2 childinfo.org/areas/vitamina/ .

3 Potrykus I. 2001. Golden Rice and Beyond. *Plant Physiology* 125:1157–1161.

4 Kryder D, SP Kowalski and AF Krattiger. 2000. The Intellectual and Technical Property Components of pro-Vitamin A Rice (GoldenRice™): A Preliminary Freedom-to-Operate Review. ISAAA Briefs No 20. ISAAA: Ithaca, NY. www.isaaa.org/kc/bin/isaaa_briefs/index.htm.

5 Paine JA, Shipton CA, Chaggar S, Howells RM, Kennedy M, Vernon G, Wright SY, Hinchcliffe E, Adams JL, Silverstone a, Drake R. 2005. Improving the nutritional value of Golden Rice through increased pro-vitamin A content. Nature Biotechnology 23(4):482–487.

6 See *supra* note 4.

7 Only within the temporal and spatial limits allowed by the patent system (note added by the authors of this case study).

8 Atkinson RC, RN Beachy, G Conway, FA Cordova, MA Fox, KA Holbrook, DF Klessig, RL McCormick, PM McPherson, HR Rawlings III, R Rapson, LN Vanderhoef, JD Wiley and CE Young. 2003. Public Sector Collaboration for Agricultural IP Management. *Science* 301(5630):174–175.

9 Current Golden Rice transformation events in the humanitarian project's development process were all designed and made by Syngenta to need access to no third party intellectual property.

Saving Forests and Creating a New Cash Crop in the Middle East and Asia: University of Minnesota

The high demand for agarwood—wood soaked with a resin produced by a small portion of Aquilaria trees in southeast Asia and Indonesia—nearly decimated the species. The trees produce the resin only when injured and, before researchers stepped in, usually when the trees were 50 or more years old.

Agarwood and its resin are highly prized in the Middle East and Asia, particularly in Islamic and Buddhist cultures, where the wood and resin are used in perfumes, ceremonial incense, traditional medicine, and other applications. Unfortunately, determining whether a particular standing Aquilaria tree contains agarwood is nearly impossible, so harvesters were felling and sawing up Aquilaria trees until they were close to extinction in much of their natural range.

Robert Blanchette, Ph.D., of the University of Minnesota, and the nonprofit organization Rainforest Project, based in the Netherlands, have jointly developed an easy and inexpensive method to induce agarwood formation in trees that are only three to six years old. Now, instead of cutting down trees found in the forest, farmers can grow stands of Aquilaria trees on plantations, induce production of agarwood in those trees, and sell them as a new cash crop.

This practice will benefit regional farmers and their local economies, reduce the threat of extinction to native populations of Aquilaria trees, and ensure a long-term supply of agarwood for centuries-old cultural and religious uses. The University of Minnesota has licensed the technology to the Rainforest Project, which is leading distribution efforts beginning in Southeast Asia. ■

AUTM. 2007. Saving Forests and Creating a New Cash Crop in the Middle East and Asia: University of Minnesota. In *Executive Guide to Intellectual Property Management in Health and Agricultural Innovation: A Handbook of Best Practices* (Krattiger A, RT Mahoney, L Nelsen et al.). MIHR (Oxford, UK), PIPRA (Davis, USA), Oswaldo Cruz Foundation (Fiocruz, Rio de Janeiro, Brazil), and *bio*Developments-International Institute (Ithaca, USA). Available online at www.ipHandbook.org.

Editors' Note: We are most grateful to the Association of University Technology Managers (AUTM) for having allowed us to adapt this case study for inclusion in this *Executive Guide*. The original was published by AUTM. 2006. *Technology Transfer Works: 100 Cases from Research to Realization* (Reports from the Field). Association of University Technology Managers, Northbrook, IL. www.betterworldproject.net.

Building Healthy Forests with Early-Stage Propagation: University of Saskatchewan

Forestry is among the world's largest industries; it has a significant impact on people's lives around the world. One of the industry's greatest challenges is increasing the efficiency of land areas designated for commercial forestry by improving their productivity. Another challenge is complying with environmental standards, which provide guidelines for reforestation, production in environmentally sensitive areas, and long-term sustainable forest management.

A crucial step toward increased efficiency is growing stronger trees. With many plant species, horticulturalists can create new varieties by taking cuttings from plants with desirable characteristics and encouraging the cuttings to root. This propagation method has yielded scores of different kinds of plants including orchids, roses, grapevines, and fruit trees. But the method doesn't work well with most forest trees because the cuttings are less likely to take root.

Researchers at the University of Saskatchewan developed a technology called somatic embryogenesis (SE), a complex propagation process that relies on the splitting of one embryo into two or more identical embryos. The method allows scientists to grow two or more plants that have the same genetic makeup. With SE, propagation occurs earlier in the plant's lifecycle and rooting is more likely to be successful.

SE offers several economic benefits to the forestry industry including greater success in propagating desirable trees and the ability to grow seedlings year-round. The University of Saskatchewan licensed the patent-protected technology to CellFor based in Vancouver, British Columbia, Canada. In 2003, the company began working with timberland managers to plant loblolly pine seedlings propagated from fast growing, disease-resistant varieties in the southeastern U.S. states including Georgia and Mississippi.

Today the company maintains more than 3,000 unique genetic lines and has an extensive network of field trials aimed at testing and further refinements. The technology allows CellFor to produce seedlings that grow faster, generate a higher yield, and produce superior wood, while reducing production costs and enhancing resistance to disease and pests.

Read more about SE at www.cellfor.com. ■

AUTM. 2007. Building Healthy Forests with Early-Stage Propagation: University of Saskatchewan. In *Executive Guide to Intellectual Property Management in Health and Agricultural Innovation: A Handbook of Best Practices* (Krattiger A, RT Mahoney, L Nelsen et al.). MIHR (Oxford, UK), PIPRA (Davis, USA), Oswaldo Cruz Foundation (Fiocruz, Rio de Janeiro, Brazil), and *bio*Developments-International Institute (Ithaca, USA). Available online at www.ipHandbook.org.

Editors' Note: We are most grateful to the Association of University Technology Managers (AUTM) for having allowed us to adapt this case study for inclusion in this *Executive Guide*. The original was published by AUTM. 2006. *Technology Transfer Works: 100 Cases from Research to Realization* (Reports from the Field). Association of University Technology Managers, Northbrook, IL. www.betterworldproject.net.

DNA Hepatitis B Vaccine: International Vaccine Institute, Korea

Intellectual property as a barrier to market entry is examined through a study of the development and introduction of recombinant DNA (rDNA) hepatitis B vaccine (HBV) in developing countries. The most widely used vaccines in the mid-1980s were produced by Merck and GlaxoSmithKline, which were the first two companies to introduce the rDNA HBV. Almost a decade later, Korean and Indian manufacturers entered the rDNA HBV vaccine market. However, the price remained relatively high (>US$7 per dose) until the Global Fund for Children's Vaccine (today amalgamated with the GAVI Alliance) was established with seed funding from the Bill and Melinda Gates Foundation. With this funding the price dropped to less than US$0.30 per dose. This study sought to identify factors that affected supplying low-cost vaccine to the public sector.

Merck and GlaxoSmithKline licensed three key patents assigned to Institut Pasteur, Biogen, and the University of California. These patents were filed in the United States, Europe, and a few other developed countries. The companies stated that licenses to more than 90 other patents relating to manufacturing processes such as isolation and purification were also needed.

The Korean companies pursued collaborations or joint ventures but chose not to focus on the United States and European markets mainly due to regulatory and market entry costs. These companies sought World Health Organization prequalification for their production facilities and approval for the vaccine from several governments in Asia and other countries in the developing world.

A Korean company, LG Chem, formed a joint venture with Chiron. Chiron had a license from the University of California (key scientists at Chiron were inventors on the University patent). Through the joint venture, LG scientists could learn how to make the vaccine. Korea Green Cross entered into a joint venture with Rhein Biotech, which had developed and patented its own method for making the vaccine. Having surveyed globally for a partner to exploit its technology, the German company chose Korea because of the low cost of production achieved by Korea Green Cross. The Korean company Cheil Sugar also sought to enter the market for the vaccine and attempted to develop its own technology. After nearly 20 years of effort, Cheil Sugar (now CJ Corp.) abandoned the effort.

These LG Chem and Korea Green Cross alliances were formed in an environment that was supportive of biotechnology innovation. The Korean government accorded high priority to R&D in biotechnology and provided strong support for overseas training and domestic research. The biotech industry received the backing of private sector investment, and domestic and export markets were encouraged by the government. High priority was given by the Korean government to hepatitis B immunization thereby ensuring an initial market for the companies.

This case study concludes that intellectual property was not a major barrier to market entry. Korean companies took several years to enter the market because of lack of resources, including a small cadre of scientific staff, the need to improve national regulatory systems, and, importantly, the small size of the global market. The international public sector market remained underdeveloped in part because of its low priority for large pharmaceutical companies, lack of demand by

Mahoney R. 2007. DNA Hepatitis B Vaccine: International Vaccine Institute, Korea. In *Executive Guide to Intellectual Property Management in Health and Agricultural Innovation: A Handbook of Best Practices* (Krattiger A, RT Mahoney, L Nelsen et al.). MIHR (Oxford, UK), PIPRA (Davis, USA), Oswaldo Cruz Foundation (Fiocruz, Rio de Janeiro, Brazil), and *bio*Developments-International Institute (Ithaca, USA). Available online at www.ipHandbook.org.

Editors' Note: This case study was originally presented at the MIHR conference Using Intellectual Property for Improved Health in Developing Countries: An Evidence-Based Approach to Good Practice, Bellagio, Italy, June 14–18, 2004.

developing countries, and little procurement by international donor agencies.

Each company sought to secure intellectual property in order to bring its vaccines to market, but patents did not hinder developing the vaccine because the companies focused on markets in countries where the three key patents were not filed. Intellectual property had some affect on access but was much less important than regulatory and manufacturing issues, and market development. However, the situation might be different post-2005 when most developing countries are required to be TRIPS compliant. In the TRIPS era, patents may be routinely filed in many countries such as Brazil, China, India and Korea thereby making it more difficult for second comers to produce in and sell to those large and important markets.

FEATURES OF THE CASE

Types of agreements

Merck and GlaxoSmithKline obtained licenses to three key patents assigned to Pasteur Institute, the University of California, and Biogen. These patents were filed in the United States, Europe, and a few other developed countries. Both companies obtained licenses to numerous other patents having to do with manufacturing processes, including isolation and purification. The Korean companies took three different routes. Cheil sought to develop the technology on its own. LG Chem (previously Lucky Gold Star) formed a joint venture through which it obtained know-how for the production of the vaccine. Korea Green Cross entered into a joint venture with a foreign company, Rhein Biotech of Germany, which had developed an alternate production method.

Patent and IP rights decisions

Merck and, to a lesser extent, GlaxoSmithKline were primarily interested in markets in developed countries and obtained all necessary licenses to patents filed in those countries. The Korean companies opted not to pursue the same markets as Merck and GlaxoSmithKline because of the costs of obtaining regulatory approval and establishing a market presence associated with those markets. LG Chem decided to proceed simply by obtaining know how and relying on its low cost of manufacture and aggressive marketing skills. Korea Green Cross and Rhein Biotech formed a joint venture in which they exploited the Rhein Biotech patent for a manufacturing method different from that used by Merck and GlaxoSmithKline. Cheil sought to develop its own proprietary technology but eventually abandoned this effort.

Policy implementation

All five companies complied with the laws and regulations applicable in their legal jurisdictions. Each company sought a clear IP path to marketing the vaccines. To the author's knowledge, no infringement lawsuits were brought against any of the companies.

EXTERNAL FACTORS THAT AFFECTED DECISION MAKING

Key factors that affected decisions made by the Korean manufacturers were the costs of regulatory compliance with respect to and market entry into the United States and Europe. In addition, the Korean Food and Drug Administration had been undertaking certain improvements, and until those improvements were completed, the Korean manufacturers could not supply United Nations agencies. The Korean manufacturers also had to obtain World Health Organization prequalification for their production facilities, which LG Chem and Korea Green Cross succeeded in accomplishing in the late 1990s. The key factor in allowing the Korean manufacturers to supply low-cost vaccine to the public sector was the establishment of a market through the Global Fund for Children's Vaccine, initially funded by the Bill and Melinda Gates Foundation.

LESSONS LEARNED AND HEALTH-ACCESS ISSUES

Intellectual property was an important issue for all the companies involved in the DNA hepatitis B vaccine project, but IP issues did not significantly impede the pace at which the Korean manufacturers were able to enter the market. The key factors were (in approximate order of importance):

- requirement for a global market
- need to meet international regulatory standards
- need to undertake in-house R&D or obtain know-how from a joint-venture partner
- time it took to construct and improve production facilities that would meet WHO requirements

Further, the ability of Rhein Biotech and Korea Green Cross to exploit the Rhein Biotech patent on an alternate production method provides support for the argument that it is easier to develop and market vaccines in a complex IP environment than it is to develop and market new defined chemical entities that have been patented. Vaccines are complex biological products that can be made through a diversity of procedures while defined chemical entities are single molecules that may be easy to produce only through one process. ■

For further information, please contact:
RICHARD T. MAHONEY, *Director, Vaccine Access, Pediatric Dengue Vaccine Initiative, International Vaccine Institute, San Bongcheon-7dong, Kwanak-ku, Seoul 151-818, Republic of Korea. rmahoney@pdvi.org*

HIV/AIDS Vaccine: Indian Council of Medical Research

This HIV/AIDS initiative is a collaborative venture between the Indian Council of Medical Research (ICMR), New Delhi, the International AIDS Vaccine Initiative (IAVI), New York, National AIDS Control Organization (NACO/Indian Ministry of Health), New Delhi and Therion Biologics, Cambridge, Mass. The project aims to develop a safe and effective HIV/AIDS vaccine—such development has been mandated by the Indian government—for India and other developing countries. The vaccine has now been developed by ICMR in collaboration with Therion and is undergoing clinical trials.

Under the terms of this public-private partnership (PPP), ICMR will provide technical expertise, obtain all necessary permissions and permits, conduct R&D to develop the vaccine in collaboration with Therion, prepare the community (in India) for clinical trials, and conduct the trials. ICMR will select an Indian partner for the manufacture of vaccine and has overall responsibility for ensuring that the project is executed according to its objectives. NACO will facilitate the execution of the project. IAVI will support the project, facilitate development of an appropriate vaccine through transfer of technology from Therion, engage in capacity building and advocacy, and facilitate technology transfer for the local manufacture of the vaccine. Therion will assist ICMR with the vaccine development and help transfer technology to the selected Indian manufacturer.

The project involved an overall agreement between ICMR and IAVI, a patent and technology transfer agreement between ICMR and IAVI, and an IP (intellectual property) rights and confidentiality agreement between ICMR and Therion Biologics. A project management committee was set up, comprising representatives from ICMR and IAVI, to coordinate and monitor all activities and assessments of the R&D programs. The committee is also responsible for strategic IP management.

All new intellectual property generated will be jointly held by IAVI and ICMR, and the Indian government shall have the exclusive right to use all patent and other new IP rights to inventions arising out of the program to benefit India and its neighboring countries. The ICMR will grant nonexclusive royalty-free and sublicensable licenses to all new intellectual property arising out of the project to selected third parties in order to make, use, sell, and import the HIV/AIDS vaccine in countries other than those indicated in the agreement (to the extent ICMR has the right to permit this use). The IAVI shall have IP rights for rest of the world.

Initially, the program was to be implemented only in India, but the Government of India, realizing that the program could benefit other developing countries as well, asked for licensing rights. In arriving at this realization, policymakers (bureaucrats) of the government needed to be educated about intellectual property and its role in technology transfer. This case has highlighted the importance of keeping government officials involved in order for an international PPP to be successful.

Although no patents were filed in India, a significant amount of clinical trial data was generated. From an IP perspective, it was crucial to recognize private sector interests. Therion has global rights for the technology needed for the vaccine construct, but India

Satyanarayana, K. 2007. HIV/AIDS Vaccine: Indian Council of Medical Research. In *Executive Guide to Intellectual Property Management in Health and Agricultural Innovation: A Handbook of Best Practices* (Krattiger A, RT Mahoney, L Nelsen et al.). MIHR (Oxford, UK), PIPRA (Davis, USA), Oswaldo Cruz Foundation (Fiocruz, Rio de Janeiro, Brazil), and *bio*Developments-International Institute (Ithaca, USA). Available online at www.ipHandbook.org.

Editors' Note: An earlier version of this case study was presented at the MIHR conference Using Intellectual Property for Improved Health in Developing Countries: An Evidence Based Approach to Good Practice, Bellagio, Italy, June 14–18, 2004.

will have rights to improvements made to the vaccine. Therion's stringent IP regulations meant confidentiality agreements were imposed on collaborating scientists, which the Government of India appreciated.

The recruitment process for the vaccine trials envisages serious ethical concerns as well as potential liability issues, as the vaccine is for HIV/AIDS. It was recognized that clinical trials must be conducted in a fair and transparent manner and the interests of participants protected through informed consent as per the ICMR's Ethical Guidelines and that all necessary safeguards to protect subjects of the study had to be built into the system.

The case study recognizes (1) the role of "honest broker" that international nongovernmental organizations like IAVI can play in a PPP, providing funding and access to high technology from a private company; (2) the need to educate policymakers (bureaucrats) from the beginning of a project to ensure smooth progress; and (3) the equally crucial need to involve policymakers, lawmakers, politicians, women's associations, and other civil society organizations in the execution of such projects that envisage clinical trials. This project is offered as an example of productive North–South collaboration and broad capacity building and a partnership in which the strengths of the partners complement each other.

TYPES OF AGREEMENTS

As part of the HIV/AIDS project, ICMR entered into the following types of agreements:

- an overall agreement between the ICMR and the IAVI for the entire project including provisions for development, upscaling, manufacture, and distribution of the vaccine in India, neighboring countries, and the rest of the world
- a separate technology transfer and manufacturing agreement between Therion and the manufacturer identified jointly by IAVI, the Government of India, and Therion

IP RIGHTS DECISIONS AND IP MANAGEMENT

The project has resulted in the following arrangements with respect to IP rights and strategic IP management issues:

- IAVI and the Government of India-ICMR will jointly hold the new intellectual property generated during the project.
- The Government of India-ICMR shall have exclusive rights to use all patent and other new IP rights to inventions arising out of the program in India and neighboring (SAARC) countries.
- ICMR grants IAVI a nonexclusive, worldwide, royalty-free sublicensable license to all new patents and other intellectual property arising out of the program that would permit IAVI or third parties selected by IAVI to make, use, sell, offer for sale, and import HIV/AIDS vaccines in countries other than those indicated in the agreement (to the extent ICMR has the right to permit the use of the same).
- Intellectual property is jointly managed by the ICMR and IAVI through the project management committee.

POLICY IMPLEMENTATION

Policy is implemented through a project management committee comprising representatives from the IAVI, ICMR and NACO, and jointly chaired by members appointed by ICMR and IAVI. The committee is responsible for the coordination and monitoring of all activities, periodic assessments and updates, and refinements and revisions of the R&D program.

EXTERNAL FACTORS THAT AFFECTED DECISION MAKING

A number of external considerations influenced ICMR's strategies and decision making. These include:

- the potential use of the vaccine(s) in India's neighboring countries
- the need to provide an effective and affordable vaccine to the people

KEY LESSONS LEARNED AND HEALTH-ACCESS ISSUES

The following items represent key lessons from ICMR's HIV/AIDS vaccine project, which may be applicable to other entities that aim to utilize intellectual property:

- Only through strategic public-private partnerships can such ventures succeed.
- Private sector's interests need to be considered.
- The role of an international nongovernmental agency such as IAVI is important and vital for the success of such a project.
- There is a need to educate government officials on issues relating to IP rights and technology transfer, as the government's role is crucial in the clearance and approval of projects of national interest.
- The importance of (1) ethics in carrying out clinical trials and (2) the need to involve policymakers, women's associations, and other civil society groups in the execution of the project cannot be overstated. ■

For further information, please contact:
KANIKARAM SATYANARAYANA, *Director, Intellectual Property Rights Unit, Indian Council of Medical Research, Ansari NagarNew Delhi-110, 029, India. kanikaram_s@yahoo.com*

Malaria Vaccine: Malaria Vaccine Institute and GlaxoSmithKline Biologicals

Malaria remains one of the world's deadliest killers. Every year, the disease takes the lives of more than one million people, mostly sub-Saharan African children under age five. Hundreds of millions more people fall ill from the mosquito-borne disease. Major hurdles to traditional prevention and treatment strategies include drug resistance by the malaria parasite and heightened resistance to insecticides by the mosquito that transmits it. Scientists have been working for decades to develop a preventive malaria vaccine. While they have successfully demonstrated that such a vaccine is possible, many challenges continue to impede progress on the road to an effective product. The complex life cycle of the malaria parasite (the most deadly being the *Plasmodium falciparum* species) represents a major hurdle. While each stage of the parasite's development offers an opportunity to attack it, the parasite's ability to evade people's immune responses has made the development of a malaria vaccine technically difficult.

PATH[1] is an international, nonprofit organization that creates sustainable, culturally relevant solutions, enabling communities worldwide to break longstanding cycles of poor health. The PATH Malaria Vaccine Initiative (MVI)[2] is a global program established in 1999 through an initial grant of US$50 million from the Bill and Melinda Gates Foundation, which has since awarded MVI an additional US$207.6 million, including US$107.6 million to complete development of the most promising malaria vaccine candidate. MVI's mission is to accelerate the development of promising malaria vaccines and to ensure that they are available and accessible in the developing world.

Among the candidates in MVI's portfolio, the RTS,S vaccine of GlaxoSmithKline (GSK) Biologicals[3] is the most advanced. Created in 1987, the pre-erythrocytic vaccine candidate's early development was undertaken by GSK Biologicals, in close collaboration with the Walter Reed Army Institute of Research. In January 2001, GSK Biologicals, MVI, and other partners—with support from the Bill and Melinda Gates Foundation—entered into an agreement to develop the vaccine for children in sub-Saharan Africa. Clinical evaluation of RTS,S began in 1992 and the results since then represent a breakthrough for malaria vaccine development. RTS,S has proved to be effective for at least 18 months in reducing clinical malaria by 35 percent and severe malaria by 49 percent. *Time* magazine highlighted this project as one of the most important health accomplishments of 2005.

PARTNERS

Partners in the malaria vaccine project are

- from academia, New York University
- from government, Walter Reed Army Institute of Research
- a nonprofit organization, PATH Malaria Vaccine Initiative
- a pharmaceutical company, GSK Biologicals
- two health-research centers, the Center for International Health (CIH), Hospital Clínic

---

MIHR/PIPRA. 2007. Malaria Vaccine: Malaria Vaccine Institute and GlaxoSmithKline Biologicals. In *Executive Guide to Intellectual Property Management in Health and Agricultural Innovation: A Handbook of Best Practices* (Krattiger A, RT Mahoney, L Nelsen et al.). MIHR (Oxford, UK), PIPRA (Davis, USA), Oswaldo Cruz Foundation (Fiocruz, Rio de Janeiro, Brazil), and *bio*Developments-International Institute (Ithaca, USA). Available online at www.ipHandbook.org.

Editors' Note: This case study was prepared by MIHR members of the Technology Managers for Global Health (TMGH), a special interest group of the Association of University Technology Managers (AUTM) (see www.tmgh.org) and adapted for this *Executive Guide*. The original version was published as part of a collection of case studies: MIHR/TMGH. 2007. *Academic Licensing to Global Health Product Development Partnerships* (ed. U Balakrishnan). MIHR: Oxford, U.K.

of the University of Barcelona and Centro de Investigação em Saude da Manhiça (CISM)

The Bill and Melinda Gates Foundation provided funding for the project.

TECHNOLOGY

This vaccine candidate is a recombinant protein that fuses a part of the *P. falciparum* circumsporozoite protein with the hepatitis B surface antigen molecule. Combined with a proprietary GSK adjuvant system, RTS,S induces the production of antibodies and white blood cells that are believed to diminish the capacity of the malaria parasite to infect, survive in, and develop in the human liver. In addition to inducing partial protection against malaria, the RTS,S vaccine candidate stimulates a protective immune response to hepatitis B, which commonly infects people in developing countries.

PROGRESS, CURRENT STATUS, AND GOALS

GSK Biologicals and MVI are currently conducting several small-scale trials in infants and young children, the groups most vulnerable to malaria and that would benefit most from an effective malaria vaccine. Working with in-country research institutions, clinical trials are ongoing in partner African countries, including Mozambique, Tanzania, Gabon, and Ghana. A variety of immunization schedules will be assessed, and the efficacy of the vaccine will be evaluated when administered with the Expanded Programme on Immunization. If these trials are successful, the partners will proceed to a large-scale Phase III clinical trial to determine the efficacy of the vaccine in the same age group. If all goes well, the RTS,S vaccine could be licensed as early as 2010.

ABOUT THE CLINICAL PARTNERS

The Center for International Health (CIH), Hospital Clínic of the University of Barcelona

The Center for International Health (CIH) is a pioneering structure within the University of Barcelona's Hospital Clínic, the leading Spanish biomedical research center.[4] The CIH is involved in health care, training, and research in global health issues. The collaborative programs in Africa, particularly the development of the Manhiça Health Research Center, which is in close partnership with Mozambican institutions, are a central component of the activities of the CIH.

THE CENTRO DE INVESTIGAÇÃO EM SAUDE DA MANHIÇA

Centro de Investigação em Saude da Manhiça (CISM) is the first peripheral health research center in Mozambique to undertake medical research into key health problems in that country. Founded in 1996, CISM was developed under a collaborative program between the Mozambique Ministry of Health, the Maputo School of Medicine (Universidade Eduardo Mondlane), and the Hospital Clínic of the University of Barcelona with core funding from the Spanish Agency for International Cooperation.[5]

MOZAMBIQUE'S MINISTRY OF HEALTH

The mission of Mozambique's Ministry of Health is to promote and preserve the health of the Mozambican population, to promote and provide quality and sustainable healthcare services, and to, with equity and efficiency, gradually increase access to sustainable healthcare for all Mozambicans. ■

For further information, please contact:
GLAXOSMITHKLINE BIOLOGICALS, *Alice Grasset, Phone: +32-2-656 8774 or +32-475-309 020.*

PATH MALARIA VACCINE INITIATIVE (MVI), *Ellen Wilson, Phone: +1-301-652 1558 or +1-301-922 4969.*

CENTRO DE INVESTIGAÇÃO EM SAUDE DA MANHIÇA, *Hospital Clínic of the University of Barcelona, Marc de Semir, Phone: +34-93-227 5700 or +34-62-794 7528.*

---

1 PATH: www.path.org.

2 Malaria Vaccine Initiative: www.malariavaccine.org.

3 GlaxoSmithKline Biologicals: www.gsk-bio.com.

4 University of Barcelona Hospital Clínic: www.hospitalclinic.org.

5 CISM: www.manhica.org.

Rotavirus Vaccine: NIH Office of Technology Transfer

The National Institutes of Health (NIH), as part of the U.S. Public Health Service (PHS), is dedicated to improving the public health of individuals worldwide through innovative research and the funding of critical medical research programs. Immunization against rotavirus disease is an important public health initiative supported by several organizations worldwide. This case study describes the partnerships between PHS and institutions in Brazil, China, India, and the United States that have been established to facilitate development of a safe, effective, and affordable vaccine for arresting the overwhelming mortality associated with rotavirus infection in the developing world.

PARTNERS

Partners in the rotavirus vaccine project are:

- from government: the National Institutes of Health/U.S. Public Health Service
- nonprofit organizations: Fundação Butantan (Sao Paulo, Brazil), Chengdu Institute of Biological Products (Chengdu, China), and Wuhan Institute of Biological Products (Wuhan, China)
- for-profit companies: Aridis Pharmaceuticals (United States), Bharat Biotech International, Ltd. (Hyderabad, India), Biological E., Ltd. (Hyderabad, India), Shanta Biotechnics, Ltd. (Hyderabad, India), and Serum Institute of India, Ltd. (Pune, India)

EPIDEMIOLOGICAL FEATURES OF ROTAVIRUS

Rotavirus is the leading cause of severe dehydrating diarrhea in infants and children worldwide. According to a report issued by the World Health Organization (WHO), each year, the disease is responsible for about 25 million clinic visits, two million hospitalizations, and between 352,000 and 592,000 deaths in children age five and under. As one can imagine, the worldwide economic burden associated with rotavirus disease is staggering, exceeding $1 billion each year in medical costs. Children in developing countries are disproportionately at risk of dying from rotavirus-related infection. In India alone, rotavirus is blamed for the deaths of approximately one out of every 250 children each year, and in China, the disease accounts for more than 34,000 deaths per year. This rotavirus-associated mortality is due in part to inadequate sanitation and to inadequate access to intravenous rehydration therapy in poor countries.

THE TECHNOLOGY

The human-bovine reassortant rotavirus vaccine is an invention of Dr. Albert Kapikian and his colleagues at the National Institutes of Allergy and Infectious Disease (NIAID) of the NIH. The invention was further developed through collaboration with Wyeth Pharmaceuticals. The vaccine technology is based on multivalent immunogenic compositions comprising four human-bovine reassortant rotaviruses and involves the insertion of the gene-encoding VP7 protein of G1, G2, G3, and G4 human rotavirus strain into a bovine rotavirus backbone. These VP7 serotypes represent the clinically most prevalent human rotavirus serotypes. Additionally, the basic quadrivalent vaccine formulation can be augmented with G9 and G8 strains (or one of these additional strains for a pentavalent formulation) to make a hexavalent formulation. Serotype 9 (G9) has emerged as an important strain in Latin America and the most important

MIHR/PIRPA. 2007. Rotavirus Vaccine: NIH Office of Technology Transfer. In *Executive Guide to Intellectual Property Management in Health and Agricultural Innovation: A Handbook of Best Practices* (Krattiger A, RT Mahoney, L Nelsen et al.). MIHR (Oxford, UK), PIPRA (Davis, USA), Oswaldo Cruz Foundation (Fiocruz, Rio de Janeiro, Brazil), and *bio*Developments-International Institute (Ithaca, USA). Available online at www.ipHandbook.org.

strain in Brazil, whereas G8 is prevalent in many African countries.

Originally, the human-bovine reassortant rotavirus vaccine was intended as a second-generation rotavirus vaccine. It was developed alongside the human-rhesus reassortant vaccine, RotaShield, an earlier invention of Dr. Kapikian that was commercialized by Wyeth following U.S. Food and Drug Administration approval in 1998. RotaShield was voluntarily removed from the market in 1999 after the vaccine was suspected of being linked to an increased risk for intussusception in children. After the withdrawal of RotaShield from the market, interest in the human-bovine reassortant technology increased, which led to multiple applications for commercial licensing as detailed below.

LICENSE AGREEMENTS

Published reports and presentations by NIH NIAID investigators generated significant interest in the human-bovine rotavirus vaccine technology from companies and institutions worldwide. In 2005, eight organizations, one in the United States and seven based in the developing world, were granted licenses from PHS to manufacture and distribute the rotavirus vaccine. The licensees are U.S.-based Aridis Pharmaceuticals; Fundação Butantan, a Brazilian government institution; Bharat Biotech International, Biological E., Ltd., Shantha Biotechnics, Ltd., and Serum Institute of India, Ltd., all India-based companies; and Chengdu Institute of Biological Products and Wuhan Institute of Biological Products, both funded by the government of China. The vaccine technology is covered by issued patents (and pending patent applications) in the United States, Europe, Canada, Japan, China, India, Korea, Brazil, and Australia, thus NIH decisions regarding the license agreements were based on thorough evaluation of the applicants and their capabilities with regard to vaccine research and manufacturing. The license agreements with all parties are based on territorial rights and include both rights for the intellectual property and to biological materials. The biological materials include all the vaccine strains, as well as the analytical reagents necessary to develop the vaccine.

Butantan was awarded an exclusive license to practice the invention for development of a rotavirus vaccine in Brazil and Latin America. In cooperation with the Brazilian Ministry of Health, Butantan plans to introduce the vaccine into Brazil's child immunization program, which provides free vaccines for all children of Brazil. Similarly, Chengdu and Wuhan will manufacture and supply the rotavirus vaccine to China's expanded program of immunization (EPI). The Office of Technology Transfer (OTT) at NIH granted to the four Indian companies licenses to the IP rights in India and rights to manufacture and distribute the rotavirus vaccine in India and other developing countries, excluding Brazil and other Latin American countries and China. Finally, Aridis was granted an exclusive license to IP rights covering the rotavirus vaccine in the United States, Europe, and Canada. By using this multipronged approach and carving out territory-specific agreements, PHS ultimately set the stage for global distribution of the rotavirus vaccine. The terms of the agreements were structured according to each licensee's mission to provide free or affordable vaccines to children in their specific territories.

PROGRESS, CURRENT STATUS, AND GOALS

The human-bovine reassortant rotavirus vaccine is expected to reach the market in developing countries in five to six years. All the licensees are currently in a stage of organization, preparing all the necessary facilities and infrastructure for manufacturing the vaccine and for clinical trials. The licensees plan to receive training in the technology involving the vaccine at the laboratory of Dr. Kapikian at NIH. It is anticipated that the Codevelopment will include collaboration with the NIH. OTT staff was recently notified by its partners and the staff of the Bill & Melinda Gates Foundation that the latter will support partial development of clinical trial procedures for screening the technology at specific institutions in developing countries. ■

For further information, please contact:
URI REICHMAN, *Branch Chief, Infectious Diseases, Office of Technology Transfer, National Institutes of Health, 6011 Executive Boulevard, Suite 325, Rockville, MD 20852, U.S.A.* reichmau@mail.nih.gov

Gastrointestinal Medicines from African Aloe: Baylabs (Pty) Ltd.

The plant species *Aloe ferox*, indigenous to the eastern and southeastern Cape regions of South Africa, has sustained an aloe tapping industry for more than 250 years. However, the industry has failed to substantially improve the economic conditions of communities in the region. Between 1,600 and 3,000 aloe tappers earn, on average, $150 per month.

In 1998, a method for producing a novel fiber in powder form from the discarded leaves of the plant was patented by South Cape Aloe (SCA). A virtual startup company with a strong emphasis on technology and intellectual property (IP) was subsequently formed in South Africa to develop a product to treat irritable bowel syndrome (IBS) and AIDS-related diarrhea (ARD).

The company, Baylabs, aims to form local partnerships to develop, manufacture, and distribute the product to both developed and developing countries. Baylabs' strategy is to focus on R&D to generate and protect intellectual property and products, while outsourcing noncore functions such as manufacturing, sales, and distribution.

SCA granted the manufacturer African Aloe exclusive rights to make the powder and gained a share hold in Baylabs in exchange for exclusive, royalty-free, worldwide rights to exploit the powder. Baylabs filed a Patent Cooperation Treaty application for the novel powder formulation, with national filings in 13 European countries and prosecutions in the United States, Japan, Australia, and China.

Baylabs has developed four over-the-counter natural remedies from *A. ferox* that are distributed to pharmacies. The revenue generated is used to file patents and obtain scientific evidence of efficacy for gastrointestinal (GI) problems. The products will continue to be marketed and regulated as a dietary supplement while scientific evidence is being gathered and until the product is registered as a medicine.

The company's value has grown through its intellectual property and clinical trials of IBS and infantile diarrhea disease (IDD). Discussions are underway with international strategic partners regarding exclusive license agreements; efforts to secure government or venture capital funding are in progress. Baylabs plans to build preprocessing field plants and a facility to manufacture the powder, with the aloe tapper community as an equity partner, which could lead to increased salaries (almost double) for aloe tappers.

There is no traditional knowledge (TK) involved in using the waste leaf but TK exists in using the *A. ferox*. A key feature of this case study is the potential for other treatments; the formulation can be used for IBS in developed countries and ARD in developing countries. Once clinical trials have been completed, Baylabs plans to register the product as a medicine. However, the advantages of registering the product as a drug rather than as a food supplement have been questioned. Such registration would require, among other things, strict manufacturing quality standards and could be fraught with regulatory difficulties. Many intended herbal remedies, if subjected to full clinical trials and toxicity (as required by regulation), would not meet these standards.

In natural products a key issue is long-term planning and supply. If the product were to become a blockbuster, arrangements would have to be made for the community to benefit, such as through a trust fund. It is important to recognize traditional harvesters and

Bunn AE. 2007. Gastrointestinal Medicines from African Aloe: Baylabs (Pty) Ltd. In *Executive Guide to Intellectual Property Management in Health and Agricultural Innovation: A Handbook of Best Practices* (Krattiger A, RT Mahoney, L Nelsen et al.). MIHR (Oxford, UK), PIPRA (Davis, USA), Oswaldo Cruz Foundation (Fiocruz, Rio de Janeiro, Brazil), and *bio*Developments-International Institute (Ithaca, USA). Available online at www.ipHandbook.org.

Editors' Note: An earlier version of this case study was presented at the MIHR conference Using Intellectual Property for Improved Health in Developing Countries: An Evidence-Based Approach to Good Practice, Bellagio, Italy, June 14–18, 2004.

traditional plant users and their stake in bioprospecting. Baylabs is set to give the aloe tapping community a stake in the project.

The Baylabs example illustrates how the development of a technology can have positive commercial and positive moral outcomes. Through the creation of strategic alliances and partnerships, there can arise opportunities for securing and developing intellectual property for the benefit of underserved communities in both developed and developing countries.

TYPES OF AGREEMENTS

As part of the GI medicines from African aloe project, Baylabs has entered into the following types of agreements:

- exclusive patent license agreement
- exclusive supply agreement

IP RIGHTS DECISIONS AND IP MANAGEMENT

Baylabs has faced key areas of IP rights decision making and strategic IP management issues including:

- securing a strong IP portfolio through international filings, scientific proof-of-concept, and rigorous clinical trials
- securing ownership of intellectual property and outsourcing noncore functions

POLICY IMPLEMENTATION

The SA Medicines Control Council (MCC) is presently formulating policy on traditional and herbal medicines. Companies are therefore able to place over-the-counter products in the market without clinical trials. These may not make any medicinal claims. This enabled Baylabs to place four elementary products (aloe gel, a high fiber tablet, a laxative tablet, and an antiarthritic tablet containing aloin as the active ingredient) on the market and to secure income from their sale. These products had to be submitted to the traditional medicines registry at the MCC to enable continued manufacturing and sales.

EXTERNAL FACTORS THAT AFFECTED DECISION MAKING

A number of considerations influenced Baylabs' strategies and decision making. These include:

- burden of disease from ARD in developing countries
- burden of disease from IBS in developed countries
- commercial opportunity from IBS
- indigenous occurrence of *Aloe ferox*
- opportunity to exploit a by-product of the aloe tapping industry
- regulatory issues relating to aloe mixture
- opportunity to alleviate IDD

KEY LESSONS LEARNED

The following items represent key lessons from the Baylabs GI Medicines/Aloe project, which may be applicable to other companies that aim to utilize intellectual property:

- have a moral as well as a commercial reason for existence (improve living standards of aloe tappers and alleviation of ARD, IBS, and IDD)
- have a global commercial opportunity, which big pharma has been unable to effectively address (IBS—a US$15 billion annual industry)
- create and protect intellectual property (registers serious intent)
- create alliances and partnerships
- a startup can be successful operating as a virtual company and securing IP ownership
- choose partners with a shared value system
- have a good IP attorney (preferably in-house)—there are always issues! ■

For further information, please contact:
TONY BUNN, *Director, Technology Development and Transfer Office, Medical Research Council, PO Box 19070, Tygerberg, 7505, South Africa.* tony.bunn@mrc.ac.za

Lapdap™ Antimalarial Drug: GlaxoSmithKline, WHO-TDR, and the U.K. Department for International Development

Lapdap™ is a new combination of two off-patent malaria drugs. The U.K. Medicines and Healthcare Products Regulatory Agency approved the drug in 2003 for the treatment of malaria caused by *Plasmodium falciparum*, which kills one to two million people every year. The combination drug was developed in response to the growing resistance among patients to malaria drugs, with failure rates in Africa as high as 40 percent.

Lapdap came out of early research funded by the Wellcome Trust and was brought to market by a public-private partnership (PPP) involving GSK (GlaxoSmithKline), WHO-TDR (a WHO/UNDP/World Bank Special Program in Research and Training in Tropical Diseases), and the U.K. Department for International Development (DfID). This was done in collaboration with scientists from the University of Liverpool and the London School of Hygiene and Tropical Medicine, African researchers and clinicians, and the Wellcome Trust.

Under the terms of a funding partnership, GSK, WHO-TDR, and DfID each paid one-third of the development costs. Their agreement covered the ownership of nonpublished data and the establishment of a product-development team to continue development and obtain regulatory approval.

Early patent applications filed on the basic biological work underlying the combination of the two existing drugs were abandoned after filing because it was later found that the work had already been published in scientific literature and so there was 'prior art.' There are currently no patents protecting the Lapdap™ product in any country.

Lapdap™ was developed to be as inexpensive as possible, with a public sector target of less than US$0.30 per dose. It is currently sold only through private sector pharmacies, with the commercial sale price varying by country. The drug is available in South Africa, Nigeria, Kenya, and Ivory Coast.

Lapdap's™ role in public health is still being assessed; Phase IV studies are ongoing and the WHO has stated that after reviewing available clinical and preclinical data, it will identify strategies for optimal and safe use. Lapdap™ has potential for future public health initiatives; a collaborative agreement was signed in April 2004 between GSK, WHO-TDR, and MMV to develop a new fixed-dose artemisinin combination-therapy drug combining chlorproguanil, dapsone, and artesunate for treatment of malaria.

Successful collaboration to ensure that developing countries benefit from the fruits of intellectual property requires an integrated approach toward networking and capacity building, involving innovation, regulatory approval, market creation, licensing, and distribution.

The lack of formal health infrastructure in rural Africa, where there are few physicians and where the drug is sold over the counter, has led to great importance being attached to the packaging and distribution, as well as education to ensure proper dosage. The establishment before registration of a public health group, under the WHO's auspices, provided a useful forum for discussing how Lapdap™ would be accessed. This case highlights the need for consensus regarding

MIHR/PIPRA. 2007. Lapdap Antimalarial Drug: GlaxoSmithKline, WHO-TDR, and the U.K. Department for International Development. In *Executive Guide to Intellectual Property Management in Health and Agricultural Innovation: A Handbook of Best Practices* (Krattiger A, RT Mahoney, L Nelsen et al.). MIHR (Oxford, UK), PIPRA (Davis, USA), Oswaldo Cruz Foundation (Fiocruz, Rio de Janeiro, Brazil), and *bio*Developments-International Institute (Ithaca, USA). Available online at www.ipHandbook.org.

Editors' Note: An earlier version of this case study was presented at the MIHR conference Using Intellectual Property for Improved Health in Developing Countries: An Evidence-Based Approach to Good Practice, Bellagio, Italy, June 14–18, 2004.

public sector use of the product between all parties involved in national malaria control.

This case study was considered 'IP neutral,' since the academic and public health mission was neither impeded nor driven by IP considerations. However, the Wellcome Trust, as part of its mission, recognizes the important role of industry and its investors (including non-commercial funders) in translating research innovations into new health products. It therefore encourages and supports the responsible use of IP rights to protect research findings where commercialization or further funding which could benefit from the existence of that underlying IP is necessary to achieve the greatest public benefit.

It could be argued that the lack of underlying intellectual property in this case, specifically patents, may have accelerated the research project and reduced transaction costs. On the other hand, the absence of patents may have slowed this process, particularly the attainment of Phase IV studies because a patent-driven time schedule did not drive the development process.

It was generally agreed, however, that intellectual property other than patents was generated in the form of regulatory dossiers (clinical trial data), know how, terms of codevelopment agreements, and trademarks. Recognizing the multiplicity of intellectual property can contribute to a more comprehensive understanding of the IP management aspects of product R&D, post-development, and manufacturing.

Lapdap's™ pursuit of WHO endorsement raised the broader policy issue of the global health body's role as a certificatory of treatment regimes. WHO approval is a vital step in products reaching developing countries and gaining public sector acceptance. However, responsibility within a PPP for securing such endorsement is not always clear.

Regulatory endorsement is but one aspect of product sustainability. Royalty streams should be examined for how their use and management can contribute to product support. Although often treated as undesirable additional costs, the generation of royalties on public sector sales is an effective IP management tool for keeping a product on the market.

The involvement of universities in this public health initiative drew attention to the role of university technology transfer offices (TTOs). It appears that TTOs are frequently given competing missions by their institutions, with no clear priority as to whether making money or delivering applications of research regardless of returns is the most important goal. Declining revenue of universities has pressured cash-strapped TTOs to increase their contribution, compelling them to turn to intellectual property. Although exploiting university research is a legitimate goal, it may be short-sighted to focus solely on patents; the transfer of know-how and trade secrets is just as important, and an overemphasis on revenue generation using IP rights may limit the potential of certain research outcomes.

In attracting commercial interest, TTOs must be mindful of overvalued patents and overestimated royalties, and must know how to manage hurdles and prevent unreal expectations. Alongside the need for flexibility in negotiations, education about technology management is required.

The challenge therefore is to use PPPs as an effective means of bringing drugs to the poor by drawing on the expertise and synergies between sectors. These partnerships afford the opportunity to segment the market in a way in which the public body can benefit from having an exclusive license for its stakeholders while satisfying commercial partners.

TYPES OF AGREEMENTS

An agreement was signed relating to establishment of the product-development team and ownership of nonpublished data. Under the funding partnership between GSK, WHO-TDR, and the U.K. DfID, each partner contributed one-third of the development costs.

PATENT AND IP RIGHTS DECISIONS

Early patent applications were filed between 1994 and 1996 by GSK (then SmithKline Beecham) on the basic biological work underlying the combination of the two existing drugs, with Dr. Bill Watkins (University of Liverpool & Wellcome Trust Research Laboratories, Kenya) as named inventor. These applications were later abandoned, because after filing it became clear that the combination had already been published in the literature and therefore was no longer novel. There are therefore no patents protecting the Lapdap™ product in any country.

POLICY IMPLEMENTATION

Lapdap™ at present is being sold only through the private sector (pharmacies). WHO does not currently recommend the use of chlorproguanil-dapsone alone as an option for national treatment policy in countries where malaria is endemic. The role of the drug in public health is still being assessed—Phase IV studies are ongoing, and pharmacovigilance activities in specific patient groups are planned. WHO has stated that after reviewing available clinical and preclinical data, it will shortly identify strategies for the optimal and safe use of Lapdap™ in malaria-endemic countries.

Because of Lapdap's™ reported efficacy, relatively short half-life, and low production cost, it has potential for future public health use in combination with an artemisinin compound. In April 2004, a collaborative agreement was signed between GSK, WHO-TDR, and MMV to develop a new fixed-dose artemisinin combination-therapy drug combining chlorproguanil, dapsone, and artesunate for treatment of malaria.

EXTERNAL FACTORS THAT AFFECTED DECISION MAKING

In the case of Lapdap™, where IP considerations did not drive the later development of the project, some external factors of relevance were:

- nature of the end market for Lapdap™ (poor countries in Africa)
- multiparty cooperation and synergy

KEY LESSONS AND HEALTH-ACCESS ISSUES

The following lessons were learned during development of the Lapdap™ drug and subsequent distribution:

- Pharmaceutical industry expertise in clinical trials, the regulatory process, and marketing are necessary to accelerate product development.
- Establishment of a public health group under WHO auspices *in advance of registration* was a useful forum for discussing how the product would be accessed.
- Consensus on the use of the product in Africa is necessary at the country level between parties involved in malaria control. ■

For further information, please contact:
DANIEL NELKI, *Head of Legal and Operations, Technology Transfer, the Wellcome Trust, 215 Euston Road, London, NW1 2BE, U.K.* d.nelki@wellcome.ac.uk

Cyclofem® Contraceptive: Upjohn, WHO, and the Concept Foundation

The case study of the development and distribution of Cyclofem® contraceptive as a project of Upjohn and the World Health Organization (WHO) is an example of innovative intellectual property (IP) management in which a collaboration between a public sector institution and a private pharmaceutical company led to the establishment of a new nonprofit organization that brought the product to developing country markets. The venture described in this case study was also new type of undertaking for WHO.

Upjohn pharmaceutical company developed the once-a-month injectable contraceptive Cyclofem®. Despite successful Phase III trials undertaken jointly by WHO and Upjohn, the drug company decided there was an insufficient market for the contraceptive and donated the clinical trial data to WHO. When no U.S. or European commercial partner could be found to take the product forward, WHO invited the nonprofit organization PATH (to which it licensed the clinical data rights) to come up with a viable solution.

PATH proposed establishing a new nonprofit organization, the Concept Foundation, which would focus on developing countries. Intellectual property and know-how was transferred via PATH to the Foundation, which licensed developing country producers on an exclusive basis in defined private sector markets and on a nonexclusive basis for public sector markets to ensure competition. A royalty stream of 4% was paid to the Foundation to support continued production and distribution. Manufacturers were expected to meet national and international (current good manufacturing practices, or cGMP) regulations. Milestones were an important part of the package, and were linked to territories, regulatory matters, and market penetration.

Production was established in Mexico and Indonesia, supplying private and public sectors with an affordable quality product that had been dropped by its developer. This was the first pharmaceutical product to result from successful WHO product R&D. The Concept Foundation is now self-sufficient and provides valuable technical assistance and introduction support, alongside economic development and technology transfer.

EXTERNAL FACTORS THAT AFFECTED DECISION MAKING

Establishing a nonprofit organization in a developing country was an appropriate option because WHO could not own, manufacture, distribute, or manage the product. PATH did not want to jeopardize its own neutral role in improving public health. Another consideration was liability. PATH, with assets in the United States, could not afford to risk its well being. Ultimately, after much discussion it was realized that the liability risk should rest in a jurisdiction that reflected the environments in which the product would be used.

The Foundation's aim was to on-license to producers and distributors in developing countries. If a government wanted to buy the product, it could go to any of the manufacturers and ask for a bid on cost prices. As time passed, the Concept Foundation identified the need to update the regulatory dossier for Cyclofem®. It carried out this updating and made the new dossiers available to current and prospective licensees.

Mahoney R. 2007. Cyclofem Contraceptive: Upjohn, WHO, and the Concept Foundation. In *Executive Guide to Intellectual Property Management in Health and Agricultural Innovation: A Handbook of Best Practices* (Krattiger A, RT Mahoney, L Nelsen et al.). MIHR (Oxford, UK), PIPRA (Davis, USA), Oswaldo Cruz Foundation (Fiocruz, Rio de Janeiro, Brazil), and *bio*Developments-International Institute (Ithaca, USA). Available online at www.ipHandbook.org.

Editors' Note: An earlier version of this case study was presented at the MIHR conference Using Intellectual Property for Improved Health in Developing Countries: An Evidence-Based Approach to Good Practice, Bellagio, Italy, June 14–18, 2004..

LESSONS LEARNED AND HEALTH-ACCESS ISSUES

This case study is an example of innovative IP management where collaboration between WHO and Upjohn led to the establishment of a new nonprofit organization with the purpose of bringing the Cyclofem® contraceptive to developing country markets. This case demonstrates that clinical trial data can be important IP that can help ensure availability of products in developing countries. Putting it simply, without clinical trial data, the product can not be marketed; thus the data are of great value. The goal of the Concept Foundation and similar ventures is to ensure availability of products to the poorest of the poor. It is not enough to ensure that the private market helps public sector distribution. As this case study shows, investing time in updating a dossier to meet the requirements of other countries and therefore helping to encourage producers to go into markets that have not been served is important. Similarly, having solid and enforceable milestones is not an indication of lack of trust; it is rather being serious about business and wanting to succeed. ■

For further information, please contact:
RICHARD T. MAHONEY, *Director, Vaccine Access, Pediatric Dengue Vaccine Initiative, International Vaccine Institute, San Bongcheon-7dong, Kwanak-ku, Seoul 151-818, Republic of Korea. rmahoney@pdvi.org*

Improved Production of a Natural Product Treatment for Malaria: OneWorld Health, Amyris, and the University of California at Berkeley

In December 2004 the Bill and Melinda Gates Foundation[1] awarded a five-year product development grant to the Institute for OneWorld Health (iOWH),[2] a nonprofit pharmaceutical company, to create a unique three-way partnership between iOWH, a university (University of California at Berkeley),[3] and a for-profit company (Amyris Biotechnologies, Inc.).[4] The goal of this project[5] is to significantly reduce the cost of artemisinin, a key precursor in the production of Artemisinin Combination Therapies (ACT), through synthetic biology, industrial fermentation, and chemical synthesis. Artemisinin is chemically converted to one of several derivatives and then combined with other drugs to make an ACT for the treatment of malaria.

Malaria is a parasitic blood disease that inflicts as many as 500 million people annually. About 1.5 million people die each year from the infection, primarily children in Africa and Asia. More than half of the deaths occur among the poorest 20 percent of the world's population. Studies in Vietnam have shown that the botanically derived medicine, artemisinin derivatives, can reduce deaths from the illness by 97 percent. However, the current cost of a three-day course of drugs containing artemisinin is US$2.40, which places it out of reach for people in many nations where the disease is most prevalent. Reducing the price would make the treatment more widely accessible.

Artemisinin is currently extracted from the wormwood plant, which is supplied by farmers in Vietnam and China (and more recently, Africa). Seasonality and availability of the plant contribute to the high price of the drug. The Gates-funded project hopes to eliminate the need for plant extraction by utilizing a platform technology of synthetic biology developed by Dr. Jay Keasling at the University of California (UC), Berkeley.[6] The goal is to lower the cost of artemisinin-containing drugs ten-fold by producing a consistent, reliable, high-quality supply of artemisinin in microbes.

The US$42.6 million grant was divided among the three partners: US$8 million to UC Berkeley for continued basic research, US$12 million to Amyris for applied research on the fermentation and chemical processes, and US$22.6 million to iOWH to perform the required regulatory work and lead the implementation of the product development strategy for the developing world. UC Berkeley's role focuses on the engineering of drug-precursor-producing microbe. Amyris' efforts span engineering of the production microbe to optimizing the semisynthesis of the drug through fermentation and novel downstream synthetic chemistry. The role of iOWH includes developing a commercialization strategy based on a thorough understanding of the worldwide regulatory requirements and an analysis of the current ACT manufacturing supply-chain and distribution models. This one grant enables activities in all three areas of development and creates an integrated team, each of the partners applying its expertise to streamline translation from bench to bedside.

MIHR/PIPRA. 2007. Improved Production of a Natural Product Treatment for Malaria: OneWorld Health, Amyris, and the University of California at Berkeley. In *Executive Guide to Intellectual Property Management in Health and Agricultural Innovation: A Handbook of Best Practices* (Krattiger A, RT Mahoney, L Nelsen et al.). MIHR (Oxford, UK), PIPRA (Davis, USA), Oswaldo Cruz Foundation (Fiocruz, Rio de Janeiro, Brazil), and *bio*Developments-International Institute (Ithaca, USA). Available online at www.ipHandbook.org.

Editors' Note: This case study was prepared by MIHR members of the Technology Managers for Global Health (TMGH), a special interest group of the Association of University Technology Managers (AUTM) (see www.tmgh.org) and adapted for this *Executive Guide*. The original version was published as part of a collection of case studies: MIHR/TMGH. 2007. *Academic Licensing to Global Health Product Development Partnerships* (ed. U Balakrishnan). MIHR: Oxford, U.K.

To ensure accessibility and affordability, the partners have committed to reduced returns in the malaria field. UC Berkeley has issued a royalty-free license to iOWH and shall grant royalty free licenses to Amyris for IP that is developed during the collaboration for the treatment of malaria in the developing world with the goal of significantly reducing the price of ACT products, and reducing the use of artemisinin mono therapies per the World Health Organization's recommendations for uncomplicated malaria.

This arrangement has benefits for all the parties. The university benefits from the research funding as well as from any royalties that may be realized on profit earned from sales by Amyris in areas outside of malaria in the developing world. As a for-profit company, Amyris can apply the innovations developed for the artemisinin project to other projects that rely on the same platform technology. As a nonprofit pharmaceutical company, iOWH is able to make malaria treatments more affordable for people in the developing world.

PARTNERS

Partners in this project are:

- from academia, the University of California, Berkeley
- the nonprofit pharmaceutical company Institute for OneWorld Health (iOWH)
- the for-profit pharmaceutical company Amyris Biotechnologies, Inc.

The Bill and Melinda Gates Foundation provided the funding for the project.

THE TECHNOLOGY

The preferred and most effective treatments for malaria today are artemisinin-based combination therapies (ACT). Artemisinin, a complex natural product known as an herbal remedy for thousands of years, is typically derived from the wormwood plant. Plant sources of the chemical are variable and crop shortages contribute to increased cost. Chemical synthesis of the molecule would require 30 to 40 steps and is therefore impractical on a commercial scale.

Dr. Jay Keasling, a UC Berkeley professor of chemical engineering, developed a process of "synthetic biology" to produce an artemisinin precursor through a multistep process in bacteria.[7] The precursor can then be chemically converted to artemisinin through synthetic chemistry developed at Amyris. Producing the drug precursor in microbes would lead to a more consistent and reliable supply and therefore reduce the cost of production.

The synthetic biology platform may also be used to produce other drugs, nutraceuticals, and flavors and fragrances.

PROGRESS, CURRENT STATUS, AND GOALS

During the five-year granting period, which began in 2005, the partners would carry out the following activities shown in Figure 1.

UC Berkeley researchers are working to identify the genes involved in the artemisinic acid biosynthetic pathway in the wormwood plant, *Artemisia annua*. Using their expertise in synthetic biology, they are inserting this biosynthetic pathway into microbes to create hosts that manufacture this direct precursor to artemisinin. Optimizing artemisinic acid production in these host cells is being achieved through cutting-edge techniques in metabolic engineering, in collaboration with scientists at Amyris Biotechnologies.

Amyris Biotechnologies is collaborating with the Center for Synthetic Biology to build a better microbe. Amyris will optimize the microbial strain developed with UC Berkeley for commercial production. In addition, Amyris will develop a fermentation and purification process for the precursor. Simultaneously, Amyris is developing a scaleable, inexpensive chemical process to convert the precursor to artemisinin.

OneWorld Health is the product development lead and has responsibility for directing this collaborative effort. In addition, the organization is leading the project's regulatory and commercialization strategies and is conducting a risk-benefit analysis surrounding the use of artemisinin derivatives in malaria-endemic regions.

DEALS

Agreements between the partners include the following:

License Grants:

- The arrangement is governed by a three-party collaboration agreement and two license agreements (from UC Berkeley to each of Amyris and iOWH).
- UC Berkeley granted iOWH a royalty-free license for the manufacture of artemisinin-based malaria treatments used in the developing world. UC Berkeley further shall grant royalty-free licenses to iOWH for IP developed under the three-party collaboration agreement for use in manufacturing artemisinin-based malaria treatments used in the developing world. OneWorld Health is to establish partnerships for ACT manufacture and distribution.
- UC Berkeley granted Amyris licenses to develop the manufacturing process for the developing-world malaria market. Amyris also has licenses for the developed-world malaria market, nonmalaria indications of artemisinin, and alternative uses of the platform worldwide. UC Berkeley further shall grant similar licenses to Amyris for IP

Figure 1: Activities of Project Partners (to be carried out by the end of the grant period)

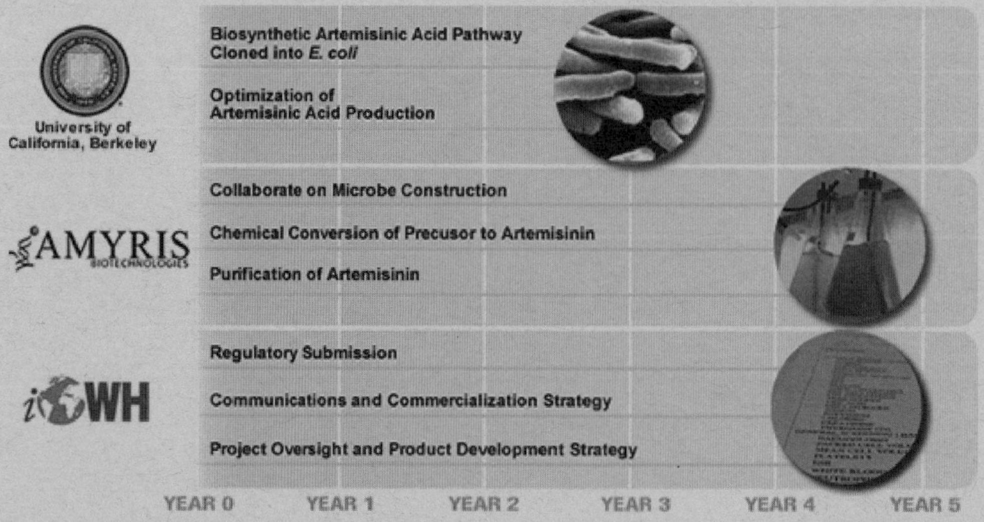

developed under the three-part collaboration agreement.

- Amyris shall grant iOWH a royalty-free license for IP developed under the three-part collaboration agreement for the manufacture of artemisinin-based malaria treatments used in the developing world.

Royalties:

- The license from UC Berkeley to iOWH is royalty free.
- The license from UC Berkeley to Amyris is royalty free for the developing-world malaria market (development for iOWH) and is royalty bearing for the developed world and nonmalaria indications in the developing world.

Patents:

- Patent costs for UC Berkeley's pre-existing patents are shared between iOWH and Amyris.
- UC Berkeley patents on IP arising from the collaborative research may be filed by UC Berkeley and licensed to iOWH and/or Amyris under the pre-arranged terms mentioned above. Costs are shared by the licensee on a pro rata basis. UC Berkeley has no obligation to file an application if it does not have a commitment by a licensee to pay patent costs.
- Patents that are the sole property of Amyris and/or iOWH may be filed by Amyris and/or iOWH, as the case may be, at their own expense.
- Logistics of filing and payment of costs on jointly owned IP will be negotiated in good faith by the joint owners when such joint IP arises. If the joint owners cannot agree and if iOWH has an ownership interest in a joint property, then iOWH may file and prosecute on behalf of the owners at its own expense.

Other:

- Amyris, as UC spinout company, is seeking venture funding to leverage applications in other markets.
- Using the process developed by Amyris and UC Berkeley, iOWH is to establish partnerships for ACT manufacture and distribution
- Similar licenses to all relevant third-party intellectual property will be obtained by iOWH as the need arises. ■

For further information, please contact:
UNIVERSITY OF CALIFORNIA, BERKELEY, *Carol Mimura, Assistant Vice Chancellor, IPIRA, 2150 Shattuck Avenue, Berkeley, CA 94720, U.S.A. carolm@berkeley.edu*

INSTITUTE FOR ONEWORLD HEALTH, *Katharine Woo, Director, Scientific Affairs, 50 California Street, Ste. 500, San Francisco, CA 94111, U.S.A. kwoo@oneworldhealth.org*

AMYRIS BIOTECHNOLOGIES, INC., *Neil Renninger, Vice President–Development, 5980 Horton Street, Ste. 450, Emeryville, CA 94608, U.S.A. renninger@amyrisbiotech.com*
.

---

1 Bill and Melinda Gates Foundation. www. gates foundation.org.

2 Institute for OneWorld Health. www.oneworldhealth.org/.

3 Office of Intellectual Property and Industrial Research, UC Berkeley. ipira.berkeley.edu/index.php.

4 Amyris Biotechnologies, Inc. www.amyrisbiotech.com.

5 The Artemisinin Project. www.artemisininproject.org.

6 Keasling Laboratory, UC Berkeley. www.cchem.berkeley.edu/%7Ejdkgrp/.

7 Martin VJ, DJ Pitera, ST Withers, JD Newman and JD Keasling. Engineering a Mevalonate Pathway in *Escherichia Coli* for Production of Terpenoids. *Nature Biotechnology*. 21 July 2003. (7): 796–802.

Nontoxic Drug Therapy for Chagas' Disease and Malaria: University of Washington and Yale University

Some of the world's most intractable diseases are predominantly in the developing world. These illnesses are known as neglected diseases because they receive little attention from the medical community and the pharmaceutical industry, even though they have a significant impact on vulnerable populations. One of these neglected killers is Chagas' disease.

According to the World Health Organization, Chagas' disease is an insect-borne, parasitic illness that infects and kills millions of people every year. Chagas' disease is endemic in 21 Latin American countries and is a major cause of heart failure in the region. Caused by the parasite *Trypanasoma cruzi*, Chagas' disease is most often transmitted by an insect known as the kissing bug, which tends to feed on people's faces. Humans, as well as wild and domestic animals, carry the parasite, and insects infected with *T. cruzi* frequently live in the thatched walls and roofs of homes, making it especially challenging to eradicate.

Controlling the disease is difficult, costly, and risky: it depends largely on treating homes in affected areas with residual insecticides and, in general, improving housing by replacing traditional thatched-roof dwellings with more modern, plastered walls and metal roofs. Management of the illness now involves blood screening to prevent transmission through transfusion. Some drug treatments are available as well.

COLLABORATING TO FIND A TREATMENT

But the standard drug treatments for Chagas' disease leave much to be desired. Most are aimed at fighting the infection, which spreads inside the heart and gastrointestinal tract of the victim. Drugs are difficult to administer and highly toxic, leading to severe side effects in many patients. And no existing medicines have consistently cured patients, according to a report from the Institute for OneWorld Health, a nonprofit pharmaceutical company the goal of which is to develop affordable treatments for neglected infectious diseases around the world.

A collaborative research effort among scientists at the University of Washington and Yale University recently brought forth a nontoxic drug therapy for Chagas' disease. The team included Andy Hamilton and Junko Ohkanda, both chemists at Yale, and Fred Buckner and Wesley Van Voorhis, infectious disease experts, and Michael Gelb and Kohei Yokoyama, chemists, at University of Washington.

"It was a wonderful collaboration between organic chemists and parasite biologists that came about through reading the literature and recognizing potential connections," said principal investigator Hamilton, who has since become a provost at Yale. "Big problems nearly always involve collaborative solutions because no one person or institution can have all the answers."

Buckner, of the University of Washington Medical School, agreed. He has worked for years with a group of chemists led by Gelb to develop compounds to treat infectious diseases caused by protozoan pathogens.

"They would make the compounds and we would test them against the parasites to see if they would do

AUTM. 2007. Nontoxic Drug Therapy for Chagas' Disease and Malaria: University of Washington and Yale University. In *Executive Guide to Intellectual Property Management in Health and Agricultural Innovation: A Handbook of Best Practices* (Krattiger A, RT Mahoney, L Nelsen et al.). MIHR (Oxford, UK), PIPRA (Davis, USA), Oswaldo Cruz Foundation (Fiocruz, Rio de Janeiro, Brazil), and *bio*Developments-International Institute (Ithaca, USA). Available online at www.ipHandbook.org.

Editors' Note: We are most grateful to the Association of University Technology Managers (AUTM) for having allowed us to adapt this case study for inclusion in this *Executive Guide*. The original was published by AUTM. 2006. *Technology Transfer Works: 100 Cases from Research to Realization* (Reports from the Field). Association of University Technology Managers, Northbrook, IL. www.betterworldproject.net.

anything," Buckner said. "Some turned out to be active against targets that were different than what we designed them to do, but we determined the mechanism of action and showed them to be active in an animal model."

APPROACHING THE PROBLEM FROM DIFFERENT ANGLES

The original patent application described "compounds and methods for treating infections caused by bacterial protozoal and fungal agents," said Aline Flower, of University of Washington TechTransfer Invention Licensing.

When asked about the potential application of the compound, Hamilton said, "we developed, in collaboration with parasitologists, compounds that target the Chagas' disease agent in animal models, and we are seeing some very encouraging data."

Buckner and his colleagues had made inroads targeting these diseases, working toward cures or vaccines. "We had discovered that protozoan parasites contain the enzyme protein farnesyltransferase," said Buckner. "This same enzyme plays an important role in cancer cells, which meant a lot of research laboratories were developing drugs against it. We were working on the hypothesis that protein farnesyltransferase inhibitors might work against parasites."

In the meantime, Hamilton and Ohkanda were working on a similar problem from another angle. "This was the result of many years of fundamental research in trying to get a novel molecular structure to target a specific enzyme," Hamilton said. "It's a question of how one synthetic molecule could recognize a biological molecule in a process called molecular recognition."

According to Hamilton, the two universities and the nonprofit pharmaceutical company developed an integrated model for drug development, perhaps just as important as the chemical compound the researchers had discovered. "We hope, as we make progress in the pre-clinical stage, OneWorld Health will help us pull together the necessary funding to allow the clinical and preclinical development of these compounds," said Hamilton.

The Yale Office of Cooperative Research senior licensing associate Alan Carr explained that an interinstitutional agreement between the University of Washington and Yale University enabled the institutions to structure a deal with OneWorld Health to license the compound affordably.

Like the drug compound, the model for drug development, borne of innovative university technology transfer, could well have a lasting impact on people around the world. ■

Diagnostic Tests for Cervical Cancer: PATH

The public sector institution PATH aims to improve global health by advancing technologies, strengthening systems, and encouraging healthy behaviors through effective collaborations with the private sector. PATH tries to reduce risks for a commercial company developing products for resource-poor countries by identifying gaps in the market that existing technology can fill, demonstrating value, and partnering in development and sustainable supply. In addition, PATH adapts products to different markets, provides training, and engages in advocacy with WHO and other public bodies. PATH is both a recipient and a provider of funding.

As a nonprofit organization that creates and manages intellectual property in house, PATH recognizes that working with private companies requires sensitivity to and awareness of commercial incentives. PATH believes that intellectual property is just one element of the economic environment of the technology. Successful collaborations with private sector companies impact positively the availability, accessibility and affordability of products in public sector health programs in developing countries.

During product development and distribution, PATH works to change behavior and to open or improve communication. It worked with India's Ministry of Health to launch a hepatitis B vaccine on a project that involved community education and communication in preparation for the vaccination program. The program's success has ensured national expansion of the program.

Diagnostics is a large field with a number of disparate groupings of intellectual property generated by scientists around the world; it is common for multiple parties to hold key pieces of intellectual property. PATH routinely conducts market and industry feasibility studies to determine the type of industry partner to pursue, to determine which is best positioned to take PATH into the target segments it is interested in, and to identify IP issues. The public sector needs to recognize that securing the necessary IP rights for diagnostic products is imperative before moving ahead with development and commercialization.

Procurement in diagnostics is not as centralized as other public health products, such as vaccines and drugs. This makes it more difficult to plan for the global public health sector. Marketing is generally on a country-by-country basis, unlike family planning products, for example, that have regional or global distribution agencies for the public sector markets.

THE CERVICAL CANCER DIAGNOSTIC TEST PROJECT

PATH is engaged in ongoing work with industry partners to develop rapid diagnostic tests for cervical cancer for use in developing countries. In addition, two major institutes, in India and China, are screening 30,000 women for cervical cancer and will then conduct the clinical trials to validate the efficacy of these simple and inexpensive tests. In addition, this work will generate useful information on viruses that have not yet been examined in detail in these countries.

Under the terms of the R&D agreements between PATH and the industry partners, PATH's obligations include funding a portion of the industry partner's direct R&D costs, conducting market and industry assessments, developing an evaluation framework

---

Brooke S. 2007. Diagnostic Tests for Cervical Cancer: PATH. In *Executive Guide to Intellectual Property Management in Health and Agricultural Innovation: A Handbook of Best Practices* (Krattiger A, RT Mahoney, L Nelsen et al.). MIHR (Oxford, UK), PIPRA (Davis, USA), Oswaldo Cruz Foundation (Fiocruz, Rio de Janeiro, Brazil), and *bio*Developments-International Institute (Ithaca, USA). Available online at www.ipHandbook.org.

Editors' Note: An earlier version of this case study was presented at the MIHR conference Using Intellectual Property for Improved Health in Developing Countries: An Evidence-Based Approach to Good Practice, Bellagio, Italy, June 14–18, 2004.

for public-health use of the new test, and conducting multicountry clinical evaluation of the new tests' performance for registration purposes. The industry partner is responsible for development of the products, management of the intellectual property (patenting costs and prosecuting infringement), manufacture and supply for clinical evaluations, and finalizing the product for registration and commercial supply.

PATH retains ownership of specimens, but data are either jointly or individually owned. A product-development committee was formed, and PATH only provides funding sufficient to reach the next agreed-upon milestone. During the R&D phase PATH can terminate, without cost, at key milestones, although industry partners terminate at a cost.

The commercialization period of the agreement runs for ten years from the first sale of a registered product. Both industry partners are required to provide preferential public sector pricing. If these specific products are sold in developed countries, PATH will earn a royalty, however PATH has forgone all royalties on developing country sales. Termination clauses covering one industry partner involve repayment of PATH's direct funding and the transfer of distribution and/or manufacturing to a third party; the other industry partner is only required to grant PATH a nonexclusive license to the product and underlying reagent.

Both companies are working on products that are different from those they will launch in the United States and Europe. Developing a product with PATH could potentially jeopardize products in other developed countries; it is therefore critical for participating industry partners to be able to segment markets.

PATH's success in being able to attract industry partners to collaborate in its effort to develop a diagnostic test for cervical cancer is an example of creating an overarching cervical cancer prevention initiative that made collaboration attractive and worthwhile—in this case, a program of cervical cancer screening including clinical work, advocacy, and policy issues. PATH does not expect to be providing the product in the future; its industry partners have the intellectual property, are developing it, and are responsible for its management.

This case study illustrates that intellectual property and technology transfer are not enough to create a broad and lasting health impact. PATH believes it is possible to attract top-tier industry partners, especially if there is a comprehensive public health initiative and not just a technology development project. Issues to consider in developing a public health initiative include determining the value of know-how, deciding whether to grant an exclusive or a nonexclusive license, dealing with key reagent IP holders, and influencing the final product price.

TYPES OF AGREEMENTS

Over the years of diagnostic-test development and commercialization, PATH has:

- in-licensed key diagnostic reagents to PATH from academic, government, and private company sources
- out-licensed diagnostic test and reagent production know-how from PATH to diagnostic manufacturers
- some with geographically defined exclusive territories
- some on global nonexclusive basis
- materials transfer agreements
- supply agreements
- confidentiality agreements
- codevelopment agreements

IP RIGHTS DECISIONS AND IP MANAGEMENT

PATH has faced key areas of IP rights decision making and strategic IP management issues including:

- managing freedom to practice risks associated with other parties' intellectual property for certain diagnostic platforms and reagents
- determining the value of know-how developed for efficient production of certain diagnostic reagents even when the know-how was not patentable
- determining whether to provide downstream licensees with a greater or lesser level of market exclusivity, or whether to license only on a nonexclusive basis
- dealing with holders of key intellectual property involving particular antigens or antibodies necessary to develop particular diagnostic tests
- deciding whether to patent incremental in-house innovations in the face of uncertain demand and usefulness
- considering how to achieve or at least positively influence final product pricing and access when third-party diagnostics importers/distributors (not the PATH-licensed diagnostic manufacturer) will be the party making the sales transaction to a developing country government

POLICY IMPLEMENTATION

On an overall policy basis PATH works under its Guiding Principles for Private Sector Collaboration, endorsed by the board of directors, which is most often relevant to PATH's intellectual property and licensing activities with diagnostics. To conform to key elements of these guiding principles, a license (and overall collaboration) between PATH and a commercial diagnostics producer must:

- exhibit a clear link to PATH's mission by improving the availability, accessibility, and affordability of important products for public health programs in developing countries
- recognize that the commercial partner must achieve commercial benefit to ensure their sustainable commitment to supplying the technology
- provide a clear definition of the roles, responsibilities, and expectations of both PATH and the commercial producer
- balance PATH's need for transparent collaboration with the commercial producer's need to protect proprietary information
- reflect a rigorous process of due diligence on PATH's part before executing an agreement

The IP elements, working relationships, and technology economics of every project or program can vary from one extreme to the other. Because of this, PATH has found it counterproductive, for the most part, to make broad institutional policies about specific individual elements of complex intellectual property and collaborative development agreements. For example, there is no PATH-wide policy that states "all licensed manufacturers must sell to public sector at cost plus 10%." In some cases that structure might be appropriate, in others it might prevent the technology from ever coming to market. In cases where PATH has developed significant technology that may have value in developed country markets, PATH maintains the flexibility to negotiate for a royalty on developed country market sales. PATH forgoes royalties on sales of licensed technologies for developing country public sector use.

EXTERNAL FACTORS THAT AFFECTED DECISION MAKING

The diagnostics arena has a number of characteristics that have historically influenced PATH's strategies and decision making. These include:

- extremely competitive nature of global diagnostic industry
- relative ease of entry into global diagnostics industry
- proprietary control (whether through formal patents or, simply, sole possession of key clones) of key diagnostic reagents by individual companies or institutions
- multilevel manufacturing and distribution channels typical for diagnostic products
- distributed nature of global public sector procurement of diagnostic reagents—no single, huge, vertical procurement mechanism as exists for vaccines and, to a degree, family planning products

KEY LESSONS LEARNED AND HEALTH ACCESS ISSUES

The proprietary control of a single key diagnostic test reagent can give some parties control and power seemingly disproportionate to their contributions to an overall diagnostic test development project. It is critical to have either IP access and/or reagent supply agreements in place early in the product-development cycle, so that access uncertainty is reduced and cost of access is fully understood. The private sector understands this well, while we (at PATH and in the broader public sector) have not always done our homework in this area.

Noncommercial development and/or stewardship of diagnostic platform intellectual property or key component intellectual property can create a positive impact. For example, PATH enhanced the local production of key rapid-test raw materials (nitrocellulose filters and colloidal gold signal reagents) in India, which created an impact beyond the transfer of technology for individual tests to specific companies. Materials suppliers are now serving additional emerging diagnostic producers.

Intellectual property and technology transfer alone are rarely enough to create a lasting impact on public health. We are all working on solutions to health problems that have fundamentally less promise as a "business opportunity," from a commercial manufacturer's standpoint, than do other health problems. To make a new diagnostic test that will deliver profit to the manufacturer and be beneficial and accessible to patients, there needs to be policy change, advocacy work, and extensive evaluations. The diagnostic manufacturer will rarely fund these types of activities, especially for price-sensitive public health markets, so it is critical to involve others who will undertake this work. Intellectual property and technology transfer are certainly important. However, for maximum lasting health impact they should be managed as components of a comprehensive public health initiative rather than as independent activities. ■

For further information, please contact:
STEVE BROOKE, *Advisor, Commercialization & Corporate Partnerships, PATH, 1455 NW Leary Way, Seattle, WA, 98107, U.S.A. sbrooke@path.org*

Regardless of which approach to technology valuation is used, the assessor should have the foresight to see where the new technology could be applied and how useful it might be. The assessor should therefore be familiar with adoption rates of the given technology in a defined market. The value of both formal (statutory, such as patents) and informal (such as know-how) intellectual property (IP) rights should also be known so that negotiation mistakes are avoided. Importantly, **there is no single best method for technology valuation**, and different methods may be used for different technologies within the same organization. Successful technology valuation depends on accurate estimates of how successful a product will be and how much it will sell for. If one can make accurate estimates, one has a good chance of building a trustworthy relationship with licensees, successfully bringing the technology to market, and increasing the chances of making more technology transfer deals in the future.

More specific methods of valuation and **methods to price** technologies are discussed extensively by Razgaitis[4] who also authored several books on the subject. He emphasizes that the *value* of a technology depends on how it is used, how much it costs to develop, how long it will take before its sales generate returns, and the probability that the technology will be commercially successful. *Pricing*, on the other hand, refers to the price a buyer and seller agree upon. This may be in up-front payments (in cash or equity) or deferred royalties, or a combination of both. Razgaitis describes six of these valuation methods.

Method I. The **Use of Industry Standards Method** looks at the range of published royalties (and other forms of payment) from technology licenses within an industry category and uses that information to guide valuation of a technology under consideration.

Method II. The **Rating/Ranking Method** looks at several license agreements for similar technologies, comparing and ranking a technology under consideration against the license agreements with respect to stage of development, scope of IP protection, market size, profit margins, and other factors.

Method III. **Rules of Thumb Methods**, such as the 25% Rule Method, apportion anticipated profits from the commercial use of the technology between the seller and buyer.

Method IV. The **Use of Discounted Cash-Flow Analysis with Risk-Adjusted Hurdle Rates Method** seeks to split expected returns but adjusts basic profit-and-loss accounting terms to account for the timing of investments and returns and the risks borne by the parties. The method introduces a discussion of some possible structures of payments, as they affect both timing and risk.

Method V. The **Advanced Tools Method** applies statistical methods, such as Monte Carlo simulations, to discounted cash-flow models in order to test the influence of various value assumptions and license terms on the possible outcomes of a deal.

Method VI. The **Auctions Method** allows interested parties to bid on a technology, based on their own independent efforts at valuing the technology, thus comparing their respective valuations, identifying the highest valuation, and striking a price based on that highest valuation.

More than one method can be used in any given valuation and, depending on the circumstances, it may be advantageous to use a combination of two or more methods. One should consider the commensurate level of valuation analysis appropriate to the magnitude of the potential licensing opportunity when choosing methods. Razgaitis provides many valuation examples, including typical royalty rates obtained by universities for software, pharmaceuticals, diagnostics, and others (see Table 1). The data illustrate a trend that appears in other examples discussed by Razgaitis: those products and industries with traditionally high operating margins (profits), such as pharmaceuticals and software, tend to exhibit higher royalty rates compared with, say, the materials industry. More specifically, and also for the purpose of establishing reasonable expectations of both licensors and licensees, Table 2 shows typical royalty rates from the medical industry. Note that the context of both tables is well defined: early-stage technologies out of research laboratories. In the second table, however, note that there is an important economic difference between the ends of the royalty ranges given: 1% versus 3% or 2% versus 10%, and so on. Unless the technology transfer manager understands where the institution's

Table 1: Example Table of Royalties Developed through Experience by a University Licensing Office

Product	Royalty (%)	Comments
Materials processes	1–4	0.1%–1% for commodities; 0.2%–2% for processes
Medical equipment/devices	3–5	
Software	5–15	
Semiconductors	1–2	Chip design
Pharmaceuticals	8–10	Composition of materials
	12–20	With clinical testing
Diagnostics	4–5	New entity
	2–4	New method/old entity
Biotechnology	0.25–1.5	Process[a]/nonexclusive
	1–2	Process[a]/exclusive

a Expression systems, cell lines, growth media/conditions

Source: L. Nelsen (M.I.T) as cited by Razgaitis[5]

Table 2: Royalty Rates for the Medical Industry

Technology/Industry	Earned Royalty (%)	Up-front Payments (in US$)	Minimum Payments (in US$)
Reagents/process	1–3	Patent costs	2,000–10,000
Reagents/kits	2–10	Patent costs	2,000–10,000
Diagnostics in vitro	2–6	5,000–20,000	2,000–60,000
Diagnostics in vivo	3–8	5,000–20,000	2,000–60,000
Therapeutics	4–12	20,000–150,000	20,000–150,000
Medical instrumentation	4–10	5,000–150,000	5,000–20,000 (yr. 1) 10,000–25,000 (beyond yr. 1)

Source: Razgaitis[6]

opportunity fits in the range identified, it is difficult to know where to begin. Further, not every opportunity falls within even these broad ranges. Some opportunities will have only negligible value; others could be unusually valuable opportunities.

Many things can be transferred between a licensor and licensee: IP rights, technical data, rights over improvements to a technology, rights to sublicense the technology, costs related to the patenting process, and so on. **The price that the licensee pays to the licensor can consist of any combination of various types of payments**, including running royalties, fixed payments, common stock (equity), R&D funding, lab equipment, consulting services, grant-backs, options, or access to other proprietary resources. The licensing contract should make allowances for the risks that a licensee will have to take in developing and commercializing the technology. **A combination of royalties and equity stakes is a particularly effective way of splitting the risk between two parties.** If a technology does not deliver, then the seller only receives the equity stake and the buyer does not need to pay any future cash. On the other hand, if the technology is highly successful, the buyer will have accessed the technology without forfeiting important cash that may have been crucial in bringing the technology to market, and the seller gains in higher royalties and higher value of the equity.

Another way to distribute the risk fairly is to discount expected returns with an appropriate hurdle (or milestone) rate. Alternatively, the schedule of payments may be adjusted as a function of milestones. Picking appropriate milestone rates or individual payments can be informed by explicitly modeling how different stages or factors in the development process contribute to the overall expected returns and to the risk of not realizing those returns.

Razgaitis provides many illustrations and offers the following broad conclusions, which are supported with examples. **There is no "right" price for a technology with each licensing transaction being unique.** To assess the future sources of value, the innate economic benefit that can be captured by using the technology in *some* market must be identified. That value is dependent upon many factors, all of which may change at any moment. But **there is often only a small window of time in which a technology can turn a profit.** Being cognizant of all information regarding the possible market value of and possible risks associated with a technology, makes it easier to arrive at an accurate valuation. This information may be difficult to collect, but it will be necessary. In fact, in order for pricing to be done properly, all of the relevant information must be gathered, preferably ahead of negotiations.

Early stage technologies may end up having little or no commercial value, but there are rare cases of immense value. As the technology ages and patents approach expiration, the bargaining position of the licensor weakens. This results in an inevitable shift in bargaining power from the licensor to the licensee, resulting in prices or royalties being renegotiated downward, not upward. Regardless of where one is in the process, there are methods practiced by technology transfer and business development professionals that can be used to guide the pricing process.

Pricing is never completely objective and always carries risk, while risk itself is subjective and each party will perceive it differently. Certain events, such as additional testing of a technology by the licensor, or by a government R&D grant awarded to the licensee, or a collaborative venture between the licensee and other R&D institutions may reduce the risk as perceived by the licensee. The important thing is to **find a price that is acceptable to both parties and that encourages the licensee to invest in the development of the technology.**

Concluding this section, Lesser and Krattiger[7] use bioprospecting examples to examine how different valuation methods have different policy implications for developing countries. For instance, a bioprospecting deal can provide a developing country the incentive to preserve its natural resources if the money it receives from the developed country institution for the right to bioprospect inside the country is greater than what the country would receive by allowing destructive activities such as logging.

The trade-offs described earlier between up-front payments and royalties are particularly relevant from a policy perspective in the context of bioprospecting. The authors show that the principal factors used in negotiating price are the uncertainty of attributing value and the uncertainty of finding marketable products.

Negotiations in bioprospecting deals strive for an appropriate balance between collection (initial) fees and royalty (delayed) payments. Lesser and Krattiger demonstrate with examples and calculations how changes in assumptions lead to different outcomes. For example, collection fees will reduce total payments except when national interest rates are very high. In-country screening, including the use of indigenous knowledge, is a potentially valuable strategy as it shifts the rise of failure to the licensee. The authors outline issues for contract negotiators and discuss the implications for biodiversity conservation.

A discount rate is what is used to adjust future income to present net value. It is often akin to an interest rate. Receiving, say, $100 today would be $110 in one year's time if an interest rate were 10%. Conversely, receiving $100 in one year's time would be worth $90.90 today if a 10% discount rate were applied. Typically, personal and corporate discount rates are greater than social rates, although the determination of the social rate is open to different interpretations. As anyone who has paid off a loan over a 10 or 20-year period recognizes, small changes in the discount rate have major implications on the outcome. Further, the concept of personal discount rate (that is, what the person on the other side of the table has internalized about risk), political and economic instability, immediate need for money, and so forth, could play a large role in the choice between collection payments and royalties.

Importantly for developing countries, when fees are "shifted forward" by increasing the collection fee and reducing royalty payments, more risk is transferred to the collecting company that develops a product, since it will have to pay the same amount of money regardless of whether successful commercial products are developed from the collected material. Shifting fees forward have particularly interesting possibilities in countries where interest rates are high (since the discount value of future payments is lower with higher interest rates). This leads to important policy considerations for national governments, nongovernmental organizations (NGOs), and development agencies. The chapter reviews these policy considerations and concludes that providing grants/loans and training/equipment for in-country screening should be given a high priority because in-country screening may be productive in the long term. With regard to national policy, Lesser and Krattiger discuss several policy considerations[8] involving both in-country screening and the allocation of payments between collection fees and royalties.

With adequate in-country funds lacking, international donors should seriously consider loans or grants for training and equipment purchases. This coupled with a series of other initiatives[8] will bring many closer to realizing the promise of bioprospecting. ■

All chapters refer to: *Intellectual Property Management in Health and Agricultural Innovation: A Handbook of Best Practices*. 2007. A Krattiger, RT Mahoney, L Nelsen, JA Thomson, AB Bennett, K Satyanarayana, GD Graff, C Fernandez, and SP Kowalski (eds.). MIHR: Oxford, U.K., and PIPRA: Davis, U.S.A. Available online at www.ipHandbook.org. The online version contains for each chapter a detailed Editor's Summary, Implications, and Best Practices.

1 Chapter 9.1 by L Nelsen titled Evaluating Inventions from Research Institutions, p. 795.

2 Chapter 9.2 by RH Potter titled Technology Valuation: An Introduction, p. 805.

3 The online version of the *Handbook* provides a spreadsheet (in Microsoft® Excel®) for the user to see how various results are obtained depending on different inputs and assumptions.

4 Chapter 9.3 by R Razgaitis titled Pricing the Intellectual Property of Early-Stage Technologies: A Primer of Basic Valuation Tools and Considerations, p. 813.

5 *Ibid.*

6 See *supra* note 4, with data adapted from an article published by Corey G and E Kahn. 1991. How to Negotiate Reasonable Royalty Rates for Licensing Novel Biomedical Products. *Genetic Engineering News* July–August 1991. p.4.

7 Chapter 9.4 by WH Lesser and A Krattiger titled Valuation of Bioprospecting Samples: Approaches, Calculations, and Implications for Policymakers, p. 861.

8 Cabrera Medaglia J. 2004. Bioprospecting Partnerships in Practice: A Decade of Experiences at INBio in Costa Rica. *IP Strategy Today* No. 11 (2004): pp. 27–40. www.bioDevelopments.org/ip lists a number of important policy conclusions stemming from experiences at INBio in Costa Rica.

FOR GOVERNMENT POLICYMAKERS

- **Determining how to translate an invention into an innovation that makes a difference in people's lives (economically or socially or both) is one of the principal reasons technology transfer offices exist.**

- Government policies ought to be **flexible and enable research institutions to customize technology transfer strategies** that align with the institutions' missions. Different approaches will serve different types of research and academic organizations working within various disciplines and cultures.

- It can be challenging to negotiate licensing agreements that are fair to everyone and conducive to making inventions become innovations. **It is often better to make an imperfect deal than no deal at all.** People do not benefit until technology is developed and distributed.

- **Public sector institutions should therefore be supported in their overall deal making efforts** rather than using individual deals as particularly good or bad examples.

- A government can **make technology transfer less risky** and more attractive for licensees by applying such policies as government R&D grants, subsidies, encouragement of clusters, financing of business incubators, and offering complementary R&D inputs or regulatory requirements that are conducive to the emergence of new technologies.

- Bioprospecting and related activities raise important issues with respect to pricing. Importantly for developing countries, when fees are "shifted forward" by increasing the collection fee and reducing royalty payments, more risk is transferred to the collecting company, which is developing the product, since the company will have to pay the same amount of money regardless of whether successful commercial products are developed from the collected material. **Shifting fees forward may have particularly interesting possibilities, as doing so allows countries to invest resources early on to capture additional value in bioprospecting activities.**

- It is important to adopt national policies that **facilitate access to biological resources under fair and equitable terms with prior informed consent**. Access mechanisms should be transparent, predictable, and managed by experts.

- There is a strong interaction between bioprospecting activity and national scientific capabilities. In countries with strong scientific capability, bioprospecting is robust. Moreover, such **capacity increases the negotiating strengths and benefit sharing** stipulated in contract agreements.

Given that IP management is heavily context specific, these Key Implications and Best Practices are intended as starting points to be adapted to specific needs and circumstances.

KEY IMPLICATIONS AND BEST PRACTICES: SECTION 9

FOR SENIOR MANAGEMENT

(UNIVERSITY PRESIDENT, R&D MANAGER, ETC.)

- ✔ **Determining how to translate an invention into an innovation that makes a difference in people's lives (economically or socially or both) is one of the principal reasons technology transfer offices exist.**

- ✔ It can be challenging to negotiate licensing agreements that are fair to everyone and conducive to "moving" inventions to innovations. **It is far better generally to make an imperfect deal than no deal at all.** People do not benefit until technology is developed and distributed.

- ✔ **Institutions need to assess whether or not patenting is the most effective way to ensure high economic and/or humanitarian impact of their technologies**.

- ✔ A public institution's **decision with regard to patenting** should depend on (1) whether such patenting would be socially responsible, (2) whether there is public interest in the technology, and (3) whether patenting would help the local economy (where applicable).

- ✔ Putting a "price tag" on an invention early on is difficult, if not impossible. Fortunately, **the full value of an invention need not be determined when the invention is transferred or licensed,** as value can be realized later through the use of running royalties, fixed payments, common stock (equity), R&D funding, lab equipment, consulting services, grant backs, or access to other proprietary resources. For public sector organizations, in-kind contributions may sometimes be particularly appealing.

- ✔ **Evaluating a new technology is difficult and the evaluation will necessarily be imprecise.** It is better to encourage a TTO to make deals creatively and expeditiously, without the imposition of minimum royalties and other restrictive terms. The important thing is to **find a price that is acceptable to both parties and that encourages the licensee to invest in the development of the technology**.

- ✔ Senior management should be **supportive of the overall deal making of its technology transfer officers** rather than be critical of individual deals. Naturally, TTO officers need to follow procedures, apply policies, and be well trained and experienced in deal making.

- ✔ **Putting pressure on TTO officers to break even or to generate revenues can constitute a perverse incentive**, almost forcing a TTO to go with up-front payments. This may drain a startup of critical financial resources and thus reduce the level of investment that is allocated to making the invention work.

- ✔ Probabilistic modeling software can aid pricing efforts. The most effective software is expensive and may not be a good investment if fewer than 100 deals are made per year. Quite **often the best approach is to get as many licenses as possible completed in a short period of time, even if an individual license does not provide the maximum possible income**. The more licenses, the higher the probability that one, or a few, will generate returns.

Given that IP management is heavily context specific, these Key Implications and Best Practices are intended as starting points to be adapted to specific needs and circumstances.

FOR SCIENTISTS

- The best approach by your TTO is usually to **disclose inventions early and disclose often**.

- You should even consider disclosing what you think might not yet be a full invention. Experience shows that **scientists are, in fact, not very good at determining when they have an invention**. In many cases, they have a dozen when they themselves think they have none!

- If your TTO officers decide not to file for patents, you shouldn't be discouraged. This is not a critique on your research, its importance, or its relevance. **The TTO has many priorities to balance, including financial.**

- It can be challenging to negotiate licensing agreements that are fair to everyone and conducive to "moving" inventions to innovations. **It is generally far better to make an imperfect deal than no deal at all.** People do not benefit until technology is developed and distributed.

- Some of the key questions TTOs address early on when an invention has been made is whether a patent application should be filed at all, how the invention would be marketed, and what value the invention might add to existing processes or products or what value might come out of a new product or process. **Determining how to translate an invention into an innovation that makes a difference in people's lives (economically or socially or both) is one of the principal reasons technology transfer offices exist.**

- Scientists must insist that the TTO **have transparent procedures for reviewing invention disclosures and making decisions**. You should not only be informed of the basis and rationale for a decision, but also, in most cases, be fully involved in the process.

- It is important to keep a detailed **record of your research procedures**. Your records may help determine inventorship and may provide clues as to the value of your inventions.

- Once your TTO patents your invention, don't expect a big revenue flow. For a TTO, quite **often the best approach is to get as many licenses as possible completed in a short period of time, even if an individual license does not provide the maximum possible income**. The more licenses, the higher the probability that one, or a few, will generate returns. Both you and senior management should be **supportive of the overall deal making of the TTO** rather than criticizing individual deals. Naturally, TTO officers need to follow procedures, apply policies, and be well trained in deal making.

- **Additional research by yourself or your group often increases both the likelihood of finding a licensee and the economic value of the license**. But this is only true if the research is specifically aimed at reducing the risks of commercializing the technology. Basic research may do little to reduce these risks. Discuss this issue with your TTO officer. Especially in an academic environment, he or she will be reluctant to provide unsolicited advice regarding this issue.

- Remember that licensing incomes reward the commercial value, and not the scientific value, of your invention. **Technology licenses may provide you with follow-on grants and other intangible incentives** to conduct further research.

Given that IP management is heavily context specific, these Key Implications and Best Practices are intended as starting points to be adapted to specific needs and circumstances.

FOR TECHNOLOGY TRANSFER OFFICERS

- ✔ A **combination of royalties and equity stakes** is a particularly effective way of splitting risk between two parties. If a technology does not deliver, then the seller receives only the equity stake, and the buyer does not need to pay any future cash. Another way to distribute the risk fairly is to **discount expected returns with an appropriate milestone rate**.

- ✔ **Many valuation approaches exist. None is perfect.** Considering that each deal is highly context specific, each technology transfer office should be able to select the best approach and adapt it to the specific circumstances.

- ✔ Licensing is always risky and no deal will be perfect. **It is often better to make an imperfect deal than none at all.**

- ✔ When devising a **patenting strategy**, you will need to make three decisions: First, should you seek patent protection? Second, what is the best patent-marketing approach? Third, what license fees or royalties ought to be levied?

- ✔ Since there is no single best way to assess the value of a technology, **all parties should agree on the valuation method to be used**.

- ✔ Probabilistic modeling software can aid pricing efforts. The most effective software is expensive and may not be a good investment if fewer than 100 deals are made per year. Quite **often the best approach is to get as many licenses as possible completed in a short period of time, even if an individual license does not provide the maximum possible income**. The more licenses, the higher the probability that one, or a few, will generate returns.

- ✔ Putting a "price tag" on an invention early on is difficult, if not impossible. Fortunately, **the full value of an invention need not be determined when the invention is transferred or licensed,** as value can be realized later through the use of running royalties, fixed payments, common stock (equity), R&D funding, lab equipment, consulting services, grant backs, or access to other proprietary resources. For public sector organizations, in-kind contributions may sometimes be particularly appealing.

Given that IP management is heavily context specific, these Key Implications and Best Practices are intended as starting points to be adapted to specific needs and circumstances.

Patents and Patenting: Balancing Protection with the Public Domain

The nature of *public* and *private* and balancing individual rights with public welfare has been a perennial concern for many societies. For more than two thousand years in the West, scholars, philosophers, and politicians have debated questions of individual rights and of a government's responsibility to protect those rights while promoting the public good. Plato argued against private property (he said it would corrupt the personality by infecting it with greed), while Aristotle essentially argued for private property stating it would enhance an individual's sense of identity and self-esteem and, in addition, allow for the optimal economic use of *the commons*.

Private goods are those over which there can be competition or rivalry, their use can be excluded from nonowners. Private goods typically are traded in markets. If a market is able to agree on a price (such as for bread), the ownership or use of the good (the bread) is transferred. Further, once the good is consumed (the bread has been eaten) others are precluded from also eating it.

Public goods are goods the use of which neither competes with nor rivals use by others (nonrival), and no person can exclude other persons from use of the goods (nonexcludable). Sunlight, traffic lights, street signs, sewer systems, and a smallpox-free world are examples of public goods. Crucially, **who provides the public good is not a factor in determining whether a good is public or private.** Governments provide public goods (such as street lights) and they also provide private goods (for example, housing and medical care). Similarly, the private sector may provide public goods, such as technical norms or streetlights. The example of the streetlight illustrates how private goods (patented, high-efficiency light bulbs, electricity produced by companies, street posts installed by local private contractors) become public goods because they are made available to all. The creation of a public good is not free of cost. Costs may have been borne by society at large (the street light will have been paid for, indirectly, by the taxpayer) but the enjoyment or use of it is free to any and all individuals who pass along that particular street.

Further, these examples demonstrate how *public* and *private* are in some respects two sides of the same coin; both are needed for the coin to exist and have value. But as Boettiger and Chi-Ham[1] show, the manner in which this plays out with information and inventions generated by science is not as straightforward as the examples above suggest. Complications are due, in part, to the fact that inventions, unlike real and tangible property, once they are disclosed are essentially nonrival and nonexcludable. Unless, of course, IP rights systems regulate ownership, access, and use.

Boettiger and Chi-Ham discuss the nexus of public and private property and provide in-depth guidance on using defensive publishing and the public domain as tools to achieve a range of IP

Krattiger A, RT Mahoney, L Nelsen, JA Thomson, AB Bennett, K Satyanarayana, GD Graff, C Fernandez and SP Kowalski. 2007. 10: Patents and Patenting: Balancing Protection with the Public Domain. In *Executive Guide to Intellectual Property Management in Health and Agricultural Innovation: A Handbook of Best Practices* (Krattiger A, RT Mahoney, L Nelsen et al.). MIHR (Oxford, UK), PIPRA (Davis, USA), Oswaldo Cruz Foundation (Fiocruz, Rio de Janeiro, Brazil), and *bio*Developments-International Institute (Ithaca, USA). Available online at www.ipHandbook.org.

management goals. The authors examine the extent to which the *public domain* can in fact be depended upon and even *leveraged* to facilitate and preserve access to technologies. The authors view **defensive publishing** and utilizing public domain research inputs as options within a broader set of IP management strategies, including the options of patenting, trade secrecy, trademark protection, and bailment contracts. All of the above can be used in various combinations to find the balance between protection and accessibility that both promotes technology development and fosters ongoing innovation. They argue that the choice of the strategic option depends on a pragmatic and realistic understanding of the nature of the *public domain*. In order to clarify and illustrate this, the *public domain* is compared to two different, but closely related, property rights concepts: *open source* and *the commons*.

Open source is defined as the body of knowledge over which owners claim property rights, but with access to that knowledge being provided systematically by the owners under terms of a license that regulates access. According to the authors, **the commons** is, by its nature, a less clearly defined concept that varies according to context, but includes a lack of private ownership, open access, or collective management.

It should be noted that defensive publishing strategies can be more viable than patenting in the following two cases:

- development of the technology will not depend on private sector investment
- the leverage ownership could provide, such as the ability to segment markets or bargain for access to complementary technologies, is not important

Defensive publishing can be less costly, eliminating patent costs and transaction costs in licensing, especially when the ultimate desired outcome is to provide broad access to a technology. When cost or infeasibility makes enforcing a patent unlikely, defensive publishing is the more sensible alternative for companies. It can be very effective when combined with patenting.

The second aspect of the chapter by Boettiger and Chi-Ham deals with the **use of public domain technologies** as research inputs. The authors outline how this approach can reduce transaction costs and mitigate potential IP access problems in downstream R&D. However, as for any IP management strategy, attention should be paid to the overlapping, web-like nature of patent claims and the ever-shifting boundary of the public domain. Even when a technology is preliminarily believed to be in the public domain, such as from an early scientific publication or an expired patent, subsequent publications or patents can still claim certain uses of the technology, such as in particular combinations, applications, or with possible improvements.

The public domain can be a vital resource to public sector institutions and also companies. Judicious defensive publishing and the careful use of public domain technologies offer IP managers everywhere effective, flexible, and less-expensive tools for exploiting these resources. Intellectual property is not a panacea for the management of innovation. Neither is open source. All have utility and limitations. **Artful management involves the creative and balanced use and handling of both public and private goods.**

If patenting is the chosen route, it must be remembered that patenting decisions need to be made well before it is clear whether or not an invention has value. It makes business and strategic sense, therefore, to minimize the initial costs of such decisions. If the invention appears to have significant market potential, then a cost-minimizing approach toward patenting is not recommended. However, most inventions have questionable or uncertain future value, and so a cost-minimizing approach is an appropriate strategy for patent application filings.

One **cost-minimizing approach is the filing of provisional patent applications**, which is possible in many countries. Importantly, foreign inventors may also file provisional applications in the United States. Provisional patents allow inventors an extra year of protection, effectively extending the patent period from 20 to 21 years. As Cruz[2] explains, the **benefits of provisional applications** include cost and simplicity. Provisional applications are not substantively reviewed by a patent office examiner, but are simply checked to ensure that they meet minimal filing requirements. A provisional application also does not require a prior art search. Since these applications are so quick and inexpensive to prepare, they offer an easy way for inventors to establish a priority date for an invention and avoid statutory bars.

But there are also **limitations associated with provisional applications.** While inexpensive to file, provisional patent applications do not reduce the costs of preparing and filing subsequent utility applications, meaning that the total cost of filing will increase, if only by a small amount. More importantly, provisional applications require a degree of disclosure, so inventors should be sure not to disclose something they wish to retain as a trade secret. Also, provisional patents may not be amended; they trigger the time line for Patent CooperationTreaty (PCT) and Paris Convention filings, and without filing nonprovisional patent applications, provisionals do not mature into patents.

Livne[3] then discusses various avenues for reducing costs in patent filings and presents a highly useful **decision tree.** He cautions that the use of patent attorney is generally an essential guide in such matters. The stakes are often high and mistakes can be costly. When publication is imminent and patent protection in foreign countries is desired, a provisional or nonprovisional application should be filed in the United States before publication. Once disclosed, filing in many countries will no longer be possible.

Although licensing is discussed elsewhere,[4] the manner in which patent applications are written, particularly the claims, can be instrumental in facilitating certain licensing strategies, particularly field-of-use licensing. The **foundation of an effective field-of-use licensing strategy is a patent application** that foresees certain licensing opportunities and accommodates unforeseen opportunities. Olson[5] discusses this using examples from the agricultural, pharmaceutical, biochemical, and chemical disciplines and illustrates how this strategy applies equally to inventions with commercial and humanitarian applications. He urges technology managers to retain control over the patent application process and to encourage creative thinking when preparing patent applications.

Applying for patents is only one element in a strategy to create IP portfolios of substantial value. For both public and private sectors, patents are a central element, but an IP portfolio should also take advantage of other forms of protection: trademarks, copyrights, and trade secrets. Dodds[6] presents various **strategies for building an IP fortress** and discusses the limitations and strengths of various approaches. For example, an *offensive* patent strategy is designed to build barriers to exclude competitors from proprietary technologies. With a *defensive* patent strategy, a company files patents primarily to ensure that innovations can be practically used. To build an IP fortress of protection, several forms of intellectual property may be used for the same invention or improvement, with different forms of IP protection serving offensive or defensive tactics.

Notwithstanding the different missions, objectives, and motivations of the public and private sectors, the central forces behind their respective IP protection strategies are identical (though the relative strength of the forces will vary significantly). **Private sector** organizations, primarily corporations, are profit-oriented and respond to the pressures imposed by the marketplace and by shareholders who expect returns on their investments. Therefore, the private sector will use defensive and offensive patenting strategies, often obtaining numerous patents with narrowly drafted claims. In this way, a series of patent portfolios is strategically used to build proprietary fortifications and the private sector organization can stake out its territory, protect its interests, and secure its profits. In the expanding world marketplace, this strategy has only become more telling, with the increasing reliance on foreign filing and patent families confirming the predominant global strategic perspective of multinational companies.

The **public sector**, on the other hand, has a very different mission, which is to serve the greater public good. Patenting strategy will focus on more broadly drafted claims that will encompass a technology or (as is often, and more importantly, the case) a key process, method, or technique, for example a technique of genetic transformation. These types of patents, when appropriately strategically licensed, enable effective development, broad dissemination, and maximum societal impact of a technological advance precisely in line with the public sector mission of providing for the general public, in contrast to the much more limited constituency of the private sector.

Patent protection is limited geographically, protecting the invention only in countries where the patent issues. Private sector companies and public sector institutions can reduce costs by focusing the patent protection to those geographic areas where there are business or humanitarian opportunities. But filing in foreign jurisdictions is not easy or cheap. For this reason, two chapters, one by Viksnins and McCrackin[7] and the other by Schneiderman,[8] review **foreign filing strategies and tactics, with particular emphasis on filing patent applications using the PCT**. The two chapters are complementary and discuss the practical aspects from different points of view.

Several key factors should be reviewed when approaching the international production, marketing, distribution, and sales of a new and innovative product or process for which patent protection will be sought. These factors include a full range of various business and legal issues that, once considered, will provide the international patent protection options that can then be evaluated and appropriately selected, according to an organization's business goals and financial resources.

Depending on an organization's goals and resources, specific patent-application options will have advantages and disadvantages. One option is to file a separate patent application for each nation or region where protection is sought. Another option is to file a patent application, in accordance with the **Paris Convention**, which establishes a priority filing date. This gives, for one year, the exclusive right to file for patents in other Paris Convention countries. This approach has advantages when filing in a very limited number of countries. It also avoids the costs associated with the intermediate steps of filing in the PCT or regional patent offices prior to filing nationally. This option has the following disadvantages: each application will be independently examined (that is, no deference is given to a prior favorable review in a different country) and government filing fees and translation costs will be due early in the patenting process.

An indispensable tool for delaying, consolidating, and minimizing international patent costs, **the PCT offers a unified and simplified procedure** for filing multiple foreign patent applications using a single initial application. The PCT has standardized the filing and preliminary evaluation of international patent applications. Consisting of over 130 member countries, the PCT is administered by the World Intellectual Property Organization (WIPO), which reviews PCT applications and then distributes them to designated member countries. The process of filing the PCT application in individual patent offices can be delayed for up to 30 months. During this time, the applicant will receive the results of the WIPO International Preliminary Examination of the PCT application. For many countries, especially those still developing capacity in patent prosecution, national patent offices give considerable deference to the PCT International Preliminary Examination Report.[9]

Using best practices in IP management involves identifying IP assets, organizing resources, building capacity, formulating options, and then pursuing strategies that will maximize the value of an organization's IP assets. Managing patent portfolios is always challenging, even more so now with the rapid globalization of technology markets. Globalization makes **best practices in patent portfolio management** more critical for effectively distributing innovations in the health and agricultural sciences, whether for commercial purposes or for facilitating humanitarian access. As public and private sector institutions increasingly work in a global context, choosing where and under what circumstances to file for patents is becoming more important, and, according to Yin and Cunningham,[10] the following factors should be considered:

- objectives of the organization with respect to its issued patents
- assertion of patents *offensively*, either as part of a licensing strategy or in litigation, if companies are unwilling to license

- assertion of patents *defensively*, as leverage in licensing negotiations or to ward off litigation by others
- identifying where potential targets are located or doing the bulk of their business, if a portfolio is to be used offensively,
- identifying where an organization may most likely encounter licensing approaches or litigation offensives by others, if a portfolio is to be primarily defensive

A global patent program should be proactive as well as preemptive in its outlook, especially regarding the potentiality of patent litigation, where knowledge of options can save time and money. Yin and Cunningham compare and contrast the **advantages and disadvantages of pursuing patent litigation** in either a federal district court or in the U.S. International Trade Commission (ITC). Although the ITC's jurisdiction is essentially limited to cases dealing with the illegal importation of alleged infringing products, there are times when it might be a good idea to pursue patent litigation in the ITC. In addition, one can also pursue litigation in both the federal district court (patent infringement action) and the ITC (unfair trade practices action) at the same time.

Obtaining patent protection and regulatory approval for biotechnology and pharmaceutical products is an extremely time-consuming and expensive process. For nonprofit organizations working with limited resources, it is especially crucial to manage the process efficiently and make the most of patent protections while they last. Fernandez, Huie, and Hsu[11] suggest that public sector entities can use private sector techniques to **maximize revenue and, in turn, provide drugs to the public at the lowest possible price.** The authors suggest that organizations carefully plan the timing of patent and Food and Drug Administration (FDA) applications to maximize the effective life of a patent and avoid unnecessary disclosures.

Nonprofits especially should note that the U.S. Patent and Trademark Office (PTO) gives special priority to certain biotechnology patent applications from small entities and nonprofits. The FDA likewise expedites approval if there are indications that the product will provide significant therapeutic benefit over existing therapies. At the other end of the patent lifecycle, after having gone through the steps of obtaining FDA approval, it is in the best interests of innovating companies to **extend the patent term** for as long as possible. This chapter is included in the *Handbook* to show the important **interface between patents and the regulatory drug approval process** and to show how this interplay affects market entry. It is not intended as an endorsement of extending effective patent life to delay the market entry of generic drugs.

As part of certain patent application filings,[12] biological resources may have to be deposited in support of a patent application. According to Harney and McBride,[13] in the United States a **deposit of biological materials** is not a requirement per se, but under U.S. patent law it can satisfy three main requirements:

- the enablement requirement, that is, that it would allow a person skilled in the art could make and use the invention
- the written description requirement, that is, it would describe the invention in sufficient detail to allow such a person skilled in the art to reasonably conclude that the applicant was in *possession* of the claimed invention at the time of filing
- the best mode requirement, that is, that it would disclose the best mode of carrying out an invention in sufficient detail to allow a person of ordinary skill in the art to practice it

Plant varieties constitute a biological resource. While in the United States while plant varieties can be protected as utility patents, the United States and many other countries also have **protection mechanisms specifically adapted to the biological and self-replicating nature of plant varieties.** Pardee[14] provides detailed and step-by-step instructions for how to obtain a U.S. Plant Variety Protection (PVP) certificate. Although the chapter focuses on PVP application procedures in the United States, the chapter is generally useful for illustrating the principles, preparations, and procedures for applying for and obtaining a PVP certificate. This is because the U.S. provisions of the PVP Act of 1970 closely follow the model developed by the Convention of the International Union for the Protection of New Varieties of Plants (UPOV). Moreover, UPOV procedures have been adopted by many countries around the world, even by many who are not members of UPOV.

In sum, the factors that drive decisions about what type of protection to seek and where to seek it are complex and will heavily depend on the context in which the decisions are made. Public and private institutions will consider the same factors but weigh them using different criteria. For example, the prospect of litigation in a foreign jurisdiction for a public sector entity will be marginally important per se but highly relevant to its potential to license. The result is that both public and private sectors will consider whether to adopt *offensive* or *defensive* protection and litigation strategies.

Because the rights accorded to the patentee are divisible (the right to exclude one from selling *or* the right to exclude another from manufacturing) one can divide the countries of the world into those where the invention can be manufactured versus countries where the invention will be sold. And even after identifying those countries where the invention might be marketed, it is often unnecessary to file in all of the identified countries. In order to best determine what strategy to pursue, an organization must know what it has and decide where it's going. The first step, therefore, in developing an IP strategy is to document what technologies already exist in the organization, what technologies are in development, and what partnerships are feasible. It will then be possible to intelligently choose the best ways to protect intellectual property and enhance its value, be it for economic or humanitarian objectives, or both. ■

All chapters refer to: *Intellectual Property Management in Health and Agricultural Innovation: A Handbook of Best Practices.* 2007. A Krattiger, RT Mahoney, L Nelsen, JA Thomson, AB Bennett, K Satyanarayana, GD Graff, C Fernandez, and SP Kowalski (eds.). MIHR: Oxford, U.K., and PIPRA: Davis, U.S.A. Available online at www.ipHandbook.org. The online version contains for each chapter a detailed Editor's Summary, Implications, and Best Practices.

1 Chapter 10.1 by S Boettiger and C Chi-Ham titled Defensive Publishing and the Public Domain, p. 879.

2 Chapter 10.2 by RL Cruz titled Provisional Patent Applications: Advantages and Limitations, p. 897.

3 Chapter 10.5 by O Livne titled Cost-Conscious Strategies for Patent Application Filings, p. 921.

4 See Section 11 in the *Handbook* and Part 11 in this *Executive Guide*.

5 Chapter 10.3 by AM Olson titled Designing Patent Applications for Possible Field-of-Use Licensing, p. 903.

6 Chapter 10.4 by J Dodds titled Patenting Strategies: Building an IP Fortress, p. 911.

7 Chapter 10.6 by AS Viksnins and AM McCrackin titled A Guide to International Patent Protection, p. 927.

8 Chapter 10.7 by AM Schneiderman titled Filing International Patent Applications under the Patent Cooperation Treaty (PCT): Strategies for Delaying Costs and Maximizing the Value of Your Intellectual Property Worldwide, p. 941.

9 A potential alternative to PCT filing or filing directly in each country of interest is to file in a regional patent office. Regional patent offices have come into existence through international treaties and include the European Patent Office (EPO), the African Regional Industrial Property Organization (ARIPO), the African Intellectual Property Organization (OAPI), and the Eurasian Patent Convention (EA). The advantages of filing in a regional patent office are that some translation costs may be eliminated, and substantive examination of the regional patent in each of the designated countries may not be required. Indeed, if protection is desired in a number of member countries, this approach is especially cost-effective. But if protection in only a few member countries is desired, it may be less expensive to file applications in each country individually and avoid the costs associated with the intermediate steps of first filing in the regional patent office. Also, not every target country may be a member of a regional convention.

10 Chapter 10.8 by R Yin and S Cunningham titled Filing and Defending Patents in Different Jurisdictions, p. 953.

11 Chapter 10.9 by DS Fernandez, J Huie and J Hsu titled The Interface of Patents with the Regulatory Drug Approval Process and How Resulting Interplay Can Affect Market Entry, p. 965.

12 Such would include filings for bacteria, fungi, eukaryotic cells and lines, hybridomas, plasmids, plant tissues, seeds, viruses, vectors, and cell organelles.

13 Chapter 10.10 by DJ Harney and TB McBride titled Deposit of Biological Materials in Support of a U.S. Patent Application, p. 973.

14 Chapter 10.11 by WD Pardee titled Protecting New Plant Varieties through PVP: Practical Suggestions from a Plant Breeder for Plant Breeders, p. 981.

FOR GOVERNMENT POLICYMAKERS

- ✔ The use of IP rights is not a panacea for the management of innovation, nor is the public domain. Both public and private goods have utility and limitations. The **art of innovation management is in using both public and private goods** and to manage the interface between them.

- ✔ Because public domain technologies play an important role in publicly funded research, **defensive publishing** can be used by public sector research institutions to help expand and reinforce the accessibility of technologies in the public domain. Academic institutions in particular should be encouraged to publish, in addition to considering IP protection.

- ✔ Because of the case-specific applicability of defensive publishing, **blanket policies that require defensive publishing by national research institutions** deny them the opportunity to develop their research results strategically in combination with IP rights protection.

- ✔ In order to realize the commercial and humanitarian potential of international markets for products and processes arising from public sector research investments, public sector research-based institutions ought to develop **strategies that judiciously balance the public domain and IP rights**. Commercial and humanitarian objectives and strategies are not in conflict, but rather are complementary aspects of best practices in IP management.

- ✔ A country's **membership in the Patent Cooperation Treaty** (PCT) can greatly help national institutions—public and private—to strengthen international technology transfer, licensing and research, and product development partnerships and can aid access to global markets.

- ✔ Membership in the **PCT can provide significant advantages** and can lead to much more cost-effective examination of patent applications.

- ✔ **Harmonizing national patent systems** across regions, as well as globally, can be a useful strategy for improving the effectiveness of the IP system and improving a national institution's ability to reach foreign markets.

- ✔ Providing for legislation, or for amendments to current statutes, that **facilitates patent filing by foreign entities** can be an important component of technology transfer and development.

KEY IMPLICATIONS AND BEST PRACTICES: SECTION 10

FOR SENIOR MANAGEMENT

(UNIVERSITY PRESIDENT, R&D MANAGER, ETC.)

- ✔ The use of IP rights is not a panacea for the management of innovation, nor is the public domain. Both public and private goods have utility and limitations. The **art of innovation management is in using both public and private goods** and to manage the interface between them.

- ✔ Because public domain technologies play an important role in publicly funded research, **defensive publishing** can be used by public sector research institutions to help expand and reinforce the accessibility of technologies in the public domain. Academic institutions in particular should be encouraged to publish, in addition to considering IP protection.

- ✔ Because of the case-specific applicability of defensive publishing, **blanket policies that require defensive publishing** deny the opportunity to use research results strategically in combination with IP rights protection.

- ✔ Few institutions anywhere in the world have transparent incentives for researchers or technology transfer officers to prepare defensive publications. Encouraging **publication with maximum inventive disclosure** through a balanced set of incentives for researchers and technology transfer officers is a useful strategy.

- ✔ Scientists should be encouraged to **use public domain technologies** as research inputs whenever feasible to reduce possible future constraints in the downstream commercialization of innovations. In many circumstances, however, **relying on patented technologies** may be the more effective way to go, particularly when the goal is to develop products.

- ✔ Building strong **institutional capacity in IP management** will enable technology managers and scientists alike to understand the complex array of options that should be considered before publishing research results or filing patent applications. Development of protocols and strategies will clarify options and retain and maximize value.

- ✔ One such capacity centers on the decision of whether patents for an invention should be filed in a manner that does **not delay publication of research results**. Provisional patent applications, where possible, offer one such avenue.

- ✔ An important component of **developing an IP strategy** is to document the technologies that already exist in the organization, plus those technologies in development (for example, through an **IP audit**). Other essential components of such a strategy are the promotion of international patent protection and the concrete steps an institution is taking to drive innovation and technology transfer.

- ✔ Management should encourage good laboratory practices and diligent **record keeping** of data to ensure that research can later be used in possible regulatory filings. Doing so could lower costs and reduce the time to market.

FOR SCIENTISTS

- ✔ Published information, or research tools provided by a colleague, may be covered by IP rights. In the life sciences, the web of patents extends far and wide. This should neither deter nor distract you from good science. An **awareness of basic IP management best practices** will minimize possible future problems.

- ✔ You can intentionally make your inventions and the associated technologies accessible to everyone by **publishing results instead of patenting them**. Publishing results, however, does not guarantee full public access. Patents can still encroach upon the technical content of the work. Speak to your technology transfer manager about publications, and ask him or her to help with performing the necessary steps for turning your publication into a readily identifiable disclosure of patentable technology.

- ✔ **If public disclosure is your goal** as a way of preventing others from patenting a particular invention, it may be valuable to consider posting online or in searchable databases, with a valid date stamp, a longer working-paper version, supporting materials, or appendices. For this purpose, consider using dedicated services, such as a university technical disclosure bulletin or a centralized registry of unpublished papers, with official date stamps posted on faculty Web sites for online searches.

- ✔ **If patenting and public disclosure are your goals**, first consult with your institution's technology transfer manager prior to disclosure. Your institution should have an effective mechanism in place to determine whether or not a patent should be filed without significantly delaying publication. But be aware that premature publication can lead to a loss of IP rights.

- ✔ Your institution's technology **transfer managers will need your input** in order to make strategic decisions about where to pursue foreign patent applications. You likely know where competitors are located and where products arising from your research are needed.

- ✔ One of the services of **PIPRA** is to advise researchers in the plant sciences about which research is in the public domain and which is available for licensing on reasonable terms. If you are engaged in the development of biotechnology crops, you may find PIPRA's Web site and services useful.

- ✔ Good laboratory practices and comprehensive **laboratory notebooks** can ensure that your research is suitable for subsequent regulatory filings. This can reduce costs and time to market.

KEY IMPLICATIONS AND BEST PRACTICES: SECTION 10

FOR TECHNOLOGY TRANSFER OFFICERS

- ✔ The use of IP rights is not a panacea for the management of innovation, nor is the public domain. Both public and private goods have utility and limitations. The **art of innovation management is in using both public and private goods** and to manage the interface between them.

- ✔ Because public domain technologies play an important role in publicly funded research, **defensive publishing** can be used by public sector research institutions to help expand and reinforce the accessibility of technologies in the public domain.

- ✔ It helps to have other tools besides patents to **get technology out of the lab and into the marketplace**. Consider first whether a technology requires investment by the private sector (and, thus, exclusivity) to be put into practice.

- ✔ Defensive publishing may run contrary to your instincts if you tend to think in terms of controlling a technology by ownership (and thus excluding others from using it). Think instead in terms of **maintaining control of the technology**—or elements of it—by casting it into the public domain and, thereby, preventing others from owning it.

- ✔ Researchers will need advice on how to craft **defensive publications**.

- ✔ It is important to understand the **advantages of provisional patent applications**. They can be very useful in controlling costs and, also, in providing additional time for weighing options as to whether it is worthwhile to pursue a full patent application.

- ✔ **Delaying patent applications** involves risk. Subsequent prior art that blocks an application might appear. Or, the same invention might be patented by a competitor.

- ✔ For any invention, evaluate whether **foreign patent rights** are truly required. This will require a combination of business, marketing, and legal analyses.

- ✔ When assembling a patent application, **attorney costs can be reduced** by providing a cohesive document containing all data and information relating to the invention, such as alternative methods, compositions and/or devices. Use attorneys, at least, to review draft patent applications and to write the **all-important claims**.

- ✔ The foundation of an effective **field-of-use licensing strategy** is a patent application that foresees certain licensing opportunities and accommodates unforeseen opportunities. It will thus be important for your office to establish and implement strategies for patent application preparation that seek to anticipate any and all licensing opportunities that can arise from an invention.

- ✔ It is essential to **retain control of patent applications**. Don't permit a licensee to gain control; their interests and your interests are likely very different.

- ✔ **Tiered or layered IP protection strategies** utilize several forms of protection for a single product or process. For example, a hybrid maize variety may be simultaneously protected by patents, trade secret, trademark, and plant variety protection.

Technology and Product Licensing

Buyers want to get the most for their money; sellers want to get the most for their products. Put differently, no one wants to pay more for an item than what is necessary. And sellers expect a fair price for their products.[1] These universal pricing concerns apply not just to food, shelter, and any other goods and services, but also to intellectual property. As a licensee, how much should you pay for a license? How much should a licensor charge for a license? And what form should payments take? Royalties on products sold? Fixed payments per year? Equity in a business? Provision of services (bartering) or some other form of remuneration? And what exactly are you paying or charging for? Answers to these questions are always complex.

The chapters in this section offer some points of reference from which to explore these and other questions that emerge during **IP licensing transactions** (a license being the transfer of certain property rights between two or more parties under a specified sharing of rights *and* obligations between those parties). These considerations apply to companies and public sector institutions alike. With a *license*, as distinguished from a sale, possession of property does not transfer but remains with the original owner.

Negotiation is one way to establish the terms under which a transaction takes place. But negotiation is just one aspect of establishing the terms of a specific transaction. Preparation for a negotiation can—or should—require at least ten times more time than the actual negotiation, since the **goal of a negotiation** is to formulate an agreement that meets the needs of both the licensor and the licensee in a manner that ensures mutually beneficial future relationships between the institutions and individuals. Anticipating the other party's needs and wants, and considering alternatives for resolving possible competing interests, is just one aspect of the preparation. Price, quite often, will not be the most difficult aspect to negotiate. Other terms can be more critical and of greater relevance, and value.

The first chapter by Freeman[2] discusses the central issues that licensors and licensees need to consider before negotiating agreements. After providing an **overview of licensing in the field of biotechnology**, he considers the main components of a license agreement, highlighting concerns specific to the field of biotechnology. A license agreement will include several key components:

- The *background section* sets out the factual predicates for the license, including the names of the parties, the effective date of the agreement, and the parties' motivations and expectations.
- The *definitions section* explains key terms used in the agreement.
- The *grant section* establishes whether only the licensee may practice the invention (an exclusive license) or whether others may

Krattiger A, RT Mahoney, L Nelsen, JA Thomson, AB Bennett, K Satyanarayana, GD Graff, C Fernandez and SP Kowalski. 2007. 11: Technology and Product Licensing.In *Executive Guide to Intellectual Property Management in Health and Agricultural Innovation: A Handbook of Best Practices* (Krattiger A, RT Mahoney, L Nelsen et al.). MIHR (Oxford, UK), PIPRA (Davis, USA), Oswaldo Cruz Foundation (Fiocruz, Rio de Janeiro, Brazil), and *bio*Developments-International Institute (Ithaca, USA). Available online at www.ipHandbook.org.

practice it. This section may stipulate rights to sublicense or rights to assign, or it may say that there are no such rights.

- The section on *fixed payments and royalties* sets out payment terms of the agreement.
- The *confidentiality section* specifies provisions for and restrictions on the disclosure of information shared between the parties.
- A section on *enforcement against infringers* specifies which party shoulders the burdens and realizes the benefits of enforcing the licensed patent against infringers. This enforcement often involves allocating the risks and rewards of the overall success of the venture.
- The *term of the agreement and termination procedures* should be fully spelled out. As with most licenses, the biotechnology license will often have a term that coincides with the patent term.

Freeman discusses a series of other points that require special attention, including the complexities of confidentiality clauses that are particularly pertinent when working with academic and public sector research institutions. Finally, to help make negotiations easier and more realistic, the chapter discusses incentives for licensors and licensees and considers some of the finer issues of developing collaborations.

Cahoon[3] addresses issues related to **agricultural biotechnology (agri-biotech) licenses.** Although these license agreements are similar in many ways to other kinds of license agreements, agri-biotech license agreements have some unique elements. The chapter explores the basic nature and purpose of a license agreement, and preferred licensing methods and terms are suggested. The chapter then turns to the complex—but highly important—singularities of agri-biotech licenses, focusing on such issues as multiple property types that cover a single technology and/or product, freedom-to-operate issues that may drive anti-royalty-stacking provisions, philanthropic and humanitarian-use clauses, and stewardship obligations. The chapter emphasizes the uniqueness of agri-biotech licenses in regard to the concept of *field of use,* which may be broadly or narrowly defined.

The complex and rapidly evolving nature of agri-biotechnology requires (at the moment) that each license agreement be tailored to the particular context in which the invention will be used. Still, such licenses do not have to be invented from the beginning, and Cahoon's chapter elucidates both the common and unique aspects of agri-biotech licensing. For practical purposes, any organization engaged in high-volume licensing will find it useful to develop its own internal template agreements[4] that are then modified and adapted to suit each special circumstance.

In addition to the more generic licensing aspects, this section of the *Handbook* contains a series of chapters that review more specific IP licensing strategies. The **in-licensing and out-licensing of plant varieties** are the two sides to the licensing equation. In-licensing of plant varieties can increase market share, building a competitive advantage by providing for customer needs. In-licensing varieties also enhances or completes a company's variety portfolio, both for in-house breeding programs (facilitating access to breeding materials) and for in-licensed varieties ready for commercial distribution. The most common reason for out-licensing varieties is for a company to maximize return on investment by allowing others to produce and sell varieties in markets that cannot be reached satisfactorily through the current marketing setup.

Importantly, the licensing of varieties is increasingly becoming more important for public sector breeding institutions. These are often funded, at least in part, by governments, and so they have a fundamental mission of serving the public interest. The same applies to the Centers of the Consultative Groups on International Agricultural Research (CGIAR). These institutions are eager to ensure that resource-poor farmers have the greatest possible access to value-added crop varieties. The central question here is how can these public sector breeding institutions provide **broad access to improved germplasm**? How can a combination of plant variety protection (PVP) and licensing accelerate the dissemination and adoption of improved varieties?

Private sector licensing expert Nilsson openly addresses plant variety licensing, sharing the experiences and approaches of a private sector entity. Nilsson[5] illustrates how plant variety licensing is a practical tool that plant breeding companies (in the private sector) or institutions (in the public sector) use to commercialize or provide access to their products (crop varieties). Licensing also facilitates technology transfer (where technology is defined as know-how, improved germplasm, a range of breeding tools, and genes) in a simple delivery mechanism:

- the seed as a vehicle for technology transfer
- the seed itself as a commodity-embedding technology

Nilsson provides practical guidance for in-licensing and out-licensing crop varieties, with a special focus on developing countries. Because of decreased funding of public sector breeding, the seed sector is gradually being served by private companies, often small-scale and local enterprises. This has created increased demand for new varieties, and seed companies are seeking to in-licence varieties while private sector breeders may desire to out-license their varieties. Capacity for negotiating and executing license agreements, therefore, is becoming all the more critical. In an organized, detailed, and understandable fashion, Nilsson's chapter presents the fundamentals of seed licensing, emphasizing how the licensee and the licensor should focus on the practical content of a license agreement: exclusivity to plant material and territory, plant variety protection, variety trials, national registration, royalty payments, and information transfer.

The best licenses are those that recognize that relationships—like markets—are not static. An agreement should thus include sufficient flexibility for evolution. The agreement should reflect changes in the market, competitors, technology, seed legislation, and PVP laws. Enabling such flexibilities is perhaps the greatest art in drafting and negotiating variety licensing agreements.

Overall, the successful licensing of varieties is contingent on the strength of PVP legislation. A PVP framework generally supports the interests of the variety owner and the farmer, facilitates the transfer of technology, and provides incentives for further investments in the development of new plant varieties. In many countries, PVP legislation is based on the Convention of the International Union for the Protection of New Varieties of Plants (UPOV). The relevance of UPOV in accelerating access to improved varieties is its harmonization and documentation of a PVP aspect that facilitates licensing by foreign seed companies and public sector institutions alike.

Due to numerous complexities in terms of geographical, cultural, and paradigmatic distances, prospective licensors and licensees frequently spend a lot of time becoming acquainted and developing a certain level of trust. Over time, they reach a point where they are speaking the same language of contracts and licenses, and they reach a satisfying agreement. This lengthy process, however, can deter or derail licensing efforts. Companies may not wish to invest such time and energy, because even commercial licenses with entities in the developing world simply take up too much valuable management time and resources; the necessary funds for extended and repeated face-to-face meetings are simply unavailable.

A complementary approach is therefore needed—a way to bridge this communication gap and more rapidly arrive at a common language. Modern computer and Web technology might provide an answer. The chapter by Krattiger, Dodds, and Bobrowicz[6] examines the **potential of a software decision tree linked to template contract language** that allows individually customized contract documents to be generated and that could ameliorate many of the aforementioned problems. Provided that some key players agree to the basic template, an appropriate software package could improve opportunities for assembling a greater array of potential partners. A test version of such a computer-generated contract template (CoGenCo) system has been developed and could be a pragmatic step toward increased licensing of proprietary and finished varieties that may or may not incorporate proprietary technologies for input or output traits. The CoGenCo system is aimed at establishing a certain international standard license, that is, a standard that all understand and agree on. In this way, a meeting of the minds is facilitated and accelerated. A standard license can be downloaded for free from the online version of the *Handbook*.[7]

The approach of CoGenCo is to facilitate the awarding of out-licenses of germplasm to developing country institutions, including by and for the CGIAR and national programs. Under the legally binding terms of CoGenCo-generated license agreements, several entities in a given country could compete against one another on price in poor countries but would not be allowed to compete against the patent holder in developed countries, in which revenues and the incentives for developing new varieties and new technologies would be undiminished. Under appropriate circumstances, the germplasm and/or traits could be licensed royalty free. Use of out-licensing in this way separates these fundamentally different markets and promotes access to improved germplasm and technologies, all by reaffirming various statutory protections as indispensable for successful agricultural research and development.

Moving on to other forms of IP licensing, **the licensing of trade secrets** presents an entirely different set of challenges. A trade secret (also called know-how in certain jurisdictions) is any proprietary technical or business information, often embodied in inventions, know-how, show-how, and tacit knowledge. The most common definitions agree on three requirements that should be met for enforceable trade secrets to exist. The proprietary information should be:

1. *Secret* in the sense that it is not generally known in the trade
2. *Valuable* to competitors that do not possess the information
3. The *subject of reasonable efforts* to safeguard and maintain the information in secrecy

Everyone knows that trade secrets are secret. Patents, on the other hand, require inventions to be publicly disclosed. But does this mean that these two forms of IP protection cannot be used together? The chapter by Jorda[8] argues emphatically that **trade secrets are complementary to patents**. By using both trade secrets and patents, the combined IP protection is stronger than if either one were used alone. But how is it possible to use both patents (which are publicly disclosed) and trade secrets (which are kept secret) to protect something? In practice, there is no conflict between the two. Patent applications are usually filed early during the research stage to get the earliest possible filing or priority date. The patent claims tend to be narrow to achieve distance from prior art, and the specification normally describes rudimentary laboratory experiments or prototypes and/or embryonic embodiments of an invention. The best mode for commercial manufacture and use are almost invariably developed later. The results of such later research need not be disclosed to obtain the patent on the early invention and can be kept as trade secrets.

As a practical matter, therefore, **patent licenses are most valuable when coupled with access to associated know-how**. A patent license alone is often inadequate for commercial development of a technology. This associated know-how is immensely important and should be part of licensing agreements; effective technology transfer requires not only patent licensing but also, and perhaps more importantly, trade secret licensing.[9]

Anyone engaged in product development, including developing countries in particular, will want to keep in mind that trade secret protection operates without delay and without undue cost. Patents, on the other hand, are territorial and thus expensive to obtain and maintain, and they can be acquired only in certain countries.

When considering the forms of IP protection available for plants, what usually comes to mind are PVP and utility patents. But as the chapter by Tucker and Ross[10] points out, **trademarks are an effective form of IP protection for plants and plant products**, either used alone or in combination with one or more other forms of IP rights protection. Furthermore, trademarks can be used to effectively protect IP rights for plant varieties internationally. Similarly, the value of trademarks for varieties and products from developing countries can be tremendous. The relative strength of trademarks is determined by how distinctive the mark is. When consumers see the trademark, they are able to easily distinguish the goods or services of the trademark owner from the related goods or services of competitors.

Two international agreements, the Madrid Arrangement and the Madrid Protocol govern international trademark registration. For plant trademarks, understanding and utilizing these provisions will become increasingly important to developing countries. Many tropical and subtropical regions are rich sources of novel fruit products, and an owner of such a product will want to adopt a strategy that both stimulates global demand for the product and maximizes commercial returns. Trademarks will be integral for such IP rights protection and global marketing strategies. In particular, three critical aspects should be considered if new branded fruit products are to be successfully launched from developing countries:

1. Determine what is to be trademarked.
2. Promptly register the trademark in the countries in which it will be used.
3. Enforce the trademark.

A successful global trademark program, built around exciting products, may be more achievable than a PVP-based strategy that relies only on licensing for returns. Instead of managers and lawyers securing licensing deals, the market itself can fuel value creation in the trademark. If successful, the returns can be tremendous.

Shifting topics once again, a very important and quite difficult aspect (especially for public sector entities) is **the granting of options and rights of first refusal**. As either a stand-alone agreement or as a clause within a broader agreement, options are a unique way of granting rights to intellectual property. The chapter by Anderson and Keevey-Kothari[11] provides a detailed discussion of the various forms of options, with tips and strategies, sample causes, and template agreements. The chapter delves deep into the legal and commercial promises and perils of granting options and concludes with a helpful section on administering options.[12]

Of special interest to university administrators and technology transfer professionals will be the sections on **incorporating options as a part of research agreements**. Universities in the United Kingdom and the United States have different approaches to handling privately sponsored research. In the United Kingdom, sponsors are often granted an option to acquire a license to develop and commercialize results, or the sponsor might in some cases own all the results. In contrast, a university in the United States normally retains ownership of any intellectual property resulting from its own research, though the university may grant rights to a sponsor to commercialize results. This emphasis on university control of research in the United States stems, in part, from provisions in the Bayh-Dole Act that prohibit universities from transferring ownership of intellectual property created from government-funded research.

Also instructive to university personnel will be the chapter's discussion of where and when not to grant a **pipeline agreement to a university spinout company**. A pipeline agreement generally refers to an option granted to a university spinout company to acquire rights over intellectual property that may, in the future, be generated by university faculty. Although a pipeline agreement may make sense, universities should be careful to stipulate how pipeline intellectual property will be identified. They will likely want to limit the agreement to intellectual property generated by specific faculty members and their labs. Universities should also recognize that in some cases spinouts may not be the licensee of choice and should therefore craft pipeline agreements with care.

Licensing is about choices, and it can be argued that no choice is more important than the field of use granted in a license. When licensing complex technologies, the licensor usually can partition patent rights based on time (duration of license grant), location (where rights may be practiced), and field of use. Shotwell[13] explains and clarifies the last of these three considerations. By partitioning a bundle of patent rights and distributing them to one or more licensees, field-of-use licensing maximizes value, optimizes delivery, and facilitates the most effective use of new technologies, whether in agriculture, biotechnology, pharmaceuticals, vaccines, or diagnostics.

With field-of-use licensing, the licensor gains greater control while maximizing the use and value of the technology. However, field-of-use licensing requires more work. The technology licensor should identify, motivate, negotiate with, and manage more than one licensee—and quite possibly many. Nonetheless, this hard work can increase royalty streams to the licensor, since multiple licensees, each with different and specialized access to the technology, can efficiently speed different types of products to market.

When using field-of-use licensing, a licensor should be flexible. For example, even if a licensor envisions only one possible field of use for an invention, it makes sense to specifically limit a licensee to just that field. This is because technology changes so rapidly that a new use for the invention has a very good chance of developing later during the life of the patent. By limiting licensees to a particular field, a licensor retains the ability to work with the best possible licensee(s) for a new use when it arises.

Shotwell recommends that the licensor retain control over patent prosecution, while seeking to fairly distribute costs over field-of-use licensees. When considering reimbursement, the field-of-use licensor should manage patent expenses creatively. For example, the licensor can cover patent expenses up front, later reimbursing them from the royalty stream, or, if costs are to be reimbursed by the licensees, language can be used to include future licensees in that reimbursement.

One of the complexities of field-of-use licensing is that it raises the important question of how to deal with patent infringement/interference problems with multiple licensees. As with patent costs, the simplest approach is for the licensor to carry interference and infringement costs alone, recovering them through royalties or settlements. This approach retains more control for the licensor and correspondingly less for the licensees. Another approach to address possible infringement and interference actions would be to work out a mechanism to share the costs and management of these activities with one or more licensees.

Possible problems with field-of-use licensing include rights that overlap across licenses. This can arise from different interpretations of the rights granted under licenses or from unexpected future technical developments. It is therefore wise to lay the groundwork for resolving disputes related to these types of potential issues.

Licenses that include **royalty stacking and royalty packing clauses** are becoming more ubiquitous because virtually all products developed now using biotechnology, genetic engineering, and chemistry are technologically complex and incorporate many different inputs. As if this were not enough, there is also the added consideration of relevant IP rights, held by third parties that may be attached to these many inputs. For example, a vaccine might be identified and tested using proprietary research tools with IP rights owned by several companies. Later, the vaccine might be produced using patented recombinant techniques and proprietary DNA sequences. Transformation vectors might be owned by others. Production of the vaccine might employ a proprietary cell line. The vaccine might be packaged with one or more proprietary adjuvants and then be delivered using a patented device. Hence, when the vaccine is ultimately "ready for use," it will likely be subject to royalty obligations to several different companies, or licensors. A dilemma results, as the various licenses involved can combine to impose aggregate royalty obligations, perhaps up to 20%, and sometimes more, of the selling price of the product. There will also be separate reporting and accounting obligations to each of the licensors. Similar problems arise in agriculture when a genetically engineered crop might be made using proprietary varieties, vectors, gene sequences, and research tools—each with IP rights owned by different entities.

Jones, Whitham, and Handler[14] discuss two scenarios that can arise when multiple royalty rates are attached to one product—royalty stacking and royalty packing:

- *Royalty Stacking.* A biotechnological product might have multiple patents attached, and thus require multiple licenses in order to make, use, or sell the product.
- *Royalty Packing.* With some biotechnologies, it usually is necessary to combine one technology with many other technologies. (In this situation, the royalties imposed on each of the proprietary products that are administered will be "packed" together.)

The chapter then presents several techniques for managing royalty stacking and packing:

- *Royalty Ceiling.* A licensee may seek a ceiling for royalties in agreements it makes with licensors.
- *Royalty Floor.* A licensor may seek a floor below which its share of the royalties may not be reduced.
- *Variable Royalties.* Licensees and licensors also might agree to have variable royalties that are conditioned on the importance of the technology in relation to the creation of the product.
- *Royalty Alternatives.* Finally, alternatives to royalty bearing arrangements can also be considered; these include lump-sum payments and patent pooling.

The last point shows that managing royalty stacking and packing does not necessarily require royalty streams. For example, a lump-sum payment for the use of a research tool may be an optimal way to disseminate and exploit a patented technology. Some technologies may be best assembled in patent pools that provide either free use of or fixed-price access to the technologies. Patent pools can thereby facilitate R&D using a variety of proprietary technologies without the need to negotiate licenses.

As all of the above chapters make clear, in our post-TRIPS environment, leaders in developing countries who seek to improve economic development and public health are advised to be well-versed in the details of global IP management. Unlike the past, today no country can comfortably remain isolated from the global IP system. Yet among many public sector institutions in developing countries, knowledge of IP licensing practices is often insufficient. To address this gap in expertise, the chapter by Satyanarayana[15] lays out several of the important features of in-licensing agreements, **common problems faced by developing countries** in constructing and implementing these agreements, and ways to avoid these common pitfalls.

In-licensing by public sector institutions is a useful, if complex, method for bringing technologies into the public sector through patent license agreements with the private sector. Although the interests of the private and public sector entities involved in these agreements will almost necessarily be in tension, it is possible for a well-crafted license to allow all parties to feel as though they have benefited from the agreement. From a public sector perspective, as Satyanarayana argues, the goal is ultimately to provide a product (be it a vaccine, drug, or new agricultural crop) to people who would not have access to it without government support. For developing countries, the good news is that legal expertise is often locally available, since many firms are already familiar with basic licensing procedures. The trick is to put this knowledge in the service of public officials to develop a comprehensive and effective plan to license and develop much-needed-technologies. These strategies include:

- developing a business strategy that balances the needs of the public sector with the needs of the private sector
- developing a marketing strategy that prices products realistically and is based on good market research to aid valuation
- forming partnerships with other suitable agencies to help manufacture and market new products
- making sure legal, business, and scientific experts are working together for optimal success
- establishing, as an important initial step, a national technology transfer office

The final chapter by Bobrowicz[16] offers a useful **checklist for negotiating licensing agreements.** For the seasoned technology transfer professional or contract attorney, the idea of preparing a detailed checklist for every licensing agreement may seem like unnecessary busy-work. Yet these same professionals could probably relate stories where a missed detail or vague contract provision led to a costly and protracted legal battle. Given the high stakes, it is certainly in the best interest of those involved in IP deal making to make sure that every last detail is checked and rechecked. When multiple deals are being negotiated at once, it is only reasonable to assume that something could get missed. To help avoid this unfortunate and potentially costly error, this chapter provides a comprehensive, yet flexible, checklist that can be deployed to help manage the details of license agreements. Although the author provides a template for most elements of the license, she is quick to note that users should feel free to alter the checklist to suit their particular business practices.

The checklist covers all the major elements of a standard IP license, particularly as used in agriculture, starting with a section detailing the most basic, yet crucial, matter of getting all the parties' pertinent contact information. Sections covering clauses, definitions, rights granted, sublicenses, improvements, warranties, and infringement, and other matters are discussed, with useful sample checklists included for each component. The chapter concludes with a consideration of boilerplate sections, including confidentiality and arbitration stipulations. Although each section is annotated, an online version can be downloaded without accompanying text.

In sum, licensing is about the development of relationships. As important as the terms of agreements are, few are more important than the long-term opportunities offered by forging good partnerships, be they between companies or between public and private sector entities. Negotiating an agreement is just the beginning of what may—or should—become a long-lasting and beneficial relationship. ■

All chapters refer to: *Intellectual Property Management in Health and Agricultural Innovation: A Handbook of Best Practices.* 2007. A Krattiger, RT Mahoney, L Nelsen, JA Thomson, AB Bennett, K Satyanarayana, GD Graff, C Fernandez, and SP Kowalski (eds.). MIHR: Oxford, U.K., and PIPRA: Davis, U.S.A. Available online at www.ipHandbook.org. The online version contains for each chapter a detailed Editor's Summary, Implications, and Best Practices.

1 In an ideal world, exchanges would take place based on "fair" and "equitable" pricing, in both the eyes of the seller and buyer. But equitable exchanges are at least in part based on perceived moral obligations that are difficult to capture in financial and market terms. Such nonmonetary and moral perceptions, often strongly personal-, cultural-, and even religion-based, are important, but there is no system, at least for the moment, for weighing different perspectives on morality or ethical grounds or for translating them and making them work in the marketplace. To be sure, it can be satisfying to the buyer and seller when each perceives a transaction to have been equitable. And we need to strive for better balance in an inequitable world. But how much is owed to whom and for what and how to achieve general satisfaction has surely eluded our modern society (and, indeed, has eluded societies for centuries, if not millennia). We can all support equity, or at least pay lip-service to it, but it is far more difficult to determine what equitable transactions, in fact, consist of.

2 Chapter 11.1 by JW Freeman titled Licensing Biotechnology Inventions, p. 991.

3 Chapter 11.2 by RS Cahoon titled Licensing Agreements in Agricultural Biotechnology, p. 1009.

4 The online version of the *Handbook* provides a large number of template agreements from myriad organizations. These templates include confidentiality agreements, material transfer agreements, contract research agreements, collaborative research agreements, and a range of licensing agreements.

5 Chapter 11.3 by M Nilsson titled The In- and Out-Licensing of Plant Varieties, p. 1017.

6 Chapter 11.4 by A Krattiger, J Dodds, and D Bobrowicz titled Potential Use of a Computer-Generated Contract Template System (CoGenCo) to Facilitate Licensing of Traits and Varieties, p. 1029.

7 www.ipHandbook.org.

8 Chapter 11.5 by KF Jorda titled Trade Secrets and Trade-Secret Licensing, p. 1043.

9 This is perhaps the principal reason why (1) most issued patents are not practiced in developed countries although their information is freely available on the Internet and (2) why compulsory licensing has tremendous limitations.

10 Chapter 11.6 by WT Tucker and GS Ross titled Use of Trademarks in a Plant-Licensing Program, p. 1059.

11 Chapter 11.7 by M Anderson and S Keevey-Kothari titled Commercialization Agreements: Practical Guidelines in Dealing with Options, p. 1069.

12 These include practical suggestions for keeping track of options, including tips for writing standard operating procedures (SOP), generating nondisclosure agreements, developing a useful database for data, and if and when necessary, calling in legal professionals for more-specialized advice.

13 Chapter 11.8 by SL Shotwell titled Field-of-Use Licensing, p. 1113.

14 Chapter 11.9 by KJ Jones, ME Whitham and PS Handler titled Problems with Royalty Rates, Royalty Stacking, and Royalty Packing Issues, p. 1121.

15 Chapter 11.10 by K Satyanarayana titled In-Licensing Strategies by Public-Sector Institutions in Developing Countries, p. 1127.

16 Chapter 11.11 by D Bobrowicz titled A Checklist for Negotiating License Agreements, p. 1133.

FOR GOVERNMENT POLICYMAKERS

- ✔ **Licensing is highly context specific**. For this reason, blanket policies on minimum requirements for licensing terms applicable to public sector institutions can discourage creative and beneficial deals and reduce the potential for national institutions to forge international linkages.

- ✔ Notwithstanding the above, public sector institutions should, as a matter of policy, consider the routine **incorporation of philanthropic use provisions** in their licenses and should always **retain research and teaching rights** to any of their inventions.

- ✔ The dual goals of **economic growth and social/humanitarian benefits** through licensing are not mutually exclusive. Indeed, they are often complementary.

- ✔ Companies regularly license their own varieties to third parties as a strategy to maximize returns on investment and reach markets that the company itself cannot easily reach. Conceptually at least, **public sector plant breeding institutions have much to gain from variety licensing** as a strategy of serving markets they do not typically reach.

- ✔ Overall, the successful licensing of varieties between and among public and private sectors is **contingent on the strength of plant variety protection legislation**. Such legislation can support the interests of the variety owner and the farmer, facilitates the transfer of technology, and provides incentives for further investments.

- ✔ Recognizing that patent applications are usually filed early in the research stage and require full disclosure, companies typically keep inventions developed later on as trade secrets. These may include the best mode for commercial manufacture. Patent licenses are most valuable when coupled with access to associated know-how. Comprehensive and enforceable **trade secret laws are thus conducive to the transfer of know-how** through licensing. The two, disclosed patents and protected secrets, are thus complementary.

- ✔ Using **trademarks as a strategy allows public and private institutions to capture more added value**. To benefit from trademarking strategies, internationally accepted legislation is important. Also important is the maintenance of high quality standards and stewardship, since trademarks (and geographical indications) provide the consumer with information on the source of the products.

- ✔ Although IP rights are governed by national statutory protection, **contract law is arguably even more important than statutory protection law**, as contracts allow institutions to exchange intellectual property in an orderly and predictable manner.

- ✔ Along with investing in a country's R&D infrastructure and capacity, it is important to sustain long-term growth. **Human and institutional capacity in IP management adds value to R&D efforts**. In- and out-licensing in particular enhance an institution's economic and social impact. Among a government's top priorities should be providing support to public sector institutions for establishing and operating effective technology transfer offices, coupled with training programs for creating capacity commensurate with the complexities of modern biotechnological products. Ideally, these capacities should reside at the institution level because of the context-specific nature of licensing.

Given that IP management is heavily context specific, these Key Implications and Best Practices are intended as starting points to be adapted to specific needs and circumstances.

FOR SENIOR MANAGEMENT

(UNIVERSITY PRESIDENT, R&D MANAGER, ETC.)

- ✔ **Licensing is highly context-specific**. For this reason, blanket policies on minimum requirements for licensing terms applicable to your technology transfer office can be counterproductive to making sound deals and can reduce the potential to forge international linkages.

- ✔ Public sector institutions should, as a matter of policy, **consider the routine incorporation of philanthropic-use provisions** in their licenses and should, as a matter of routine, always **retain research and teaching rights** to any of their inventions.

- ✔ The dual goals of **economic growth and social/humanitarian benefits** through licensing are not mutually exclusive. Indeed, they are often complementary. Much will depend on a sound institutional licensing strategy and on good relationships with licensees.

- ✔ The value of trademarks for both varieties and products, from developed and developing countries alike, cannot be understated. To **benefit from trademarking strategies**, particularly of global reach, the maintenance of high quality standards is important, since they provide the consumer with information on the source of the products. In addition, using trademarks as a strategy allows public and private sector institutions alike to capture more of any added value.

- ✔ Licensing is about choices, and it can be argued that no choice is more important than the **field of use** granted in a license. A public sector institution should have a clear policy statement on how it deals with field-of-use licensing and may even wish to consider making field-of-use licenses the preferred method of licensing. This is especially applicable to platform technologies and diagnostics. However, field-of-use licensing requires more work. A well-trained and well-staffed technology transfer office will be essential.

- ✔ A public sector institution can contribute significantly to its mission through **in-licensing intellectual property from private sector entities**. For this, it is useful to develop a set of strategies (business, marketing, partnership building, and legal) during discussions with the licensing office that balance the needs of the public sector with the needs of the private sector.

- ✔ **Business decisions, more than legal aspects, should determine licensing terms**. Nevertheless, lawyers should ensure that the contracts comply with prevailing law. This is equally applicable to private sector and public sector deals.

- ✔ **Patent licenses are most valuable when coupled with access to associated know-how**. Comprehensive staff training in the handling of confidential information from third parties is therefore critical.

- ✔ But public sector organizations should **exercise caution when accepting trade secrets** (as opposed to confidential information). In some jurisdictions there may be significant liability obligations related to trade secrets, and public sector institutions may not be in a position to cope with all such obligations.

Given that IP management is heavily context specific, these Key Implications and Best Practices are intended as starting points to be adapted to specific needs and circumstances.

FOR SCIENTISTS

- ✔ Ideally, you will leave detailed aspects of negotiations, such as collaboration or license agreements, to the relevant offices of your institutions. However, do **participate in the internal discussions** prior to in- or out-licensing negotiations. Your input will be important and should be valued.

- ✔ The dual goals of **economic growth and social/humanitarian benefits** through licensing are not mutually exclusive. Indeed, they are often complementary. Much will depend on a sound institutional licensing strategy and on good relationships with licensees. Your role in the latter may be critical.

- ✔ Make an effort to consistently document the **origin of biological and other materials** you use in your research, and keep a comprehensive record. Although it is not your responsibility to resolve IP conflicts, your detailed records will help if such a conflict arises.

- ✔ Interface with the technology transfer office (TTO) in order to understand **options** and whether you might have a role in their implementation or fulfillment. Although options are complex and a matter most appropriately addressed by your TTO officers, the granting of options may significantly impact your research options. Make sure you discuss the implications with them prior to the incorporation of options in licenses that relate to your research.

- ✔ When you disclose an invention to your TTO officers, inform them of any ideas you may have on the various fields of endeavor in which your invention could find applicability. This will help the TTO write better patent applications and, later to draw up license agreements for many different players under different **field-of-use licenses**. This approach can maximize the value of your research and may **accelerate commercial and humanitarian development** of technologies based on your research.

- ✔ **Your role in field-of-use licensing is essential.** You can provide your TTO with valuable information on licensable components for different applications and entities.

- ✔ The products arising from your program's research efforts, particularly from product development activities, will invariably embody numerous technologies, including components and processes that might have IP rights from third parties attached to the technologies. This can create complex IP management and licensing issues as these products approach commercialization. If you are engaged in product development, **maintain a good line of communication with your TTO** and ensure that early on they address IP ownership by third parties.

Given that IP management is heavily context specific, these Key Implications and Best Practices are intended as starting points to be adapted to specific needs and circumstances.

KEY IMPLICATIONS AND BEST PRACTICES: SECTION 11

FOR TECHNOLOGY TRANSFER OFFICERS

- ✔ Besides reflecting the business deal that has been made, few components are more important in a license than clear and unambiguous **definitions**.
- ✔ For practical purposes, any organization engaged in high-volume licensing will find it useful to develop its own **internal template agreements** that are then modified and adapted to suit each special circumstance. **Checklists** for different types of recurring licensing negotiations should be reviewed prior to and during negotiations.
- ✔ Recognize that relationships—like markets—are not static. Any provision in an agreement must, of course, be adhered to, but the practice of including **sufficient flexibility in licensing agreements** can be a valuable strategy in forging strong partnerships.
- ✔ The granting of options (rights of first refusal, pipeline agreements, and so forth) can be a rather controversial aspect for public sector licensing. But **options can be tremendously powerful** in forging strong and lasting relationships and in optimizing your institution's economic returns and humanitarian effects.
- ✔ **Field-of-use licensing** should be adopted as the preferred method of licensing whenever possible. It allows you to gain greater control while maximizing the use and value of your licensed technology. But be flexible and study the licensee's motivations and business model carefully as a way of conferring the highest possible incentives. Always strive to retain control over patent prosecution and infringement actions when adopting a field-of-use licensing strategy.
- ✔ Familiarize yourself with the various ways to deal with **royalty stacking and royalty packing** issues as a way of balancing risks and returns. The choice will depend on how far downstream into product development your institution stays involved.
- ✔ Negotiating about low-probability events can sidetrack progress toward agreement on core issues, so care should be taken during the negotiation to **attend to issues in a manner commensurate with their strategic importance**. It is often best to focus on the overall deal before entering into discussions about specifics.
- ✔ In a license agreement, the **rights to sublicense and assign a license** ought to be explicitly articulated.
- ✔ In research collaborations, in which employees of two or more entities share ideas and information, **confidentiality provisions** are important. Make sure the scientists in your institution understand these obligations and rights.
- ✔ Licensee agreements are contracts. Hence, a practical understanding of contract law will be fundamental to negotiating and drafting good license agreements. Many smaller TTOs use outside counsel to **ensure that agreements are compliant with national law**.

Given that IP management is heavily context specific, these Key Implications and Best Practices are intended as starting points to be adapted to specific needs and circumstances.

Dealmaking and Marketing Technology to Product-Development Partners

A licensing agreement establishes, in written form, the rules of an ongoing relationship, the success of which will depend on many factors. Mutual trust is one of the factors. Another is the development of a certain dependence based on the *value* that is being transferred between the parties. As Mahoney[1] explains, one party may have a product that can potentially have a very large market, while the other party has research, manufacturing, or distribution capabilities essential to reaching that market. The key to successful negotiation is having a clear understanding of the value each party brings to the relationship. But value is multifaceted. There is an **objective value**, represented by, for example, how many units can be sold at a certain price yielding a certain level of profit. There are also **qualitative values**, for example, the additional value assumed to exist when one company feels that a particular product, owned by a second company, would enhance or complete a particular product line.

Perhaps the most important element in a negotiation is to be clear—internally and in discussions with the negotiating partner—about the benefits that will or could be realized through a license agreement. Only with a clear understanding of the transfer of value can both parties intelligently and fairly negotiate an agreement. Mahoney discusses this along with numerous suggestions for successful licensing negotiations, including the following:

- In general, the public sector organization should consider offering the first draft of a licensing agreement (the draft needs to cover a number of topics of particular concern to public sector organizations that would probably not be addressed by a company).
- The use of a term sheet that lists the major issues expected to arise in the negotiations should be shared ahead of time indicating the outcome that the proposing party hopes to achieve. Such a sheet could also include the *needs* and *wants* (in other words, the must-have terms and the desired ones) of each party.

Furthermore, says Mahoney, negotiating such agreements requires talent, expertise, and sound tactics that cover the following areas:

- **business strategy**. The business strategist is usually the lead negotiator with considerable experience in structuring business relationships, assembling the inputs of other experts, and maximizing the benefits to all parties.
- **marketing**. Market analysis is essential to negotiating a good agreement. Failure to carry out such an analysis is dangerous because it can lead either to overestimation or underestimation of the market potential, which, in turn, can lead to a suboptimal agreement or rejection of an agreement that could have been successful.

Krattiger A, RT Mahoney, L Nelsen, JA Thomson, AB Bennett, K Satyanarayana, GD Graff, C Fernandez and SP Kowalski. 2007. 12: Dealmaking and Marketing Technology to Product-Development Partners. In *Executive Guide to Intellectual Property Management in Health and Agricultural Innovation: A Handbook of Best Practices* (Krattiger A, RT Mahoney, L Nelsen et al.). MIHR (Oxford, UK), PIPRA (Davis, USA), Oswaldo Cruz Foundation (Fiocruz, Rio de Janeiro, Brazil), and *bio*Developments-International Institute (Ithaca, USA). Available online at www.ipHandbook.org.

- **legal inputs.** A lawyer should be retained at least to review agreements and, more appropriately, to be part of the negotiating team and possess intellectual property (IP) expertise and valuation skills, understand knowledge of freedom to operate issues, and be able to access country-specific legal advice.
- **scientific and regulatory.** A negotiating team must have scientific expertise and detailed knowledge about regulatory issues, product safety, and related matters
- **production.** Staff members who can contribute their knowledge about required production equipment and good manufacturing processes as well as their understanding of time lines, cost implications of various manufacturing processes, and so forth should be involved in the licensing negotiation.
- **finance.** A careful financial assessment of the project is essential, even before negotiations. The assessment often can help the business strategist determine options for approaching a deal, to decide which new funds will be required to launch and sustain the project, and so forth.

In addition, Mahoney illustrates specific best practices for public sector entities to meet public sector goals. These are summarized in Table 1 and serve as guidelines for public sector organizations striving to widen and improve access to innovation through various licensing strategies. Price is probably the most difficult area for a licensor to get involved in.

Up to this point we have dealt with the overall strategies and best practices used to meet public sector goals. The chapter by Mongeon[3] provides a broad overview of **marketing tactics** as a way of understanding what buyers need and how to meet those needs. In essence, he invites the reader to think of marketing not as simply a way to *push* technologies into the market rather than a way of allowing the needs of buyers to *pull* them in. Indeed, marketing is not merely advertising or selling. Rather, **marketing is a multistage process**: first, the essential characteristics or benefits of a technology must be quantified; next, people who would find these characteristics or benefits desirable and therefore be willing to pay for them must be identified; and finally, the benefits of the technology must be communicated clearly and compellingly to those potential users. Mongeon offers five basic marketing questions that would-be licensors will need to address:

1. *Who* will buy the technology? Will the purchasers be producers or consumers?
2. *What* does the buyer of the technology want? What characteristics, qualities, or capabilities of the technology are valuable to the buyer, and how valuable are they?
3. *Why* would a party choose to license or purchase a technology? What is particularly compelling about it?
4. *Where* are potential users of the technology located? In which markets? Through which channels can they be reached?
5. *When* can you sell the technology to buyers? Is the technology so new that the market is not yet receptive to it?

The answers to these questions should guide a marketing plan and be supported through market research. Such research may even reveal that a technology can be used in a completely new and unexpected way in a previously unanticipated market. Perhaps the most important advice this chapter has to offer scientists and technology managers is that the "unique selling proposition" of a technology—that is, the features, advantages, or benefits that it offers the user—is rarely the science behind the technology. Good marketing makes a technology understandable and attractive to buyers, then allows their demand to draw the technology into the market.

But how does one "find" potential licensees? And how should one approach them? Marketing workshops tend to suggest a haphazard mix of different tools and strategies that may or may not work. For these reasons, MacWright and Ritter[4] offer a detailed and systematic approach to technology marketing (which is different from product marketing). The chapter contains many models for establishing contacts and prioritizing these according to specific criteria, as well as numerous worksheets that will help plan for different marketing approaches.

Table 1: Illustrations of Best Practices for Licensing to Meet Public Sector Goals

Topic	Basic concept	Public sector consideration
Areas of use	Specifies limitations on the application of the patent in developing products; simplest approach: grant an exclusive right to all applications, including not only those specified in the patent, but others that may emerge as further research and development proceeds.	The clause could grant an exclusive license only for those products that the licensor actually wishes to pursue. Also, the clause could grant an exclusive license only for those products that were unlikely to have a significant market among the poor in developing countries.
Territory	Specifies the geographic areas in which the licensee has the right to exercise a patent; simplest approach: grant a worldwide exclusive license. (A license is valid only in the countries where a patent has been issued, but rights can be granted, at licensee's expense, to file for patent protection in additional countries.)	The clause could grant an exclusive right to a set of developed countries and another exclusive license to other developed countries. Both licensees, and perhaps a third one, could receive nonexclusive licenses for an agreed list of developing countries. Then the two (or more) licensees would have to compete for sales to developing countries.
Price	In most licensing agreements, there will be no conditions with respect to price. The licensor assumes that the licensee will determine the best price to ensure the greatest return on investment.	• The price could be specified. This is feasible when the licensor has detailed technical knowledge of the production, marketing, and distribution costs. • The price could be set as cost plus (cost of production plus a reasonable markup, say 15%). This is feasible when the licensor has a reasonable expectation of being able to monitor the cost of production. • The price could be set at "no higher than the lowest price offered to any private sector buyer." This may be preferred when there are large bulk purchases by private sector buyers who are good at negotiating the very best price.
Labeling	In most licensing agreements, there will be no conditions about labeling. The licensor assumes the licensee will prepare labeling in conformity with national drug regulatory agency requirements.	The licensor can help ensure that the product is licensed properly, especially in developing countries in which national regulatory agency requirements for labeling may not be rigorous nor enforced.
White knight condition	This concept has been developed by the U.S. National Institutes of Health. It calls for the licensee to undertake some specific actions that will benefit the public sector.	This could include a donation of products for clinical evaluation in public sector research programs, joint efforts to develop markets, free supply under specified conditions to developing countries, and so on.
Royalties	Usually a licensee will negotiate the highest royalty in order to maximize revenue from the license.	The licensor can specify that royalties apply to sales only in developed countries and zero royalties in developing countries. The impact for the licensor would normally be minimal.

Source: Modified from Mahoney[2]

In essence, the marketing approach comprises four basic steps:

1. Collect information about the invention from the inventors.
2. Collect information from potential clients.
3. Review and prioritize your prospective client list.
4. Make contact with potential clients.

It is relatively easy to sell a finished product, such as shoes, and more difficult to sell a technology to make better shoes. It is even more difficult to sell (or license) the intellectual property for making better shoes, especially if the intellectual property has not yet been proven in a productive process. For this reason, university technology managers in particular often find it difficult to license individual patents. Burdon[5] thus argues that universities could gain a lot by pursuing a **portfolio approach, or rather, an integrated intellectual property management** (IPM) approach that blends sophisticated IP data search-and-analysis techniques with continuous product improvement.

At the highest level, an integrated IPM approach is differentiated into strategic and tactical decision-making. Strategic decision-making is a broad analysis; tactical decision-making analyzes specific products or technologies in a known competitive landscape. Each approach to managing IP portfolios requires different types of tools, searches, and analyses, ranging from very broad technology scans to very specific patent infringement or validity searches. Importantly, attention should be paid to how data analysis can be integrated with a product innovation process, how to identify new opportunities or resolve old problems (that may also lead to the amendment of patent applications). Perhaps the most important reason for an IPM approach is that it enhances understanding of the processes in which licensees are engaged and how a licensed technology would support their endeavors, thus reinforcing Mahoney's earlier points on the importance of integrating business strategy, marketing, scientific and regulatory expertise, and so forth.

Unfortunately, inventions by universities are generally not developed in response to market needs, which presents challenges for technology transfer offices (TTOs). Keiller[6] addresses this challenge and stresses the importance of having a clear sense of the IP goals and IP strengths of one's own institution. **An IP audit[7] is a useful way to improve an institution's marketing prowess**, because it identifies and classifies an institution's intellectual property, whether it is owned, licensed, or simply possessed. Unless the technologies, their IP status, and their respective levels of development are known, at least to some extent, it will be difficult to persuade others to pursue a license deal. Keiller describes a range of marketing approaches and shares persuasion techniques. In short, marketing packages should be tailored to accommodate customers' needs, the benefits of the invention should be emphasized, effective time management must be adopted, and above all, contacts must be followed up on.

It is important with any IP management activity to be clear about the context in which it occurs. For example, dealing with a small company will require a fundamentally different approach than would be taken with a large one. Dealing with an agricultural company will require a different approach than would be taken with a pharmaceutical company. Neagley[8] describes **in-licensing strategies (and typical terms) as they apply to small agri-biotechnology companies** that typically depend on strong IP portfolios. IP portfolios are the foundation for their R&D, encouraging outside investment and making product commercialization possible. In-licensing is especially important as it allows a company to obtain IP rights without having to invest in research.

Neagley discusses the entire range of provisions in a typical license agreement, including:

- exclusive versus nonexclusive
- enabling technologies versus traits versus plant materials
- rights granted to the licensee (covering such topics as sole licenses, coexclusive licenses and territoriality)
- compensation due to the licensor
- liability, diligence terms, and milestones,
- the licensee's responsibilities vis-à-vis the patent
- license term and termination
- issues of assignability

Importantly, compensation may be a combination of fixed fees, which can be paid up-front and/or periodically, and earned royalty fees. Both the level and timing of compensation are important to the company with respect to its planning and budget. In determining what compensation it is willing to pay, the company will need to estimate the potential value of the licensed technology and assess the potential value of any commercialized products that might be developed under the license. But compensation may also take nonmonetary forms: stock in the licensee company, an exchange of license grants, a cross-license arrangement, or a grant-back to the licensor (grant-back is compensation that involves the licensee granting the licensor rights to future inventions made by the licensee using rights received from the licensor).

Dunn, Lund, and Barbour[9] share **the approaches of a multinational agri-biotech company** with emphasis on market and policy factors that influence and constrain agricultural companies regarding how to market technologies to them, and on what these companies look for in terms of license agreements.

Early-stage agricultural technologies, whether they are genetically modified technologies or conventional ones, can be risky because they may not have commercial applications or they may fail to receive regulatory approval in the necessary markets. Gaining regulatory approval can be a slow and costly process. In addition, a low marginal revenue is made on agricultural inputs, and there are only a handful of crop species with sufficient acreage to generate the necessary returns to warrant significant investments in regulatory clearances. For these reasons, a few large corporations develop most transgenic technology; only they have the necessary capital and can assume the high risks involved.

For a university to market its technology successfully to a large company, **it is useful to have good contacts inside the company**, someone who is willing to accept the risk and "sell" the deals internally. Indeed, the value of networking cannot be overstated. For any company, though, the value of a new technological opportunity is determined by the risk involved, the additional investments required to develop the technology (and the corresponding opportunity costs), and the type of technology in question.

These concepts are further explored by Edwards[10] who surveyed deals made by biotechnology and pharmaceutical companies during the last ten or so years and analyzed the types of alliances and their terms. **The four characteristics of an alliance that generally defines the allocation of value of a technology between an originator and a commercial partner** are:

- the stage of development of the technology
- the role retained by the licensor in product supply or other ongoing activities
- the size of the market opportunity
- the scope of the market granted to the development partner

Because biotechnology companies have become highly specialized, it is no longer necessary, or even possible, for any one company to be involved in every stage of the R&D process. Up to half of the product candidates in pharmaceutical companies' R&D pipelines originate from elsewhere, and 60 to 80 percent of the leading therapeutics on the market were developed or distributed through some form of alliance.

Edwards shows that universities and research institutes are a significant source of early-stage technology, drug leads, and, occasionally, more mature technologies. **A biotechnology company with the appropriate business model is most likely to find early-stage technologies and drug leads attractive.** Once smaller biotechnology companies have developed technologies and drugs, they will probably need to enter into alliances with larger pharmaceutical companies in order to conduct clinical trials on, commercialize, and then market these products. A university developing a more mature technology might ally itself directly with larger pharmaceutical companies. Empirical evidence shows that **the more mature the technology is when an alliance agreement is assigned, the more profitable that technology is for the technology provider**.

Edwards goes on to discuss some of the fundamental terms found in biotechnology alliance agreements such as fixed fees, reimbursement of expenses, development milestones, equity investments, and royalties, as well as the terms for other, more specialized, types of postcommercialization payments. No matter whether a university wants to join a commercialization alliance itself or license an innovation to a biotechnology company that is allied to other companies, it is essential for university TTOs to understand and influence the terms of the alliance agreements in order to protect the value of their intellectual property.

Finally, Shotwell[11] integrates the ideas of this section in a discussion of a core theme of the *Handbook*: **how public sector and nonprofit efforts can utilize intellectual property to achieve their goals in serving society**. To illustrate this important point, the chapter focuses on product development partnerships (PDPs) and their innovative IP strategies. PDPs, in essence, facilitate and accelerate the flow of public and philanthropic investment through the innovation pipeline, to a far greater extent than has been typical of universities alone.

With a two-pronged approach of product specialization and taking advantage of the efficiencies of the larger marketplace for technologies, PDPs strategically mobilize intellectual property. Investments are made in a new product technology to advance it through the stages of development. This happens within the overall marketplace through the selective targeting of projects based upon their risk–reward profile. Using this approach, the measure of "reward" is not returns to the organization, but rather the potential impact on social welfare that the new drug or vaccine might have.

There are certain **similarities between PDPs and biotechnology companies**. Both occupy a similar niche in the innovation pipeline. Both share many IP goals. Both seek to maintain an appropriate mix of access and exclusivity to innovations, in order to have sufficient freedom to operate and sufficient bargaining power to implement the overall strategy of their organizations. There are also similarities between the IP strategies of PDPs and public research institutions. Both PDPs and public research institutions use intellectual property to entice or leverage private investment, enhance access to other intellectual property, build partnerships, and cultivate political goodwill to advance their missions.

Just as there are several business models used by the biotechnology industry, so there are several business models used by PDPs. The business model that a PDP chooses will depend on the technologies it deals with, the stage of development of the technologies, and the nature of the market. One factor that determines which kind of business model a PDP or any other entity will adopt will depend on whether or not the product being developed is potentially profitable and can therefore attract the interest of for-profit companies. Sound IP strategies and product development partnering also will uncover opportunities to use new technologies to benefit those who are traditionally excluded from markets. ■

All chapters refer to: *Intellectual Property Management in Health and Agricultural Innovation: A Handbook of Best Practices*. 2007. A Krattiger, RT Mahoney, L Nelsen, JA Thomson, AB Bennett, K Satyanarayana, GD Graff, C Fernandez, and SP Kowalski (eds.). MIHR: Oxford, U.K., and PIPRA: Davis, U.S.A. Available online at www.ipHandbook.org. The online version contains for each chapter a detailed Editor's Summary, Implications, and Best Practices.

1 Chapter 12.1 by RT Mahoney titled Negotiating an Agreement: Skills, Tactics, and Best Practices, p. 1155.

2 *Ibid.*

3 Chapter 12.2 by MD Mongeon titled An Introduction to Marketing Early-Stage Technologies, p. 1165.

4 Chapter 12.3 by RS MacWright and JF Ritter titled Technology Marketing, p. 1173.

5 Chapter 12.4 by J Burdon titled IP Portfolio Management: Negotiating the Information Labyrinth, p. 1195.

6 Chapter 12.5 by TS Keiller titled The IP Sales Process, p. 1203.

7 See Chapter 5.6 by M Blakeney titled Conducting IP Audits, p. 515.

8 Chapter 12.6 by CH Neagley titled Patent Licensing for Small Agricultural Biotechnology Companies, p. 1213.

9 Chapter 12.7 by M Dunn, B Lund, and E Barbour titled Business Partnerships in Agriculture and Biotechnology that Advance Early-State Technology, p. 1221.

10 Chapter 12.8 by MG Edwards titled Biotechnology and Pharmaceutical Commercialization Alliances: Their Structure and Implications for University Technology Transfer Offices, p. 1227.

11 Chapter 12.9 by SL Shotwell titled Product Development and IP Strategies for Global Health Product Development Partnerships, p. 1247.

FOR GOVERNMENT POLICYMAKERS

- ✔ **Technology marketing** is a process by which owners of a technology create relationships, between themselves and potential users that will enable the technology to be developed and made widely available, through commercialization, alliances or other methods.

- ✔ Policies that **encourage alliance building between the public and private sector** have been particularly successful in bringing innovation to market.

- ✔ **Product development partnerships** (PDPs) facilitate and accelerate the flow of public and philanthropic investment through the innovation pipeline. The ultimate measure of success is not maximum profit but maximum social benefit.

- ✔ In addition to the IP legislative and capacity framework, other **determinants of innovation need to be addressed** by governments to ensure a vibrant and innovative technology industry. In agricultural biotechnology in particular, many current **regulatory approaches and frameworks** significantly increase regulatory costs, causing years and years of delays. The result often is that only multinational companies can afford to introduce new technologies, thus stifling national innovation significantly.

- ✔ Negotiating between public and private sectors ought **not be confrontational** and should be seen as an opportunity to forge a long-lasting and mutually beneficial relationship. Put differently, **negotiating a fair licensing agreement** should not be seen just as a process of "bargaining" toward a win-win outcome.

- ✔ For the private sector party, a well-tested and successful approach to negotiating an agreement is to **offer initial terms that the public sector organization would be willing to agree to** if it were on the other side of the negotiating table.

- ✔ **Negotiation and technology marketing skills** are fundamental for successful licensing and technology transfer. People working in the public sector need to be well qualified and have strong negotiating skills, thereby enabling institutions to take advantage of their own R&D efforts and to realize broad public sector and commercial goals.

- ✔ Policies and legislation that are beneficial to **small biotechnology companies and startups,** in general, can be instrumental in accelerating the pace of innovation in a country, particularly when it comes to commercializing public sector–generated inventions.

Given that IP management is heavily context specific, these Key Implications and Best Practices are intended as starting points to be adapted to specific needs and circumstances.

KEY IMPLICATIONS AND BEST PRACTICES: SECTION 12

FOR SENIOR MANAGEMENT

(UNIVERSITY PRESIDENT, R&D MANAGER, ETC.)

- ✔ For private sector companies, the ultimate purpose of IP management is to **enhance competitiveness and reduce risk**. For public sector institutions, the ultimate purpose of IP management is to **serve the greater public interest**. These are not mutually exclusive goals, and they can be reconciled through sound technology marketing and licensing practices.
- ✔ The **four characteristics of an alliance** that generally define the allocation of value between an originator and a commercial partner are (1) its stage of development, (2) the role retained by the licensor in product supply or other ongoing activities, (3) the size of the market opportunity, and (4) the scope of the market granted to the development partner under the alliance agreement.
- ✔ The key to successful negotiation is having a clear understanding of the value each party brings to a relationship. **Value may be objective and quantitative, or of a more qualitative nature.**
- ✔ Perhaps **the most important element in a negotiation** is clear communication—also internally—with the negotiating partner about the benefits that will or could be obtained through a license agreement.
- ✔ In general, the **public sector organization should consider offering the first draft of an agreement** to cover a number of topics of particular concern to public sector organizations that would probably not be addressed by a company.
- ✔ Negotiating between public and private sectors ought **not be confrontational** and should be seen as an opportunity to forge a long-lasting and mutually beneficial relationship. Put differently, negotiating a fair licensing agreement should not be seen just as a process of "bargaining" toward a win-win outcome.
- ✔ For the private sector party, a well-tested and successful approach to negotiating an agreement is to **offer initial terms that the public sector organization would be willing to agree to** if it were on the other side of the negotiating table.
- ✔ Specific **best practices and terms that allow public sector entities** to meet public sector goals (ensuring broad access to innovation) include area of use, territory, price, labeling, white-knight conditions, and royalties.
- ✔ Senior management can **set a positive tone for negotiation** that will ensure that deals made with others are a vehicle for building strong relations and trust between parties.
- ✔ **Integrated IP management** (IPM) considers the critical role of IP management throughout the entire innovation life cycle. IPM allows managers to intervene, change course, amend or enhance patent applications, and in-license useful patents or technologies.
- ✔ **Networking** is important, if not essential, for successful technology marketing. Technology transfer officers and scientists particularly should be encouraged to network.

Given that IP management is heavily context specific, these Key Implications and Best Practices are intended as starting points to be adapted to specific needs and circumstances.

FOR SCIENTISTS

- ✔ The **"unique selling proposition" of your technology**—in other words, the features, advantages, or benefits it offers—is probably not the science behind the technology. The science behind an invention is usually not its selling point.

- ✔ **Technology marketing** is a process by which owners of a technology create relationships, between themselves and potential users, that will enable the technology to be developed and made widely available, through commercialization or other methods.

- ✔ Negotiating between public and private sectors ought **not be confrontational** and should be seen as an opportunity to forge a long-lasting and mutually beneficial relationship. Put differently, **negotiating a fair licensing agreement** should not be seen just as a process of "bargaining" toward a win-win outcome.

- ✔ You should **think about the practical applications of your inventions**. Or dream about them! And share your ideas about what applications you think your invention might have. Good marketing makes a technology understandable and attractive to buyers.

- ✔ As much as your science may be interesting and fascinating, when you speak to potential licensees or investors, it is often best not to place emphasis on the science. Rather, in extremely simple language, **stress the potential applications** of your invention.

- ✔ Remember to keep an eye on newly **published patents** and patent applications. They can help inform R&D decisions and keep you abreast of the latest technical developments in your field.

- ✔ Collaborations create contacts. Contacts build networks. **Networks provide opportunities.**

- ✔ **Your contacts and network** can help your technology transfer office's marketing efforts. For example, private sector colleagues may facilitate licensing deals with their organizations.

Given that IP management is heavily context specific, these Key Implications and Best Practices are intended as starting points to be adapted to specific needs and circumstances.

KEY IMPLICATIONS AND BEST PRACTICES: SECTION 12

FOR TECHNOLOGY TRANSFER OFFICERS

- ✔ One of your responsibilities will be to bring together individuals with different backgrounds and experiences before negotiating agreements. Ideally, a team should include **business strategy, marketing, legal, scientific, regulatory, production, and finance expertise**.
- ✔ Marketing inventions should not simply be a push of technologies; rather, it should be an approach that allows the needs of **buyers to *pull* inventions**.
- ✔ Marketing is not merely advertising or selling but a **multistage process** that addresses the who, what, why, where and when of an invention.
- ✔ Marketing inventions should use a systematic approach (which is fundamentally different from product marketing). **Particularly useful are portfolio approaches to marketing** (also called integrated intellectual property management which blends sophisticated IP data search-and-analysis techniques with continuous product improvement).
- ✔ **Public sector institutions should pay particular attention to the following terms or aspects of a license** when negotiating with companies: exclusive versus nonexclusive; enabling technologies versus traits versus plant materials; rights granted to the licensee (covering topics such as sole licenses, coexclusive licenses, territoriality, duration, field of use, and retained rights issues, as well as options or rights of first refusal, and favored-nation clauses); grant-back clauses; compensation due to the licensor; liability, diligence terms, and milestones; licensee's responsibilities vis-à-vis the patent; license term and termination; and issues of assignability.
- ✔ Some people believe that licensors should stay away from **conditions on final product price** because of its complexity; this is one reason why the public sector needs to develop its skills.
- ✔ You can help ensure that a licensee will fulfill the terms of the agreement if you require milestone payments or certain reporting conditions when milestones are reached, minimum annual royalties, or research funding-level commitments. It is particularly important to **ensure diligence** for exclusive licenses.
- ✔ There are three key ways that a license grant, either nonexclusive or exclusive, can be limited or defined: **territorial limitations, field-of-use limitations, and limits on duration**.
- ✔ Conduct **comprehensive IP audits** to determine where your IP assets are, when IP protection is needed, whether there are potential IP liability issues, whether there are licensing needs or opportunities, and whether there are inventions to be harvested.
- ✔ The key implications and best practices listed for senior management are also pertinent for TTO officers.

Given that IP management is heavily context specific, these Key Implications and Best Practices are intended as starting points to be adapted to specific needs and circumstances.

The Public Sector and Entrepreneurship

Is it better to license a technology to a start up, a spin-out, or an existing company? This is one of the questions facing public sector technology transfer offices (TTOs) around the world. There is no simple answer. The choice can depend on whether an entrepreneurial spirit characterizes the institution that developed the technology. The more entrepreneurial, the more likely the institution will wish to set up a new company. However, if an incubator infrastructure exists, then a spinout becomes more feasible.

This section considers the **benefits and risks of dealing with spinouts** by reviewing experiences, with continual reference to the situations of developing countries. The associated factors of **venture capital, technology transfer intermediaries, and the formation of business incubators** are also discussed. Anyone engaged in dealing with spinouts, venture capital companies, and incubators will want to read the entire section; it covers a range of issues from licensing considerations, to the use of milestones, to compensation, and offers plenty of elucidating analyses about realistic expectations, based on a series of real examples from the United States and the United Kingdom. All of these chapters show in one way or another that while there is certainly an "art" to entrepreneurship, reliable recipes are available.

Brown and Soderstrom[1] present the rationale for, and a comprehensive practical overview of, the creation of university spinout companies. Based on the successful experiences of **Yale University**, the authors **advocate a hands-on approach,** through which the university actively and directly manages the creation of new companies and invests in the human and physical resources needed for their success. Such an approach may provide a greater chance for success than licensing the university's technology to a start-up company, which would likely be completely separate from the university. The approach of establishing and licensing to a spinout does, however, introduce a number of significant risks. Brown and Soderstrom identify these risks and demonstrate how each of them can be mitigated in order to increase the potential for success.

The strategy of creating university spinouts, as opposed to simply licensing technologies to existing companies, is particularly likely to appeal to universities in developing countries for several reasons. Licensing is often the "preferred" option for university technology transfer. Simply because it is less complex, it requires an acceptable licensee who is both interested in and capable of developing the technology. In many countries where the biotechnology industry, for example, is in the early stages of development and where there is a smaller chance of finding an acceptable licensee, creating a spinout may hold more promise. To the extent that the goal of commercializing university technologies is to generate economic growth, the creation of new companies can have

Krattiger A, RT Mahoney, L Nelsen, JA Thomson, AB Bennett, K Satyanarayana, GD Graff, C Fernandez and SP Kowalski. 2007. 13: Public Sector and Entrepreneurship. In *Executive Guide to Intellectual Property Management in Health and Agricultural Innovation: A Handbook of Best Practices* (Krattiger A, RT Mahoney, L Nelsen et al.). MIHR (Oxford, UK), PIPRA (Davis, USA), Oswaldo Cruz Foundation (Fiocruz, Rio de Janeiro, Brazil), and *bio*Developments-International Institute (Ithaca, USA). Available online at www.ipHandbook.org.

a greater impact close to home by generating jobs, attracting additional investment, and facilitating the growth of a biotechnology cluster. Because universities and public sector research institutes are often the giants of R&D within a developing economy, they need to be relied upon as sources for human capital and investment in entrepreneurship, since there may be no other sources.

Despite the promise spinouts may hold, they may not always be appropriate. Garner and Ternouth[2] address the question of what **realistic expectations** universities and research institutions should have concerning the risks of investing institutional resources in creating and spinning out new technology-based companies and they hold almost opposite beliefs to those of Brown and Soderstrom. The authors conclude that publicly funded institutions should consider how best to achieve their primary missions of delivering social and economic benefits, and they caution policymakers against exerting too much pressure on their region's universities to create new companies, because the process is difficult, consumes limited institutional resources, and is risky. The authors recommend that universities and research institutions should, as a rule, favor licensing-out to existing companies and third-party start-up companies and get involved only in the higher-risk strategy of investing the institution's own time and resources to create a spinout with measured and informed caution.

The process of creating a spinout is essentially one of providing the right social/professional environment, legal/financial framework, and resources for something new to grow and succeed and—given the risks—to fail "gracefully" if need be, without causing harm beyond the loss of opportunity and the initial investment. A very important element of creating spinout companies is to channel the enthusiasm and commitment of those who believe in the technology, want to see it succeed, and aspire to a positive outcome (for example, by providing products that improve the wellbeing of under resourced populations). Finally, a key point is to incubate a spinout long enough to enable it to run once it is "out there."

Pragmatic information about how organizations can transfer their intellectual property (IP) rights to a spinout company (normally through a licensing agreement) and then **convert the intellectual property into products or services for the public's benefit** can be found in the chapter by Sandelin.[3] Based on three decades of experience at **Stanford University**, the chapter identifies some key issues related to negotiating such transfers. These include:

- the general attitude toward spinouts held by a public research organization's senior administration and governing board
- various licensing considerations
- the use of milestones
- the amount and kind of compensation that should be received for licensing a technology

The chapter provides guidance on how to best reach a successful agreement. The definitions of particular terms in a contract, such as infringement responsibilities, sublicensing, and warranties and indemnities, are all carefully considered. In addition, the chapter covers conflict of interest (COI) and conflict of commitment (COC) issues that arise when employees of public research organizations become engaged in spinout companies. The authors provide clear examples to help policymakers and administrators better deal with the issues involved in a licensing agreement.

Governments everywhere are encouraging public research organizations to use their inventories of IP rights to create spinouts. Successful spinouts create new jobs, contribute to economic development, and potentially grow into large multinational corporations. TTOs are key players in this effort, but they should balance the interests and mission of public research organizations with the objectives of the spinout and the needs of society.

One surprising way that public research institutions can more effectively use their intellectual property is by attracting venture capital. Wyse[4] advances the premise that, **rather than venture capital driving the creation of new companies, it is the creation of new companies that attracts venture capital.** Research institutions and government policies are able to constructively influence the creation of new companies. The chapter seeks to inform those in public research institutions, and government policymakers, about the role that venture capital can, and does, play in technology-based entrepreneurship, and the types of environments that can encourage entrepreneurship and thereby attract venture capital.

Venture capital is a specific sector of the financial industry that channels investment from institutional and private investors, corporations, pension funds, and government agencies into venture funds that in turn invest in portfolios of equity in new companies. The model essentially **spreads out and shares the technology risks involved in each of the individual companies**. It also seeks—to the extent possible—to reduce the risks involved by specializing in a certain field of technology where the venture fund's management has expert knowledge. Venture capital may also be actively involved in the management of the companies, participating on the board and even providing business services. In return for bearing and managing such risks, venture capitalists expect to achieve sufficiently high internal rates of return, typically between 20 and 40 percent.

The availability of financial capital is not, generally, the limiting factor. In 2005, US$34 billion was invested in U.S. biotech companies from all sources (Table 1) with nearly US$4 billion in investment capital coming from venture capital. While venture capital is concentrated geographically to a few locations, individuals and institutions with interests in investing in growth opportunities can be found worldwide—including in developing countries. The fundamentals of success are straightforward: **the formation of new companies creates an environment that increases their probability for success**. Thus, the two essential pieces that need to be provided are:

1. Planting the "seeds" of new companies—encouraging skilled people with new ideas to develop those ideas.
2. Creating an environment favorable for entrepreneurship and success. Universities and research institutes can plant the seeds, while government policies can shape the environment.

A **favorable environment for creating and growing new companies** consists of an encouraging business culture (one that rewards success and treats failure as a learning opportunity), access to intellectual capital (such as that flowing from universities), sufficient financial capital, and reliable physical capital (facilities, laboratories, communications). All of these are enhanced if a region enjoys a low cost of living and a high quality of life.

While governments cannot legislate entrepreneurship, they can encourage it by providing a favorable environment. Once enough companies exist, they will themselves further transform the environment, attracting or creating the skills and capital that can **develop into a technology cluster**. Ultimately, the practice of investing venture capital is a skill that can be imported to a region or country, where it can be mastered by local investors. Wyse clearly implies that the next stage in the growth of the venture capital industry will involve spreading into new regions across the globe.

He also encourages the public sector to provide more funding for translational research, that is, research that moves a technology or product further up the value chain and closer to market, thus reducing both the investment needed for commercialization and the risk (Figure 1). The point of the figure by Wyse is that knowledge-based biotech industries in agriculture require a greater emphasis on translational research, compared to the pharma industry, to be able to attract the venture capital and corporate investment necessary to commercialize new products and technologies

It is also important to know what other forces discourage or encourage the commercialization of inventions. Cook[5] focuses on **the barriers created by cultural differences between academic institutions and business.** He contends that these barriers can be overcome by motivated technology transfer intermediaries. Inventors are usually creative, self-motivated, flexible individuals, but this does not mean that they naturally pursue the commercial potential of their discoveries. Whether or not an inventor ever shows his or her invention to the outside world actually depends on two variables:

1. whether he/she wants to disclose it

TABLE 1: SOURCES OF CAPITAL IN THE BIOTECH INDUSTRY

SOURCES OF CAPITAL		TOTAL INVESTMENTS (US$, MILLIONS) 2005	2006 (1ST Q TO 3RD Q)
Public	IPO[a]	819	567
	Follow-ons[b]	4,194	3,032
	PIPES[c]	2,376	1,817
	Debt	5,565	12,241
Private (venture capital)		3,518	3,186
Other		1,114	303
TOTAL CAPITAL		17,586	21,146
Partnering		17,268 (50%)	12,463 (37%)
TOTAL		34,854	33,609

a. IPO – initial public offering: a private company files to have a portion of its shares sold to the public on a regulated stock exchange, such as NASDAQ.
b. Follow-ons – When public companies sell additional shares on the stock exchange to raise additional cash.
c. PIPES – Private investments in public entities: the sale of public shares to private financial institutions that may take public shares off the public market as a way for companies to raise cash.

Source: Wyse[6]

FIGURE 1: COMMERCIALIZING KNOWLEDGE-BASED BIOTECH INDUSTRIES IN AGRICULTURE AND PHARMACEUTICALS

Agriculture requires much translational research.

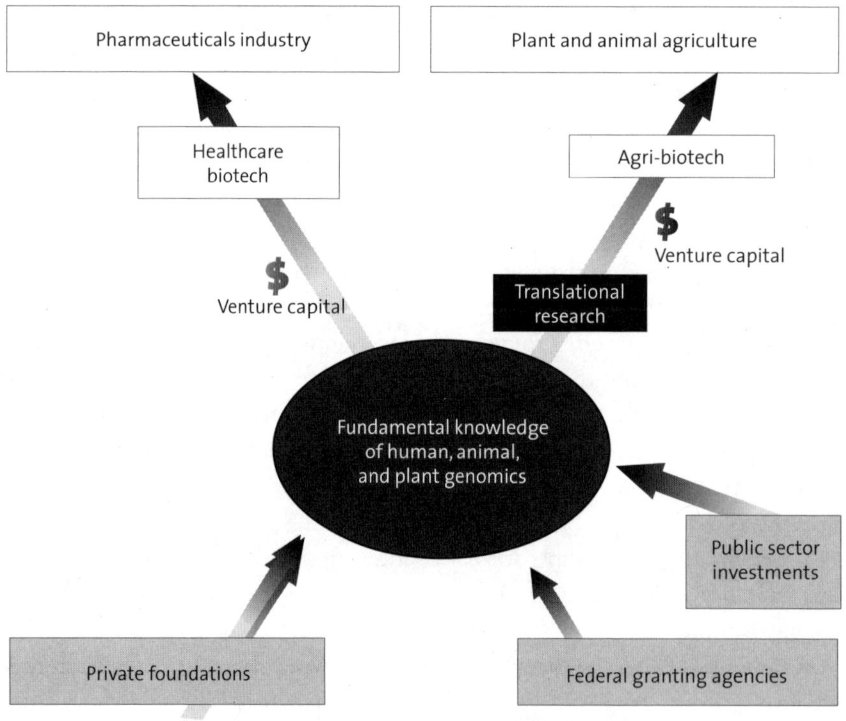

Source: Wyse[7]

2. whether the environment in which the inventor operates encourages or discourages disclosure

Much can be done to improve the environment surrounding an inventor. If an environment promotes creativity and is receptive to invention disclosure, it will not matter as much if an inventor has less self-confidence or is less of a risk taker. The manager of an institution's technology transfer effort should make every attempt to create an environment that fosters disclosure.

On the other hand, new companies operate in a very different environment. They generally have no established market position, are trying to convince potential investors that the company will succeed, and are usually understaffed and under-resourced. Such companies are most in need of effective leadership and of professional technology transfer intermediaries with the ability to translate a pioneering invention into a successful product. Such intermediaries should:

- understand the value systems that drive the inventor and the market
- be fluent in the vocabulary in both situations, so they are able to translate while retaining all linguistic nuances
- appreciate the various types of risks and how to mitigate them
- be credible to inventors as well as investors

These same qualities are valuable for those who are working to establish partnerships between the public and private sectors of developed and developing countries. Identifying, motivating, and retaining individuals with the capability to be intermediaries should be an important element of any effort to commercialize intellectual property.

The basic message of the chapter (along with the other chapters in this section of the *Handbook*) state that the role of the university is to channel its limited public resources into activities that create new opportunities that can be taken up by the market, but not to intentionally supplant or engage directly in the market. This means leaving venture creation to the market whenever possible. Indeed, most universities do not have sufficient professional resources or experience to manage spinout companies.

This and other chapters explore when and how universities can work with the market to channel investments into high-risk, high-return opportunities. This suggests a further question: when and how might universities work with other nonprofit and philanthropic funding sources to create spinout product development partnerships (PDPs) around high-risk, high-social-return opportunities? This is a question will be explored in the future by leading universities that have begun to master the process of spinning-out successful for-profit companies.

A more methodical approach has been gaining popularity over the last 15 years: **the creation of business incubators as tools for stimulating local economic development**. The concept of an incubator is simple and appealing: it provides a facility and services (for example, business planning and legal, accounting, and marketing support) to catalyze small business growth. Incubators have proven very effective. Incubated companies have a dramatically higher rate of survival than the average spinout. Additionally, companies that "graduate" from incubators provided an average of 85 full-time jobs per incubator. Used to promote the growth of entrepreneurial ventures of every imaginable type, small business incubation is now entrenched in both urban and rural areas throughout the United States.

Zablocki[8] discusses in detail the **six steps for setting up and operating successful incubators:**

1. Conducting a *feasibility study*. For a proposed incubator, such a study can achieve a number of important objectives and, if properly done, can provide a solid basis for judging the economic and political viability of the proposed project.
2. Identifying and *securing stakeholders*. While each incubator's circumstances are unique, anticipated stakeholders would likely include local and state governments and a variety of public and private sector organizations (universities, major corporations) interested in fostering new-business development in the region. Stakeholders might also include economic-development organizations that could fund the rehabilitation of a facility or the operation of the

incubator program. The support of these stakeholders is critical to initiating an incubator program.

3. Identifying a *market niche*. This requires much attention to detail. Successful businesses carefully attend to the work of defining the market position of their products and services relative to their competitors, as well as to modifying their market position in response to changing customer preferences.
4. The *formation process*. The basic structure of an incubator facility is determined by owner attributes and regional demographics (it could be private, local-government led, or university led or it could be a nonprofit company).
5. *Services*. As the incubator concept has evolved, the range of services offered by incubators has greatly expanded. Early incubators provided access to a photocopier and a conference room, clerical support, and perhaps switchboard services. Today, incubators themselves provide, or provide access to, a broad spectrum of office support, business consulting, and professional services. Business consulting services may include business-plan preparation, financial planning, advertising and marketing, strategic planning, technical and commercial communications, relocation planning, capital development (equity and debt services), business taxes, employee relations, R&D, and government procurement.
6. *Strategic Planning*. Strategic planning compels incubator management to confront tough issues. How will the incubator continue to operate if revenue projections from rental income are not achieved? How will major facility repairs (for example, a ruptured boiler) be paid for? Addressing these worst-case scenarios through strategic planning can provide both a clear course of action if things go as planned and, if they do not, the necessary contingency plans to navigate what may be a difficult beginning.

Economic development programs for small businesses proliferated in the 1980s. These programs have been referred to as **incubators without walls.** Well-managed incubators often distinguish themselves by serving as a focal point for access to the broad spectrum of available business services. Incubator managers thus provide the point of contact for entry into various programs. Many efforts to assist small business are, by contrast, programmatic in nature and limited by the scope of their intent. A well-positioned incubator, on the other hand, will help its tenants to access the range of existing programs and, in addition, provide access to informal networks for business and financial advice and assistance. For example, a retired executive may agree to help out a struggling firm, or a business angel may appear, discretely looking for new investment opportunities.

Successful incubator programs are marked by foresight, focus, and leadership. Successful incubator programs also know how to identify, organize, and maximize talent and resources, making the most of community support and entrepreneurial networks. A core group committed to starting a business incubator must recognize that its efforts cannot be pursued in a vacuum. As Zablocki puts it: "*The dream of a few must become the dream of many.*" ■

All chapters refer to: *Intellectual Property Management in Health and Agricultural Innovation: A Handbook of Best Practices*. 2007. A Krattiger, RT Mahoney, L Nelsen, JA Thomson, AB Bennett, K Satyanarayana, GD Graff, C Fernandez, and SP Kowalski (eds.). MIHR: Oxford, U.K., and PIPRA: Davis, U.S.A. Available online at www.ipHandbook.org. The online version contains for each chapter a detailed Editor's Summary, Implications, and Best Practices.

1 Chapter 13.1 by A Brown and J Soderstrom titled Creating and Developing Spinouts: Experiences from Yale University and Beyond, p. 1253.

2 Chapter 13.5 by C Garner and P Ternouth titled New Companies to Commercialize IP: Should You Spinout or Start-up? p. 1295.

3 Chapter 13.2 by JC Sandelin titled Dealing with Spinout Companies, p. 1271.

4 Chapter 13.3 by R Wyse titled What the Public Sector Should Know about Venture Capital, p. 1281.

5 Chapter 13.4 by T Cook titled The Role of Technology Transfer Intermediaries in Commercializing Intellectual Property through Spinouts and Start-ups, p. 1289.

6 See *supra* note 4.

7 *Ibid.*

8 Chapter 13.6 by Edward M. Zablocki titled Formation of a Business Incubator, p. 1305.

FOR GOVERNMENT POLICYMAKERS

- ✔ Much of **the success of a spinout or start-up will depend on the entrepreneurial spirit** at the institution. The more entrepreneurial, the more likely it will be that someone wants to set up a new company.

- ✔ Governments should **encourage public research organizations to use their inventories of IP rights to create spinouts** because they create new jobs, contribute to economic development, and potentially grow into multinational companies. But governments should recognize that setting up a new business is a **high-risk activity** and should allow new companies to fail gracefully.

- ✔ Rather than venture capital driving the creation of new companies, it is usually the creation of new companies that attracts venture capital.

- ✔ The formation of new companies creates an environment that increases the probability of success for other companies. Thus, the public sector should (1) plant the seed, encouraging skilled people with new ideas to develop those ideas and (2) create an environment that favors entrepreneurship and success. **Universities and research institutes can plant the seeds, while government policies can shape the environment.**

- ✔ While a government cannot *legislate* entrepreneurship, it can *encourage* entrepreneurship by providing a **favorable environment for creating and growing new companies**. This would be an environment with (1) an encouraging business culture that rewards success and treats failure as a learning opportunity, (2) access to intellectual capital (such as that flowing from universities), (3) access to sufficient financial capital, and (4) reliable physical capital (facilities, laboratories, communications).

- ✔ Public sector institutions may be the largest economic entities present in a developing country. Hence, they can contribute much by **taking the lead in developing and fostering the establishment of spinout companies** that are seeded with technologies generated in the public sector and protected and managed as IP assets.

- ✔ To successfully commercialize intellectual property, a country ought to have a **stable economic and institutional environment, available investment capital, commercializable intellectual property, a commercial environment that can develop intellectual property, and competent technology transfer intermediaries**.

- ✔ Technology transfer of any sort is likely to succeed only if there is **sustained commitment** at the most senior levels of both government and research institutions.

- ✔ Governments can encourage regional economic development by fostering and financing business incubators. Ideally, they ought to be located in strategically selected regions and **build on potential synergies of existing institutions**. Small business incubators in particular have proven to be effective economic development tools.

Given that IP management is heavily context specific, these Key Implications and Best Practices are intended as starting points to be adapted to specific needs and circumstances.

FOR SENIOR MANAGEMENT

(UNIVERSITY PRESIDENT, R&D MANAGER, ETC.)

- ✔ Experts are divided as to what approach should be taken by public institutions with regard to creating companies. Some advocate **a "hands-on" approach** in which the institution actively and directly manages the creation of companies and invests in the resources needed for their success. Others argue that the university should **channel its resources into activities that may result in marketable technologies**, but not engage directly in marketing activities.

- ✔ **The creation of business incubators as a tool for stimulating local economic development** should not be underestimated. Incubated companies have a dramatically higher rate of survival than the average spinouts.

- ✔ **Spinouts often create enhanced opportunities for its faculty.** If spinouts remain in the region, faculty inventors can remain active as consultants. Also, a university's success with spinouts can attract new talent.

- ✔ Much of the **success of a spinout or start-up will depend on the entrepreneurial spirit** at the institution. The more entrepreneurial, the more likely it will be that someone wants to set up a new company.

- ✔ The formation of new companies creates an environment that increases the probability of success for other companies. Thus, the public sector should (1) plant the seed, encouraging skilled people with new ideas to develop those ideas and (2) create an environment that favors entrepreneurship and success. **Universities and research institutes can plant the seeds, while government policies can shape the environment.**

- ✔ When engaging in entrepreneurial activities, **risks to the university** include potential impact on tax-exempt status, liabilities for the actions of the company, conflicts of interest and/or commitment, and conflicts with the university's mission.

- ✔ To be an **effective entrepreneurial university**, representatives of senior administration should routinely review company-founding and business-maintenance activities.

- ✔ Clear policies are needed for **disposing of equity in spinout companies**, both for the sake of the university's integrity—to prevent conflicts of interest—and for the sake of the company—to prevent the university's divestment from sending a damaging signal to the market about the value of the company or its technology.

- ✔ To demonstrate the importance of technology transfer, the **TTO should generally report directly to upper-level administration**.

- ✔ In order to **attract venture capital in agriculture**, public sector institutions need to take steps to reduce the risk of investing in agricultural projects.

- ✔ Rather than venture capital driving the creation of new companies, **it is usually the creation of new companies that attracts venture capital**.

Given that IP management is heavily context specific, these Key Implications and Best Practices are intended as starting points to be adapted to specific needs and circumstances.

FOR SCIENTISTS

- ✔ Not all university inventors are **entrepreneurs** nor are they interested in being company founders, and not all spinout company founders from a university are the technology's inventors.

- ✔ While inventors are treated equally under university patenting and licensing policies, **involvement as a company founder entails a greater degree of risk and commitment** to move an invention to commercialization. You may be valuable as an active partner of a spinout company to prevent the repetition of unsuccessful experiments (blind alleys) and to add needed creativity with respect to problem solving as development and commercialization proceeds.

- ✔ **Participation in a spinout** can be a particularly rewarding experience, financially as well as personally, as it involves the practical application of your ideas.

- ✔ **Venture capital investors** combine a broad view of the market with solid technical expertise. You will need to be prepared to convince investors not so much of the technical merits of your research, but of how your ideas lead to economic returns.

- ✔ Rather than venture capital driving the creation of new companies, **it is usually the creation of new companies that attracts venture capital**.

- ✔ Venture capital investors can be great allies, but will impose, for good reasons, distinct conditions on the project. **Be open, patient, and willing to work with investors.**

- ✔ Much of **the success of a spinout or start-up will depend on the entrepreneurial spirit** at the institution. The more entrepreneurial, the more likely it will be that someone wants to set up a new company.

- ✔ There are many factors that determine the feasibility and success of a spinout company. The technology's intrinsic value and **your commitment to your invention** are only part of the picture. If you can find an existing partner with market penetration, the chances of success increase. If you are still convinced, even after failing several times to find a willing licensee for your technology, then it may be time to consider creating a company. As these matters arise, seek the guidance of your institution's technology transfer office.

Given that IP management is heavily context specific, these Key Implications and Best Practices are intended as starting points to be adapted to specific needs and circumstances.

KEY IMPLICATIONS AND BEST PRACTICES: SECTION 13

FOR TECHNOLOGY TRANSFER OFFICERS

- ✔ **Spinouts carry a number of risks**, but with certain factors in place they can represent the best opportunity for developing early-stage technology. This is particularly true because the inventor, and other university participants, will have a vested interest in, and commitment to, the success of that technology.

- ✔ **Potential investors** in a spinout will ask two major IP questions. Could previously existing intellectual property block the technology? Could your intellectual property dominate the market and prevent entry by others? Other key questions involve the characteristics of the market opportunity and the financial bottom line of revenue and expense projections over the life of the technology.

- ✔ **Solid, long-term support from your institution** will be required to: (1) operate the technology transfer office efficiently, so that it can evaluate invention disclosures, obtain IP protection when appropriate, coordinate the search for people or companies that will develop the invention into products and services, and negotiate and prepare the necessary legal agreements (for example, license agreements for IP rights); (2) cover the costs of obtaining IP rights; and (3) provide funding to convert good ideas into working prototypes. (A good idea not put into use is wasted.)

- ✔ Your job is complex and challenging because you have to **balance the needs and expectations of many parties** with divergent interests: Remain responsive to such needs and interests; keep people informed of progress and developments; effectively utilize available resources.

- ✔ When licensing to or creating new ventures, several key attributes are essential for **attracting venture capital investment**: a strong management team, a viable technology, a strong IP position, a large potential market, and location in an environment favorable for entrepreneurship.

- ✔ New ventures in developing countries have much to gain by **attracting and building on international investor networks**. They have the potential to open new markets and bring in new alliances.

- ✔ Much of **the success of a spinout or start-up will depend on the entrepreneurial spirit** at the institution. The more entrepreneurial, the more likely it will be that someone wants to set up a new company.

- ✔ It is necessary to strike a balance between reliance on **licensing-out to existing companies** and investing time and resources in **creating new companies**.

- ✔ Rather than venture capital driving the creation of new companies, **it is usually the creation of new companies that attracts venture capital**.

- ✔ When creating spinout companies, always remain focused on your **institution's primary mission**, such that the spinout will be consistent with, and even serve, that public sector mission.

Given that IP management is heavily context specific, these Key Implications and Best Practices are intended as starting points to be adapted to specific needs and circumstances.

Freedom to Operate and Risk Management

Freedom to operate (FTO), a simple and straightforward concept, means that for a given product or service, at a given point in time, with respect to a given market or geography, no intellectual property (IP) from any third party is infringed. But to translate this concept into a productive strategy for companies and for public sector institutions alike is not so straightforward.

For public sector organizations, the **opportunities presented by incorporating FTO considerations** into product development strategies are numerous. These may include benefits through higher competitive intelligence, the ability to bring about culture change, and the forging of strong partnerships with providers of intellectual property and technology. An FTO strategy, therefore, is a *plan* that begins with research into the IP landscape of a potential product and evolves into an *attitude* through a product's R&D and commercialization/distribution cycle. Krattiger discusses these policy and strategy elements in detail.[1] It is useful, however, to first consider how FTO analyses are conducted before reviewing the principal FTO strategies.

At the beginning of a research project, a company would typically consider scientific feasibility, the effect of the research on the organization's business position (whether or not the research and the product would eventually strengthen its competitive position), the project's impact on financial status in terms of costs and potential rewards, and legal aspects (such as infringement risks). That is where an FTO analysis comes in as an initial, cursory, or quite possibly detailed, overview of the patent landscape and competitors' positions. Hence an **FTO analysis need not be a costly legal FTO opinion** (note that an FTO opinion is rendered by patent counsel whereas an FTO review can have any level of legal review, or none). Rather, an FTO analysis is an assessment of the set of patents and other IP rights that are or would be connected to the product and/or method under consideration. Kowalski[2] outlines in detail how different levels of FTO analyses are conducted in practice. Addressing scientists, business people, and legal staff, he describes how products and/or methods are broken down into fundamental components, processes, and combinations thereof, and then how each component is carefully analyzed for attached IP rights of third parties. An FTO analysis, irrespective of the level of detail, requires good preparation, systematic review, and rigorous record keeping.

Although FTO is often viewed as simply a legal issue, when approached from a more practical product-development perspective, **FTO is a strategic risk-management tool;** it relies on a synthesis of scientific and legal expertise, business development, and strategic planning. FTO for a given product in a given market is difficult to achieve because it can never conclusively be established. Obtaining FTO, therefore, becomes a

Krattiger A, RT Mahoney, L Nelsen, JA Thomson, AB Bennett, K Satyanarayana, GD Graff, C Fernandez, and SP Kowalski. 2007. 14: Freedom to Operate and Risk Management. In *Executive Guide to Intellectual Property Management in Health and Agricultural Innovation: A Handbook of Best Practices* (Krattiger A, RT Mahoney, L Nelsen et al.). MIHR (Oxford, UK), PIPRA (Davis, USA), Oswaldo Cruz Foundation (Fiocruz, Rio de Janeiro, Brazil), and *bio*Developments-International Institute (Ithaca, USA). Available online at www.ipHandbook.org.

strategy, or even a position, mindset, or culture. This is because the patent landscape is dynamic: new patents issue; old patents expire; some patents are abandoned. Therefore, *freedom to operate* does not imply absolute freedom from the risk of infringing another party's intellectual property. Whether or not FTO exists is an assessment based on the analysis and knowledge of IP landscapes for a given product, in a given jurisdiction, at a given point in time. This statement underscores a critically important principle: there can be no risk-free decision.

FTO is thus a concern that remains throughout the R&D process, to commercialization and even afterward. By setting a goal of having *reasonable* FTO, a set of ten FTO strategies for managing potential IP infringement are proposed and discussed in detail by Krattiger in Chapter 14.1 (Table 1). In practice, typically two or more of these strategies will be adopted, with the specific mix of strategies varying. Which strategies will be appropriate depend on, for example, how advanced the product is, the type of organization that develops the product, and relevant market dynamics. And not all of the listed strategies are feasible for public sector institutions.

How much attention should a public sector organization give to FTO? Since some public sector research is not directly intended for commercial use, the answer is sometimes quite simple: Not much. This condition certainly applies to a great deal of university research. However, if the project is specifically aimed at product development, a goal that is becoming more prevalent in the public sector, then FTO becomes a concern. For example, through collaborations with product-development partnerships (PDPs), the primary reason for funding the research is to develop products to help the poor. Such is the case also for the research centers of the Consultative Group on International Agricultural Research (CGIAR) and for many national agricultural research systems (NARS). Universities, too, are shifting their research focus; some manage their innovations in novel ways and aim to bundle technologies and intellectual property in order to license "solutions" rather than individual patents.

A relevant example of the importance of the public sector managing FTO is the development of **Golden Rice**.[4] No attention was given to FTO until the first material was nearly ready for transfer to developing countries. The Rockefeller Foundation then commissioned an FTO analysis that demonstrated, first of all, how many inventions from scientists around the world enabled—or accelerated—the development of Golden Rice. Although a large number of patents—and patent applications—were identified, the FTO analysis also demonstrated that licenses to only a few would be required for the transfer to and use of Golden Rice in developing countries. The FTO analysis provided a list of primary owners of patents (and of materials that went into Golden Rice under material transfer agreements) for which licenses were needed. With the leadership of the Rockefeller Foundation and Syngenta, a large agro-chemical company headquartered in Switzerland, the relevant intellectual property was quickly assembled (or in-licensed) into a package. That package then was licensed, essentially royalty-free, to public sector institutions in developing countries. This approach, in essence, represents the various aspects of FTO, from analysis to strategy, to action.

Each of the ten strategic approaches to obtain FTO listed in Table 1 presents certain risks and opportunities. Any action—including the decision not to take action—carries risk. Delaying the licensing of third-party intellectual property, for example, could lead eventually to expensive licensing terms, the inability to obtain a license, or the possibility of being sued for patent infringement. For some organizations, such as those developing genetically modified crops (GMOs), the opposite may be the case (where it might be advantageous to delay in-licensing) due to stewardship issues that are the main concern with biotechnology crops. Krattiger concludes his discussion, in Chapter 14.1, by urging the public sector to:

- judiciously evaluate whether and when FTO concerns should be considered
- build in-house capacity to conduct patent searches and cursory FTO analyses (as opposed to legal opinions)

Table 1: The Ten Strategic FTO Options

Option	Pros	Cons
LEGAL/IP MANAGEMENT STRATEGIES		
1. License-in	Is relatively straightforward	May not foster in-house R&D initiatives and may be costly
2. Cross-license	Involves give and take	In certain cases, antitrust issues may arise
3. Oppose third-party patents	Can be cost effective	Can be expensive and result might be undesirable (stronger or broader patent)
4 Seek nonassertion covenant	Is cheap and effective	Rarely allows for the in-licensing of valuable know-how
5. Seek compulsory license	Allowed under TRIPS under certain circumstances	Will not allow for the in-licensing of know-how and brings many constraints and complexities with it
R&D STRATEGIES		
6. Modify product	Can be fairly simple if planned early in R&D stage	May not be possible due to lack of readily available alternatives; incurs opportunity costs
7. Invent around	Could lead to cross-licensing position	Could lead to delays in product launch and might be costly; incurs opportunity costs
BUSINESS STRATEGIES		
8. Wait and see	Gives time for strategic positioning	Could lead to litigation and jeopardize investment already made
9 Abandon project	Is simple and effective	May be costly (need to write off R&D investments already made; incurs opportunity costs)
10. Merger/ Acquisition	Is highly effective	May distract from main business focus

→ **In Practice** **A combination of several options implemented concurrently**

Source: Krattiger, 2007.[3] The original table in Chapter 14.1 also includes a column listing the key challenges for the public sector for each of the ten strategies.

One of the underlying "technologies" for conducting an FTO analysis is the **patent search.** Such searches are also relevant when an institution is deciding whether to file a patent on a new invention (meaning when one is searching for prior art) or when scientists want to review patent literature. Fortunately, many IP search tools are accessible online at no cost. The chapter by Thangaraj, Potter, and Krattiger[5] provides a "guided tour" of online patent search engines, including a description of the major ones:

- the European Patent Office (EPO's esp@cenet)
- the U.S. Patent and Trademark Office (PTO)
- the Patent Cooperation Treaty (PCT) Web interfaces of the World Intellectual Property Organization (WIPO)

In addition, the chapter reviews subscription-based services and other paid services, such as Delphion and Derwent World Patent Index (DWPI). Although the discussion is by no means exhaustive, the chapter lists links to many useful sites.

The chapter by Fenton, Chi-Ham, and Boetiger[6] provides an examination of **how the private sector thinks about FTO**. Offering a comprehensive overview of the FTO process, the chapter sheds light on when to invest in FTO analysis and highlights the important role played by law firms in obtaining FTO. As mentioned earlier, for the public sector the strategic relevance of FTO is quite different from that of the private sector. Even when the public sector intends to commercialize products, its mission and goals differ from those of the private sector, and deciding when to pursue FTO becomes a very different question. Nonetheless, given the growing number of public-private partnerships, it is important for the public sector to understand how private companies approach FTO issues. This chapter discusses both private- and public-sector considerations for deciding **whether, when, and how an FTO analysis should be conducted.**

FTO analysis defines options; it provides a map of the relevant IP landscape. Hence, an FTO analysis presents the most viable options for achieving institutional goals. Fenton and colleagues conclude their discussion with a case study of an FTO analysis initiated by the Public Intellectual Property Resource for Agriculture (PIPRA). The case study explains the process used by PIPRA, from defining the scope of the investigation, to the delivery of the findings.

In the last chapter of Section 14, Boadi[7] looks at the aspect of FTO that includes legal liabilities beyond intellectual property and, appropriately, considers **stewardship as the central tool to managing liabilities.** The legal framework for dealing with liability relies on the country, or legal jurisdiction, in which the intellectual property is being exploited. Even so, GMOs (and indeed non-GMOs) have the potential to cross national borders. This has led to intense debate about whether a liability regime specific to such organisms and crops should be created. Providing an overview of current common law and statutory theories of liability, the chapter considers liability issues facing public sector efforts to develop and disseminate agricultural biotechnologies.

While debate rages about liability and redress issues contained in the Cartagena Protocol on Biosafety, developing countries need to think carefully about how to manage liability today. Referencing the African Agricultural Technology Foundation (AATF), Boadi provides several liability-management practices that can be of great value, including:

- ensuring compliance with intellectual property, license, and regulatory requirements
- including indemnification provisions in technology transfer agreements
- using warranty disclaimers
- obtaining letters of nonassertion
- adhering to appropriate technology stewardship best practices

Already, innovative developing countries (or IDCs), including India, Korea, China, Brazil, South Africa, and others are embracing novel opportunities provided by the new global IP regime. Having established technology transfer offices (TTOs) for organizations in both the public and private sectors, these countries have overseen the controlled, streamlined transfer of crucial technologies, often with clear public benefits. Such efforts, of course, require investments in both infrastructure and personnel to in-license, out-license, and ensure that investments in product development are accelerated through appropriate FTO considerations during the R&D process and beyond. ■

All chapters refer to: *Intellectual Property Management in Health and Agricultural Innovation: A Handbook of Best Practices*. 2007. A Krattiger, RT Mahoney, L Nelsen, JA Thomson, AB Bennett, K Satyanarayana, GD Graff, C Fernandez, and SP Kowalski (eds.). MIHR: Oxford, U.K., and PIPRA: Davis, U.S.A. Available online at www.ipHandbook.org. The online version contains for each chapter a detailed Editor's Summary, Implications, and Best Practices.

1 Chapter 14.1 by A Krattiger titled Freedom to Operate, Public Sector Research and Product-Development Partnerships: Strategies and Risk-Management Options, p. 1317.

2 Chapter 14.2 by SP Kowalski titled Freedom to Operate: The Preparations, p. 1329.

3 See *supra* note 1.

4 See Case Study 3 by A Krattiger and I Potrykus titled Intellectual Property Management and Freedom to Operate for Golden Rice in the insert of this *Executive Guide*, p. 11-14.

5 Chapter 14.3 by H Thangaraj, RH Potter, and A Krattiger titled How and Where to Search for IP Information on the World Wide Web: The "Tricks of the Trade" and an Annotated Listing of Web Resources, p. 1345.

6 Chapter 14.4 by GM Fenton, CA Chi-Ham and S Boetiger titled Freedom to Operate: The Law Firm's Approach and Role, p. 1363.

7 Chapter 14.5 by RY Boadi titled Managing Liability Associated with Genetically Modified Crops, p. 1385.

FOR GOVERNMENT POLICYMAKERS

- ✔ As intellectual property becomes more prevalent in health and agricultural research, public and not-for-profit institutions may increasingly need to consider the intellectual property of **third parties**. This may allow for efficient in-licensing of intellectual property and accelerate the development of products. For such purposes, a good knowledge of "who owns what" is needed. That is what a freedom to operate (FTO) analysis provides.

- ✔ Translating an FTO analysis into an effective strategy requires some **shifts in culture and thinking** by those public sector institutions that are engaged in the development of products. Although a legal opinion by an attorney may be based on a solid FTO analysis, the use of such an analysis is strategic. National governments have a great responsibility to encourage the establishment of best practices in IP management, through sound national policies and funding allocations.

- ✔ Taking FTO into consideration as one element of any product development strategy allows for a more judicious use of resources that can often lead to stronger and more effective partnerships, can increase **opportunities for international collaboration,** and may underpin effective public-private partnerships.

- ✔ Governmental policies and programs that support capacity building in IP management should include the **training of senior management** in FTO strategies, including institutional boards. A dialogue between boards, which are responsible for policy, and senior managers that are more concerned with implementation is essential since an FTO analysis is a **risk-management tool.** This approach increases efficiency in the handling of products for further development and/or commercialization, even if the goal is to address the needs of the poor.

- ✔ **High speed Internet** access and patent databases are valuable tools that can assist research-based institutions in the undertaking of meaningful patent and information searches that are necessary to conduct FTO analyses.

Given that IP management is heavily context specific, these Key Implications and Best Practices are intended as starting points to be adapted to specific needs and circumstances.

KEY IMPLICATIONS AND BEST PRACTICES: SECTION 14

FOR SENIOR MANAGEMENT

(UNIVERSITY PRESIDENT, R&D MANAGER, ETC.)

- ✔ As public sector and nonprofit institutions increasingly move in the direction of **product development,** whether they do so independently or in partnership with other organizations, freedom to operate (FTO) will contribute increasingly to sound IP management strategy. As such, an FTO analysis is a management tool for assessing and managing certain types of risks.

- ✔ Some public sector institutions need not be concerned with FTO. For example, a typical university that mainly licenses patents or occasionally forms a spinout company can **leave FTO concerns to others.**

- ✔ Public sector research institutions should not necessarily assume that they are exempt from IP infringement liability due to their nonprofit (or governmental or parastatal) status. Although government institutions per se may be shielded from liabilities, FTO rarely ends with these institutions. Eventually other institutions taking on the products may need to be able to deal with FTO (such as commodity exporters). Hence **FTO analysis is just one tool** for making the technology transfer process more effective, and FTO is particularly warranted as an institution expands its mission into product development and distribution.

- ✔ FTO opinions do not eliminate risks related to third-party intellectual property. Instead, they allow for the development of sound **risk-management strategies** (which may be of a legal/licensing nature, involve business approaches, or be research based). Implementing the strategy requires clear pathways of communication and dialogue between science managers, product development, licensing personnel, and senior management.

- ✔ Obtaining FTO includes the review of the patent landscape (FTO analysis) and may include a formal legal FTO opinion. But, in essence, obtaining FTO is a process to be "managed" as an **interdisciplinary endeavor** and considered within the context of the institution's overall mission (as such it involves senior management), business development, research and technology transfer, and tolerance for risk.

- ✔ Institutional policies that support capacity building in IP management should include the training of senior management in FTO strategies, including institutional boards. A **dialogue** between boards (responsible for policy) and senior management (more concerned with implementation) is essential, since FTO analysis is a risk-management tool.

- ✔ Commitment to the principles of FTO will demonstrate that a given institution is committed to respecting, and of **building upon, the intellectual property of others**.

- ✔ The more downstream a research-based institution operates, the more important FTO considerations become. A system should be in place to help decide **whether, when, and how** a public sector institution should conduct an in-house FTO analysis or commission a legal FTO opinion.

Given that IP management is heavily context specific, these Key Implications and Best Practices are intended as starting points to be adapted to specific needs and circumstances.

FOR SCIENTISTS

- Collaboration among scientists and the professionals who conduct freedom to operate (FTO) analyses is essential. **The scientist is the most important person** to explain the science behind technology, to help others understand the materials and the scientific approach, and sometimes to explain what specific patent claims mean. A scientist is the expert in his or her area of research and can provide important leads to other scientific groups, publications, and terms of art.

- Teams conducting FTO analyses will also need to understand precisely **what the product is, how it was developed, what materials were used,** and what reports were prepared. The purpose is not to check on the work, but to ascertain that all relevant information has been considered in the FTO analysis. It is important also to know what tangible property from a third party contributed to a product. A scientific member of an FTO team will need to provide this type of basic information for the FTO analysis. One of the best ways to manage that information is through careful record keeping, including rigorously kept laboratory notebooks.

- The results of an FTO analysis may **allow you to make better use of technologies in the public domain** and inform your choice of research tools or vector constructs. The analysis also may alert you to scientific discoveries and inventions related to your work.

- An FTO analysis is a foundation of IP management, but it is also something more. It is a way to demonstrate to your colleagues that you respect their property rights and understand that, when properly managed, intellectual property leads to the greater sharing of technology and related information. In a very real sense, it is a way of **building relationships** based on trust.

- Both patent search engines and scientific search engines are available at no cost on the Internet (such as the U.S. Patent and Trademark Web site and Google® Scholar).

- Knowledge of how to access, manipulate, and mine these tools for valuable information will serve you and your program well. Hence, you should encourage your staff to become **well versed in Internet database search skills,** and do not hesitate to ask your technology transfer office to organize short patent search workshops.

Given that IP management is heavily context specific, these Key Implications and Best Practices are intended as starting points to be adapted to specific needs and circumstances.

FOR TECHNOLOGY TRANSFER OFFICERS

- ✔ The management of patent infringement risks requires a good knowledge of the strategic options available. These options include **legal/licensing, business strategies, and R&D strategies.**

- ✔ Unlike at a private company, where business/legal/financial conditions often determine R&D strategies, licensing officers in public sector institutions rarely influence research projects and institutional policy. **The role of the technology transfer officer as communicator** in the public sector is therefore much more important for bringing about an IP management "culture" throughout the organization.

- ✔ A freedom to operate (FTO) analysis is an interdisciplinary endeavor best executed through FTO teams. These teams, made up of legal, business, and scientific professionals, are in themselves useful for **strengthening intra-institutional dialogue** and communications.

- ✔ The role of the technology transfer officer, and that of attorneys who may produce legal FTO opinions, is generally **to advise senior management**. It is a manager's purview, based on your input, to decide how to deal with the risks identified in your FTO analysis.

- ✔ Much work leading to a legal FTO opinion can be done in-house, working with scientists, technology transfer professionals, business people, and others. **The role of patent counsel** is important for formal legal FTO opinions, but this expense may not often be required or justified in public research settings.

- ✔ Evaluate the pros and cons of free versus subscription-based **patent search sites.** Quite often, free services are limited in content and scope and do not allow for myriad search capabilities of paid services. But many free sites, such as WIPO's PatentScope, are increasingly adding extremely valuable features.

- ✔ For an academic or public institution, legal **FTO opinions are unlikely to be needed** for the majority of technology transfer functions. They might be applicable if the institution is engaged in downstream product development and commercialization.

- ✔ One way to cut costs is to conduct the background research for an FTO analysis in-house. The compiled file of relevant art can then be provided to patent counsel, who can then further analyze, conduct additional searching to fill in suspected gaps, and render an FTO opinion. **Universities with law schools** might be able to give law students valuable internships in this manner.

- ✔ Through **good licensing practices** (including appropriate indemnification provisions and warranty disclaimers), much of the risk associated with IP infringement can be transferred to licensees who take over products from the public sector.

Given that IP management is heavily context specific, these Key Implications and Best Practices are intended as starting points to be adapted to specific needs and circumstances.

Monitoring, Enforcement, and Resolving Disputes

Intellectual property is not a static asset. It is dynamic, requiring ongoing attention and management practices that will allow an institution to protect its value and maximize its utilization. Treating intellectual property as such is a fundamental best practice, regardless of whether an institution is public or private, whether located in a developed or developing country or whether its mission is directed toward commercial or public interests. Intellectual property, if it is to be an asset, cannot be simply be shelved and left alone, or even licensed and then left alone. Intellectual assets, with patents, as a particularly cogent example, must constantly be managed, monitored, maintained, and policed as part of a continual "cultivation" of IP rights.

The larger the IP portfolio, the greater the likelihood that disputes of one sort or another will arise. Few disputes end up in litigation, as there are many options and strategies for resolving disputes. Particularly in the context of partnerships between entities in developing and developed countries, litigation would be a complicated, time consuming, expensive, and risk-laden process. As Part 7 on contracts and Part 11 on licensing demonstrated, good contracts and good licensing practices anticipate that disputes arise with partnerships and licenses. But the best way to avoid disputes is to manage agreements in a manner that leads to effective resolution of disputes.

Feindt[1] details how the Office of Technology Transfer (OTT) of the National Institutes of Health (NIH) administers its technology licenses. Licensing is an important part of NIH's operation and an important part of any technology transfer endeavor anywhere. The portfolio of licenses at OTT includes over 1,400 active technology licenses, 750 of which generate about US$100 million in revenues.[2] These licenses represent five types of technology licenses:

1. Commercial evaluation licenses (also known as options), which enable companies to provisionally evaluate whether a new technology suits their needs
2. Patent commercialization licenses, which provide access to rights in patented or patent-pending technology. These licenses can be either:
   a. Exclusive, providing a single licensee with the right to practice the patent
   b. Nonexclusive, providing patent rights to, potentially, multiple licensees
3. Nonexclusive patent licenses for internal use, which provide access to tools or processes, useful for research purposes
4. Biological materials licenses, which provide access to nonpatented biological materials
5. Software licenses, which provide access to nonpatented software

Krattiger A, RT Mahoney, L Nelsen, JA Thomson, AB Bennett, K Satyanarayana, GD Graff, C Fernandez and SP Kowalski. 2007. 15: Monitoring, Enforcement, and Resolving Disputes. In *Executive Guide to Intellectual Property Management in Health and Agricultural Innovation: A Handbook of Best Practices* (Krattiger A, RT Mahoney, L Nelsen et al.). *bio*Developments-International Institute (Ithaca, USA), MIHR (Oxford, UK), PIPRA (Davis, USA), and Oswaldo Cruz Foundation (Fiocruz, Rio de Janeiro, Brazil). Available online at www.ipHandbook.org.

Although the above licenses differ as to the types of technology licensed, the specific terms of the license, and the reporting obligations, one aspect remains consistent: every license is written with well-defined financial terms and clearly delineated reporting obligations.

As a licensor, the NIH OTT administers, monitors, and enforces its technology licenses. It accomplishes this by monitoring licensee compliance with royalty payments and reporting obligations throughout the term of the license. Typically this is not a confrontational relationship. Instead, the NIH OTT seeks to build cooperative relationships with its licensees that, in turn, facilitate problem-solving discussions, resolve outstanding issues, and identify possible opportunities for advancing commercialization of products and/or services pursuant to the technology license.

In practical terms, OTT maintains licensee contact lists (people within the licensee organization) as they are important when royalty payments or issues of noncompliance need to be addressed. Such a list, and a certain level of personal rapport, greatly facilitate communication and save much time. Another essential operational procedure is maintenance of a well-organized filing system, possibly with archival, working, and computer files. A computer filing system can be structured as a searchable database for license administration, with integrated interactive modules organizing data on contracts, inventions, patents and license applications, royalties, receipts, and reporting. (Such a database is available for download free of charge from the online version of the *Handbook*.)[3] Many of the best practices listed and discussed by Feindt relate to licensees' reporting obligations, to amendments to license agreements, and to sanctions for noncompliance. Amendments reflect mutually agreeable changes in the expectations of licensor and licensee that occur with the passage of time and changing circumstances; they might involve term extensions, royalty adjustments, or changes in the field of use. Sanctions, on the other hand, are unilateral actions by the licensor that are triggered by licensee noncompliance (and may include a threat of license termination and even legal action in order to enforce lapsed financial obligations).

A different type of dispute is that of patent infringement. Patent infringement is, at least conceptually, analogous to trespass in that it is an invasion and/or misappropriation of another party's exclusive property right. Hence, identifying and taking action to remedy infringement is an essential part of IP asset management. There are four main categories of patent infringement:

1. *Literal infringement,* in which each and every element of a patent claim is found in the alleged infringing product or process
2. *Doctrine of equivalents infringement,* in which the alleged infringing product or process is substantially the same as the patented product or process
3. *Contributory infringement,* in which a party contributes to infringement of a patent by selling a component that has no use other than as part of a patented product
4. *Inducement to infringement,* in which a party actively and knowingly aids and abets another who is directly infringing a patent

Maintaining the integrity and value of public sector intellectual property is, as Haeussler[4] points out, a strategic process, which will vary somewhat depending on the category of infringement. First, it is important for a patent applicant, public or private, to consider claim structure and scope when drafting and filing patent applications. Unless the claims of a patent are sufficiently broad so as to confer clear economic potential, prospective licensees will be reluctant to enter into commercialization agreements or partnerships. Second, it is important to stay vigilant, establishing surveillance protocols **for possible infringement of patents**. For example, inventors should be contacted, on a regular basis, and asked if they know of anyone who is, or might be, infringing their patents; in addition, technology transfer staff members should regularly review key media related to the technology in order to watch for potential infringers. Surveillance is not only sound business practice but is essential for maintaining and preserving patent rights. For example, lack of enforcement can lead to a loss of patent rights.[5]

Patent infringement can also lead to a lawsuit. And litigation is expensive, risky, uncertain and often protracted. With good negotiation, a settlement through modified license terms often can be amicably reached. If litigation becomes inevitable, then a series of questions need to be addressed including whether to use in-house or external counsel, whether to file suit based on patent infringement or breach of contract, and where to file the lawsuit. Importantly, credible communication that the IP owner is serious about protecting its IP assets will go a long way to bringing infringers to the table to discuss the issues and to negotiate. Haeussler concludes that early communication with potential infringers, and good license and licensee diligence, are the foundations for policing and maintaining intellectual property, irrespective of whether the intellectual property is owned by a public or a private entity. Feindt, in his chapter, illustrates the importance of early communication with infringers using examples from NIH.

Due to the costs—and risks—associated with litigation, alternative dispute resolution procedures, such as mediation and arbitration, should under many scenarios be attempted first. These forms of dispute resolution do not work through formal legal systems, but are instead set up by the parties involved. They are established by dispute resolution clauses articulated when a partnership is set up or a license granted—before any problems have arisen. The goal is to have an already agreed-upon system when difficulties arise.

There are several elements unique to arbitration and mediation that can help parties resolve disputes as Min discusses in detail.[6] Figure 1 shows the two processes side-by-side. For arbitration, the parties have the power to decide on the number of arbitrators, the type of arbitration (ad hoc or institutional), the place of arbitration, the language of arbitral proceedings, and the applicable substantive law. Unlike judges, whose powers are defined by national laws, an arbitral tribunal's powers are limited to those conferred by the parties. Mediation involves the same kinds of choices, although unlike a judge or an arbitrator, whose mandate is to issue a binding decision or award, a mediator does not have the power to impose a settlement on the parties. Instead, the mediator serves as a catalyst for party negotiations. The advantages of arbitration and mediation include the following:

- Through arbitration or mediation, the parties can, using a single procedure, resolve disputes involving intellectual property in a number of countries.
- In both arbitration and mediation, parties may resolve a transnational dispute on neutral territory.
- Arbitration and mediation are based on the consent of the parties.
- Parties can select arbitrators or mediators.
- Parties to arbitration or mediation can keep the proceedings and any results confidential.
- The protracted nature of litigation, which pushes parties into multiple rounds of appeals, is a common problem when litigating transnational disputes. The end result of arbitration, however, is a final, binding award.
- The mediator's role is to broaden dispute resolution options, allowing the parties, with the help of the mediator, to craft innovative, common-sense solutions that (preferably amicably) settle the dispute.
- Mediation involves low risk. If a party feels that it is not making any progress, that the procedure is becoming too costly, or that the other party is not acting in good faith, the party may withdraw from a mediation process at any time and seek to resolve the dispute through litigation or arbitration.

Min provides a list of the kinds of concerns addressed by dispute-resolution clauses. She also highlights the usefulness of arbitration and mediation for developing countries, but public sector entities in many parts of the world will find the discussion, and these tools, useful. Such entities often lack the resources to pursue extended litigation, a process that also frequently places them on unfamiliar cultural and legal ground. By formulating dispute resolution policies, institutions in developing countries can place themselves in a fairer, less expensive, and less antagonistic forum for resolving disagreements.

FIGURE 1: PRINCIPAL STEPS IN A TYPICAL MEDIATIONS AND (WIPO) ARBITRATIONS

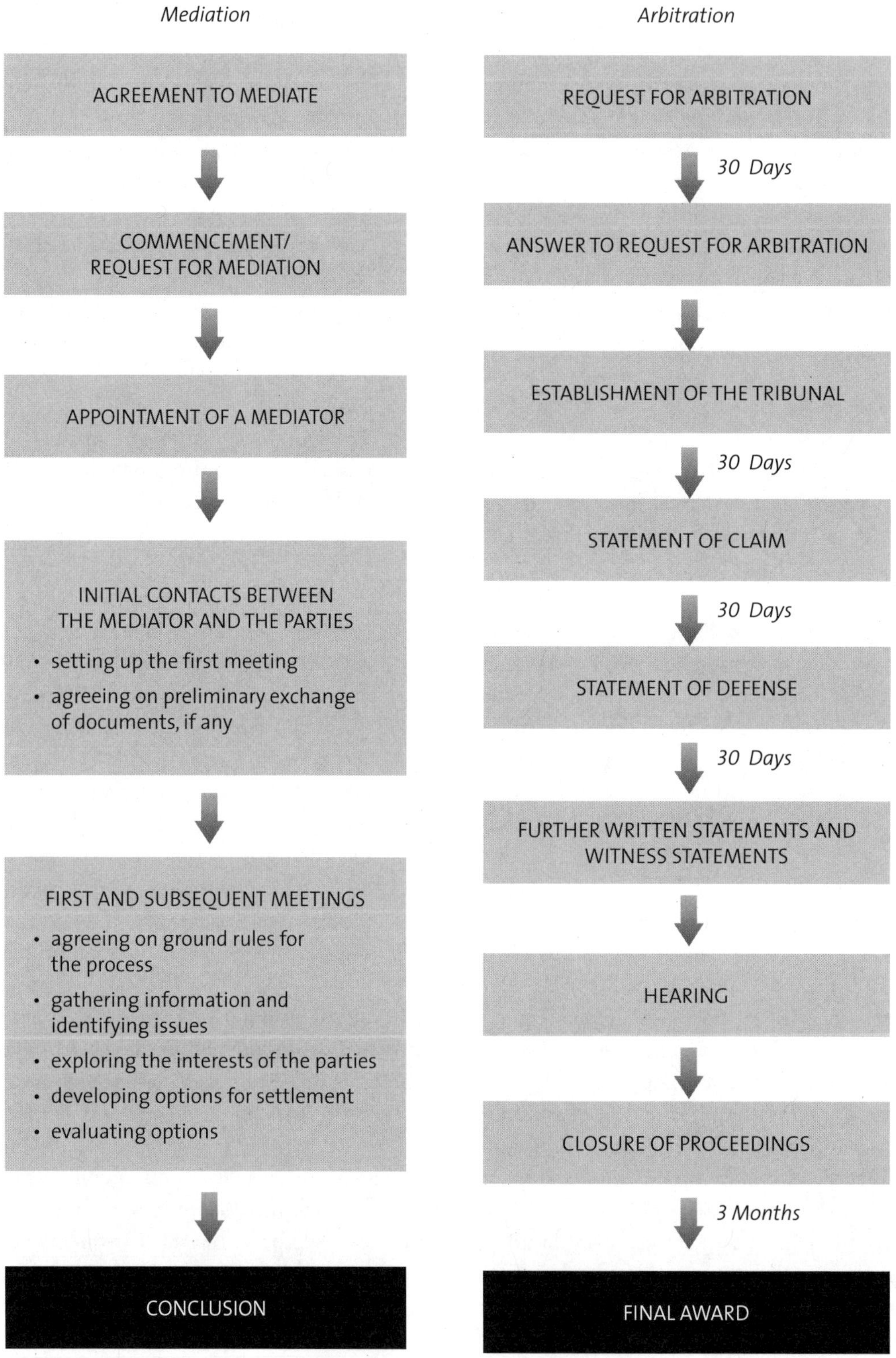

Source: Min[7]

Finally, the chapter discusses extensively the activities and services of the **Arbitration and Mediation Center**, which functions under the World Intellectual Property Organization (WIPO).

A different type of dispute may emerge under **parallel trade** practices (involving "gray market" imports). As Matthews and Munoz-Tellez write,[8] parallel trade occurs when products produced under the protection of a patent, trademark, or copyright in one market are subsequently exported to a second market and sold there without the authorization of the local owner of the IP right. Often, the local owner of the IP right will also be a local dealer who, through a license or other exclusive agreement, has been authorized by the patent, copyright, or trademark holder to market the protected product. Naturally, when the licensed dealer has an exclusive agreement, he or she expects to be the only party supplying the product in the local market. Importantly, parallel trade does not refer to the trade of pirated or counterfeit products. These are unauthorized versions of products that infringe an IP right. Parallel imports, on the other hand, are imports of genuine, often branded, products that do not violate an IP right per se, but importing the product will not have been authorized by the right holder.

Engaging in parallel trade is a legal option provided within the Agreement on Trade-Related Aspects of Intellectual Property (TRIPS). The Doha Declaration reaffirmed that developing countries can use parallel imports to support public health. Such countries can obtain access to lower-priced patented and/or branded products, such as medicines and basic agricultural inputs, by incorporating legislation to allow for parallel imports. TRIPS permits member states of the World Trade Organization (WTO) to design their own *exhaustion of patent rights* regimes. Hence, the state's legal framework for parallel trade is based on its own exhaustion of patent rights doctrine:

- *national exhaustion*, whereby the exclusive rights of patent holders cease only after the first sale of a product within the national borders (parallel imports can be blocked at the border)
- *regional exhaustion*, whereby the exclusive rights of patent holders cease after the first sale in the regional market (parallel trade permitted within the regional group)
- *international exhaustion*, whereby the exclusive rights of patent holders cease after the first sale in any market (parallel trade permitted)

The chapter focuses on how parallel trade can provide developing countries with greater access to medicines and to basic inputs for agricultural production (such as pesticides and fertilizers) at lower prices.

Thus, developing countries can incorporate into their national laws the principle of international exhaustion of rights, thus allowing for parallel imports on an international scale. In other words, developing countries can decide whether or not to allow parallel importation for all or particular IP rights.

Although parallel trade has obvious benefits for developing countries, there are also potential disadvantages. For example, the chapter notes that parallel trade might:

- reduce incentives for investment in the pharmaceutical and agricultural sectors
- reduce the incentives for rights holders to donate products at low cost or free of charge to developing countries due to fear of re-importation elsewhere
- reduce the willingness of rights holders or licensed local owners to supply particular markets

When implementing measures to facilitate parallel trade, developing countries should ensure an effective system by adequately regulating the quality and safety of parallel imports. At the same time, they need to prevent low-priced patented products in developing countries from entering high-priced developed country markets. Otherwise, patent holders, particularly in the pharmaceutical industry, could be discouraged from pricing their products differently in different markets in order to benefit developing countries. The chapter offers model legislative provisions to enable parallel imports and concludes by urging policymakers in developing countries to promote access to medicines and to support poor farmers by fully utilizing the parallel trade options available under TRIPS. ■

All chapters refer to: *Intellectual Property Management in Health and Agricultural Innovation: A Handbook of Best Practices*. 2007. A Krattiger, RT Mahoney, L Nelsen, JA Thomson, AB Bennett, K Satyanarayana, GD Graff, C Fernandez, and SP Kowalski (eds.). MIHR: Oxford, U.K., and PIPRA: Davis, U.S.A. Available online at www.ipHandbook.org. The online version contains for each chapter a detailed Editor's Summary, Implications, and Best Practices.

1 Chapter 15.1 by HH Feindt titled Administration of Technology Licenses, p. 1395.

2 Although impressive, such revenue flow is not the OTT's principle mission. Rather, by instituting and running an organized and professional office, the NIH furthers its mission of a timely introduction of new products and technologies into the marketplace. In this way, the fruits of NIH research and development are made commercially available, fostering economic development and serving the greater public good through the introduction of critical advances in health care. Furthermore, the NIH OTT fully recognizes that potential licensees will be from both developed and developing countries, such that the range of beneficiaries is truly global in scope.

3 This database has been provided by the Whitehead Institute for Biomedical Research. See also Chapter 6.12 by A Hamzaoui titled WIIPS™: Whitehead Institute Intellectual Property System (A Relational Database for IP Management and Technology Transfer), p. 649.

4 Chapter 15.2 by HW Haeussler titled Policing Intellectual Property, p. 1405.

5 This may happen pursuant to two defenses in equity: (1) *laches*, when the patentee waits too long (an inexcusable delay) before taking action against a presumed infringer and (2) *equitable estoppel*, when the presumed infringer, relying on actions or communications from the patentee, reasonably believes that he or she can practice the patented product or process.

6 Chapter 15.3 by EJ Min titled Alternative Dispute-Resolution Procedures: International View, p. 1415.

7 *Ibid.*

8 Chapter 15.4 by D Matthews and V Munoz-Tellez titled Parallel Trade: A User's Guide, p. 1429.

FOR GOVERNMENT POLICYMAKERS

- ✔ A fundamental best practice in IP management, regardless of whether an institution is public or private and whether located in a developed or developing country, is to view **intellectual property as an evolving and dynamic asset** requiring ongoing attention, management, monitoring, and policing. Only such an "IP cultivation" will allow institutions to protect the value and utility of the intellectual property.

- ✔ A country's statutory code, combined with a reliable system of **fair adjudication and judicial enforcement**, is the requisite basis for enforcing institutions' IP rights. Supporting policies that promote this legal infrastructure is essential.

- ✔ **Court action** is often stymied because of cost, length of procedure, legal uncertainty, the decision maker's lack of expertise, confidentiality/publicity, the difficulty of seeking action in foreign jurisdictions, and the negative impact on existing business relationships. But public and private institutions alike should always have the flexibility to opt for court action if this seems to be in their best interests.

- ✔ Policymakers should strive to promote policies and advocate for laws that encourage **alternative dispute resolution procedures** as the best alternatives for settling differences between parties to an agreement. These procedures are particularly important in international contract dispute resolution.

- ✔ Governments and public institutions can help make arbitration or mediation procedures accessible and available by identifying and supporting neutral institutions that can provide cost-efficient, timely dispute-resolution services. The World Intellectual Property Organization offers such services through the **WIPO Arbitration and Mediation Center**.

- ✔ Pursuant to the TRIPS Agreement and the Doha Declaration provisions on parallel trade, countries can **implement patent rights exhaustion regimes** that either permit or restrict **parallel importation**. As a result, developing countries can decide whether or not to allow parallel importation for all or for particular IP rights. Despite the evident benefits of parallel trade, there are also disadvantages, and both the benefits and the risks should be carefully considered. (Drawbacks of broad parallel importation practices include the reduction in incentives for investment in the pharmaceutical and agricultural sectors and the reduction in incentives for rights holders to donate products at low cost or free of charge to developing countries due to fear of re-importation to lucrative developed country markets. Re-importation hinders the ability of governments in different countries to maintain price controls on pharmaceutical products within their territory and reduces the willingness of rights holders or licensed local owners to supply particular markets.)

Given that IP management is heavily context specific, these Key Implications and Best Practices are intended as starting points to be adapted to specific needs and circumstances.

KEY IMPLICATIONS AND BEST PRACTICES: SECTION 15

FOR SENIOR MANAGEMENT

(UNIVERSITY PRESIDENT, R&D MANAGER, ETC.)

- ✔ A fundamental best practice in IP management is, regardless of whether an institution is public or private and whether located in a developed or developing country, to view **intellectual property as an evolving and dynamic asset** requiring ongoing attention, management, monitoring, and policing. Only such an "IP cultivation" will allow institutions to protect the value of intellectual property and maximize its utility.

- ✔ Your institution's technology transfer office should have systematic procedures to **administer, monitor, and enforce** its technology licenses. This includes compliance with royalty payments and reporting obligations in a nonconfrontational manner.

- ✔ Public and private institutions alike should always have the flexibility to opt for **legal action** if this seems to be in their best interests. But legal action is often stymied because of cost, length of procedure, legal uncertainty, a decision maker's lack of expertise, confidentiality/publicity, the difficulty of seeking action in foreign jurisdictions, and the negative impact on existing business relationships.

- ✔ Encouraging **alternative dispute resolution procedures** can be a viable strategy and, indeed, often a preferred one, for settling differences between parties to an agreement. These are particularly important in international contract dispute resolution.

- ✔ Public sector institutions should have an **institutional policy on the use of arbitration and mediation**.

- ✔ Public institutions can help make arbitration or mediation procedures accessible and available, by identifying and supporting neutral institutions that can provide cost-efficient, timely dispute resolution services. The World Intellectual Property Organization offers such services through the **WIPO Arbitration and Mediation Center**.

- ✔ Where permitted by national legislation, **parallel importation** may provide universities and public sector research institutes with lower-cost access to legitimate imports produced in other markets.

- ✔ For universities and research institutes in particular, **parallel importation may have substantial benefits** as it allows for the lower-cost import of copyrighted products (books, computer software, periodicals, and related products). Hospitals may also benefit from parallel-trade imports by access to cheaper, patented pharmaceutical products. Sometimes, however, the final cost of the parallel-imported product is higher than locally supplied goods, while quality and warranty may be lower.

- ✔ But **parallel importation also has drawbacks**. These include the reduction in incentives for investment in the pharmaceutical and agricultural sectors and the reduction in incentives for rights holders to donate products at low cost or free of charge to developing countries due to fear of re-importation to lucrative developed country markets.

Given that IP management is heavily context specific, these Key Implications and Best Practices are intended as starting points to be adapted to specific needs and circumstances.

FOR SCIENTISTS

- ✔ A fundamental best practice in IP management, regardless of whether an institution is public or private and whether located in a developed or developing country, to view **intellectual property as an evolving and dynamic asset** requiring ongoing attention, management, monitoring, and policing. Only such an "IP cultivation" will allow institutions to protect the value of intellectual property and maximize its utility.

- ✔ As a scientist, you should **regularly review** all of the agreements that relate to your projects. This specifically includes ensuring that milestones are met, royalties paid, and that any other obligations are taken care of.

- ✔ Your institution should continuously monitor **patent infringements** through various surveillance protocols. A lack of patent enforcement can lead to a loss of patent rights. Your role in this is important, since you are well connected in the area of your research and can indicate to the technology transfer office which companies might be practicing your inventions.

- ✔ Keep laboratory records and notebooks organized, ideally consistent with your institution's **laboratory notebook policy**. These can be essential for drafting patent applications, prosecuting patents and, if necessary, pursuing litigation.

- ✔ If your institution conducts alternative dispute resolution procedures such as **mediation or arbitration**, you might be called upon to participate, particularly if aspects of your research program are involved in the ongoing discussions.

- ✔ If your university or institution is in litigation with a partner you have been collaborating with, **do not let disputes interfere with your research or your relationships with colleagues** at the other institution. Many companies litigate with other parties while, at the same time, negotiating on other licenses or joint ventures with that party. **Litigation is nothing personal** and should never influence your research collaboration. Notwithstanding this, you should always be cautious when speaking about matters related to the topic of dispute. It is best never to comment on ongoing litigation matters.

Given that IP management is heavily context specific, these Key Implications and Best Practices are intended as starting points to be adapted to specific needs and circumstances.

KEY IMPLICATIONS AND BEST PRACTICES: SECTION 15

FOR TECHNOLOGY TRANSFER OFFICERS

- ✔ A fundamental best practice in IP management, regardless of whether an institution is public or private and whether located in a developed or developing country, is to view **intellectual property as an evolving and dynamic asset** requiring ongoing attention, management, monitoring, and policing. Only such an "IP cultivation" will allow institutions to protect the value of intellectual property and maximize its utility.

- ✔ A technology transfer office must have systematic procedures to **administer, monitor, and enforce** its technology licenses. This includes compliance with royalty payments and reporting obligations in a nonconfrontational manner.

- ✔ A TTO should **regularly review** active license agreements. This specifically includes ensuring that milestones are met and royalties paid.

- ✔ Potential **patent infringements** should be monitored continuously through sound surveillance protocols, and action taken to remedy infringement is an essential part of IP asset management. The lack of patent enforcement can lead to a loss of patent rights.

- ✔ If **litigation** seems to become inevitable, credible communication that the IP owner is serious about protecting its IP assets will go a long way to bringing infringers to the table to discuss the issues and negotiate a mutually beneficial outcome. Importantly, early communication with potential infringers and good license and licensee diligence, are the foundations for policing and maintaining intellectual property, irrespective of whether the intellectual property is owned by a public or a private entity.

- ✔ Essential to contract management is a **well-organized electronic filing system**, possibly with archival, working, and computer files with integrated interactive modules organizing data on contacts, inventions, patents and license applications, royalties, receipts, and reporting. A TTO should establish such a system as early as possible and before the number of agreements and licenses becomes large.

- ✔ The online version of the *Handbook* allows users to **download an electronic contracts-management system** free of charge.

- ✔ Most **IP disputes should not end up in litigation**, as there are many options and strategies for resolving disputes. Good contracts and good licensing practices anticipate that disputes arise with partnerships and licenses.

- ✔ **Mediation and arbitration** can be effective dispute-settlement procedures, provided they have been agreed upon and established in contract clauses at a time when a license or partnership is being negotiated—and before any problems arise.

- ✔ The success of an arbitration or mediation depends largely on the **"quality" of the arbitrators and mediators**. The challenge is often to find candidates who have arbitration/mediation skills, have experience with the specialized knowledge of the disputed subject matter, and are acceptable to both parties.

Given that IP management is heavily context specific, these Key Implications and Best Practices are intended as starting points to be adapted to specific needs and circumstances.

Bioprospecting, Traditional Knowledge, and Benefit Sharing

Humankind has always been—and will always be—completely dependent on the Earth, therefore our treatment of it is paramount to our survival. We have relied particularly on its wealth of biological resources and its biodiversity. For millennia, a balance has existed between the production and consumption of resources. The impact of people on the environment has made relatively few irreversible changes over this time. That is, until recently. Suddenly, the impact of these environmental changes on human activities (such as agriculture, increasing populations, industrialization, and rising rates of consumption and standards of living) has became clear. The root of biodiversity loss and environmental degradation is the notion that biodiversity is the "common heritage of mankind" (*sic*) and must be preserved for future generations. This means that, while the environment belongs to no one, it is entirely our collective responsibility.

Beginning some 50 years ago, biodiversity losses began to increase at an alarming pace. Desertification became a recognized problem in many regions of the world with ensuing biodiversity loss. By the late 1970s, biodiversity loss, desertification, and even climate change, had begun to receive significant international attention as more and more people began to recognize that the Earth's resources were finite and that our activities were unsustainable. Due to a accelerating depletion of resources, these resources began to have increasing economic value. Deep concern over an environmental crisis was widely expressed for the first time in an international forum at the United Nations Conference on Environment and Development, commonly known as the Earth Summit, held in Rio de Janeiro, Brazil, in 1992. Concurrent to these activities, biotechnology emerged and with it came the promise of creating life-saving new drugs from genetic resources. Modern biotechnologies allowed new and novel uses of biological resources, giving additional value to biodiversity. At the confluence of these world events new concerns emerged over ownership, over the contributions of generations past, and over traditional knowledge (TK) held by indigenous populations. In short, equity concerns arose.

Equity is a moral issue that has repercussions with respect to the distribution of benefits and environmental conservation. However, equity is in the eye of the beholder; different individuals come to different conclusions about what is equitable and about how to achieve equity. Unfortunately, market systems created to place a *price* on equity do not work because market systems are constrained in what they measure. Furthermore, with regard to indigenous knowledge, because its products are intangible, once the knowledge or information is disseminated, control over the knowledge is lost. From an objective standpoint, that knowledge has no

Krattiger A, RT Mahoney, L Nelsen, JA Thomson, AB Bennett, K Satyanarayana, GD Graff, C Fernandez and SP Kowalski. 2007. 16: Bioprospecting, Traditional Knowledge, and Benefit Sharing. In *Executive Guide to Intellectual Property Management in Health and Agricultural Innovation: A Handbook of Best Practices* (Krattiger A, RT Mahoney, L Nelsen et al.). MIHR (Oxford, UK), PIPRA (Davis, USA), Oswaldo Cruz Foundation (Fiocruz, Rio de Janeiro, Brazil), and *bio*Developments-International Institute (Ithaca, USA). Available online at www.ipHandbook.org.

direct monetary value unless the knowledge can be *translated* into a market-based commodity (or service), whereby the value of different contributions (knowledge, technology, labor, capital, and so forth) can be quantified and traded.

In addition to these problems, the western system of IP **(intellectual property) rights, particularly patenting, is based on the premise that anything that is already known cannot be protected**. Indigenous knowledge is often communal, has been disclosed, and has been passed on from previous generations. The very nature of indigenous knowledge, therefore, does not meet the criteria for intellectual property in today's IP system. Not surprisingly, some people view the use of TK in modern science as a form of biopiracy, which is the unfair acquisition of biological resources and/or associated know-how. Some even argue that the modern IP rights system has harmful effects on indigenous peoples.

Karjala[1] breaks down these arguments into two distinct issues:

- **Biopiracy**: to what extent do patent systems exploit traditional indigenous knowledge?
- **Patenting of living organisms**: how can we justify patenting gene-sequence and gene-product information taken from living organisms (especially humans) when these are naturally occurring substances? And if patented, how do we answer the ethical questions surrounding such patents?

Karjala argues that the core of the biopiracy problem is not patenting inventions derived from traditional indigenous information, but rather **biopiracy is unfair acquisition (misappropriation) of knowledge and the inequitable distribution of benefits derived from developing such information into valuable commercial products.** But he cautions against exclusive information rights outside the patent and copyright regimes for indigenous peoples, pointing to the need for incentives for product development. Provided that traditional information is given voluntarily and that fair compensation is paid to the group who owns the information, it is not the use of TK in a patent that is inherently wrong. Therefore, the question becomes one of how to provide for equitable benefit sharing of TK that finds its way into patent applications and is subsequently commercialized.

Policymakers ought to formulate methods for equitable access to the TK held by indigenous societies and for compensating its owners. However, this issue involves a delicate balance: access should be granted only via authorized permission, yet the price that is assessed for permission to bioprospect should not be so high that it dissuades companies and individuals from seeking access.

Although nothing in extant patent statutes or international IP/trade agreements requires that naturally occurring chemicals (such as DNA sequences and genes and their natural products) be treated as patentable subject matter per se, these can be patented once utility and novelty requirements are met. Patenting natural products, however, is not the unique concern of indigenous peoples. It is also a concern of policymakers in developing and developed countries.

Furthermore, patents on upstream "inventions," (in this context, isolated genes) might inhibit subsequent downstream research and development. This is because the upstream patenting of natural products (such as specifically isolated gene sequences) would effectively eliminate downstream incentives for inventive activity. Also, such patents would inhibit information flow, thus promoting over investment in the search for genes and under investment in the utilization of genes for advanced applications.

The important ethical issues raised by gene-related patents include whether:

- private control over genes or their products monopolizes the "common heritage of mankind"
- patents denigrate human life by reducing it to a commodity
- patents may be inconsistent with individual or collective privacy
- patents promote or inhibit distributive justice when they are concentrated in a few economically developed countries. A related concern is that patents on crop varieties might threaten biodiversity.

Importantly, Karjala notes that **these issues affect both indigenous and non-indigenous**

populations. In addition, where there are differences in how costs or benefits are distributed, patenting is not necessarily the problem. In negotiating technology transfer and access, the author proposes that careful cost/benefit analyses should guide decisions.

Interpreting the concept of the common heritage of mankind broadly, one can include nearly everything (in other words, the common heritage is not limited to indigenous peoples). Therefore the concept does not represent an ideal paradigm for building a legal strategy. Hence, traditional patent law is a better approach. The real question is not whether a gene or a gene product should be protected as the common heritage of humankind, but whether or not it is even an invention within the well-established strictures of patent rules and regulations.

As for the commoditization of genes, it is difficult to see how this would impact most indigenous societies that, for the most part, are far removed from the commodity markets of developed countries. Furthermore, the human genes at issue would most likely confer some sort of positive advantage and would therefore not implicate either privacy concerns or stigmatization. Once again, patent law would likely most effectively address genes with potential commercial value. Nevertheless, freely available information should not be protected by IP rights. If IP protection is appropriate, possibly other forms of statutory protection would be more suitable, such as breach of confidence or privacy rights.

One critical concern is whether patenting conflicts with indigenous knowledge and value systems. In a theoretical sense, patents can significantly add costs to new inventions and thereby act as barriers. However, when one balances the costs and benefits of patent law in developing countries, there may be little correlation between access and patent status. Furthermore, as Karjala points out, **there would be essentially no financial loss to owners of patented biotechnology products if they were to sell at cost in such countries**, assuming no redirection of such biotechnology to more lucrative markets. However, the prevention of product "leakage" would entail enforcement capacity, and this sort of distribution is not feasible without strict market segmentation.

Costanza, Christofersen, Anderson, and Short[2] add to this analysis of bioprospecting by presenting **practical examples of how indigenous peoples and companies can reach agreements** that are fair by most standards and conducive to further collaboration. The authors explain that international agreements such as the Convention on Biological Diversity (CBD) and the International Treaty on Plant Genetic Resources (ITPGR) provide a broad framework for protecting and utilizing genetic resources.

For bioprospecting activities, companies choose countries that have unique and protected ecosystems, a solid legal framework, sufficient political will, fair and equal treatment for all access seekers, and strong science experts or institutions to partner with. Countries will seek partnerships with foreign companies and universities that adhere to international conventions and best practices, and that have an established track record. Guiding principles for a successful partnership between collaborators in the host country and a company include a commitment between parties to maintain a fair, trusting, long-term relationship, with an efficient and reasonable authorization process, and equitable sharing of benefits between partners.

However, international agreements do not provide detailed guidance on structuring the relationships between parties involved in commercial bioprospecting activities. Companies involved in the exploration, screening, and use of genetic resources have begun to accumulate experience with building such relationships, including selecting countries with rich biodiversity, selecting partners, and drafting terms in biodiversity access agreements (BAAs) that govern these relationships.

In order to be successful, these BAAs must have a clear definition and assignment of legal rights to all genetic resources involved. Informed consent from all domestic parties affected by the bioprospecting, including landowners and managers, must be attained prior to partnership. There must exist a clear delineation of rights to patent and commercialization of the products derived from these endeavors. Each BAA is a confidential document, which supports a lack of competition among the partners to the agreement, and does not allow the transfer of proprietary technologies or technical capacity to third parties or exclusivity.

Identification of the parties to the BAA can be complicated because there may be multiple agencies within a country that have authority over access to genetic resources. There may also be multiple parties, such as landowners or company managers who could legally prevent access to or receive compensation for the resource if and when they are affected by the biodiversity prospecting. Each country that is a signatory to the CBD has a responsibility to establish a *national focal point for access and benefit sharing*,[3] a designated individual and national office that is able to identify all necessary authorities and potential claimants for the partnership.

The rights that need to be spelled out in a BAA include rights to retain or distribute samples, rights to intellectual property under different scenarios, (such as conditions of discovery and invention) and rights to publish discoveries and inventions. Responsibilities, such as the handling of reporting, communications, and administrative filings also need to be spelled out.

The parties should come to an understanding about the relative importance or value of each of their contributions (such as carrying out sampling, cleaning, or analyzing). This will directly affect the equitable sharing of any benefits arising from collaborative activities. Given the nature of bioprospecting and the regions where bioprospecting is often conducted, the full scope of returns is understood to include both financial and nonfinancial components (that is, various sources of potential value to the individual parties). The possible returns can also be divided roughly into short-term, medium-term, and long-term time frames. Thus, a BAA has enormous flexibility for structuring the terms of compensation to the parties. While advanced payments, sample fees, running royalties, and milestone payments—terms typical of many technology agreements—are available for financial benefit sharing, there are many more possibilities, including the provision of equipment and infrastructure, sharing of IP rights or rights to product sales, funding of related research, and assistance with conservation services.

Despite progress on the technical side, a BAA almost always creates controversy. The natural response of governing authorities is to move slowly, fearing criticism from competing domestic interests and international groups that watch out for cases of undervaluing biodiversity and non-support for economic development. Many such groups consider the private sector to be inherently corrupt; thus, no matter what benefits are offered the arrangement is perceived to be inequitable. Ironically, this reaction reflects negatively on those companies taking the lead in supporting the CBD and creates strong disincentives to engage in bioprospecting or to share information about such endeavors. This in turn, decreases the very value of biodiversity resources. In the end, the **commitment of both parties to a sustainable and rational use of biodiversity in a way that both encourages commercial development and protects the unique resources of the Earth** is as important as the technical aspects of deal making.

The technical aspects of technologies, however, must still be mastered. Indeed, there is an emerging new regime, Thornström[4] calls it a "world order," regarding biological matter: an international regime which govern access to genetic resources and the sharing of benefits arising from their use. The chapters by Thornström and by Thornström and Björk,[5] explore the *what*, *why*, and *how* of this new regime. The authors provide the reader with a comprehensive road map for understanding the details and finding the correct path to compliance with the laws, rules, and regulations that cover access in a given country.

The new regime is driven by **access and benefit-sharing (ABS) systems**, which apply to research carried out for either scientific or commercial purposes. ABS involves accessing organisms, or parts thereof, and related TK, that are obtained (accessed) from a country that is party to the CBD. In addition, other international treaties, accords, and agreements have also added new legal ABS regimes legislation through the acquisition and use of biological material and related information.

Everyone (tourists, nature conservationists, scientists, photographers, journalists) is subject to these new ABS regulations, but the ABS system especially affects scientists and researchers who seek to access and use proprietary genetic resources, other biological matter, and related information, such as TK and farming know-how. In national legislation, such knowledge may be treated as intellectual property or confidential trade secrets, putting it outside the public domain and not subject to *any* form of unauthorized appropriation. Violation by foreign parties (such as scientists conducting unauthorized collection activities) of the new ABS regimes may result in a range of negative and stringent consequences; fines and/or imprisonment, denial of future visits to the collection site or country, increased transaction time for obtaining formal access permits, and/or denial of access to colleagues of the violator. Obviously, it's important to know how to properly proceed.

To understand the fundamental principles of ABS, one needs to know the relevant rules, regulations, laws, customs, and conditions for benefit sharing in the country where one intends to conduct research and/or collect samples. Basic questions to ask before collecting include:

- Under which conditions may I, as a scientist, *enter* another sovereign state's territory in my scientific capacity?
- Under which conditions may I, as a scientist, *collect* biological material and related information?
- Under which conditions may I, as a scientist, *carry out or export* biological material and related information from that sovereign state's territory?
- Under which conditions may I, as a scientist, make *further use* of collected biological material and related information?

Thornström and Björk present a practical overview of the principles and procedures underlying ABS regimes that will be useful to various types of research and access situations. The authors also provide a series of template documents as illustrative examples of what might be necessary, depending on the specific requirements of the collection activities. To assist in understanding the various ABS scenarios and the documents, potentially applicable letters and agreements are presented as examples, such as letter of intent, research permit, prior informed consent (PIC), mutually agreed terms (MAT), model or material transfer agreement (MTA), and confidentiality agreement.

Although all of this might seem daunting initially, **the documents are necessary, and in a growing number of countries are required by law.** Careful planning and management will pay off in the long term, since they minimize the possibility of misunderstandings and other problems and, in turn, can reduce the chance that legal problems will arise. Perhaps most importantly, these ABS regimes are in place to facilitate the building of solid, equitable, and sustainable networks for future partnerships.

Drawing on the experiences of exemplary partnerships, Soejarto and colleagues[6] explain an organizational model for the responsible governance of bioprospecting arrangements between institutions in developed and developing countries based upon the **International Cooperative Biodiversity Groups** (ICBG) program of the U.S. National Institutes of Health (NIH). The model assumes that resources and expertise from both the North and the South are required for bioprospecting to succeed. Incentives need to be properly aligned for both regions to be fully engaged and committed. To align incentives, the ICBG model offers a clear definition of the benefits that might arise from a project, a clear recognition of all parties involved, negotiation guidelines for the parties, and a formal structure for the resulting agreement. The agreement contains the scope and objectives of the project, the long-term benefit-sharing scheme, and milestones, as well as terms for IP ownership, informed consent, and royalty distribution. Details of how the ICBG model works in practice are illustrated with an example of one such bioprospecting arrangement between the University of Illinois at Chicago (UIC), research institutes in Vietnam and Laos, and GlaxoSmithKline.

Informed consent was another critical issued to be covered. In this case, informed consent offered provisions for the collection and use of plant/genetic materials and for individuals and their communities regarding traditional medicinal use or uses of a plant. In addition, prior informed consent was to be secured before the implementation of the work. The governments of Vietnam and Laos were acknowledged as the owners of the genetic materials and their derivatives in their respective countries.

Fundamentally, the ICBG model recognizes and emphasizes the importance of several parties and the outcomes they seek. Often overlooked in typical international research consortia and business agreements, these additional parties include poor communities and the regional authorities in locations where biodiversity prospecting is to be conducted. The additional objectives include biodiversity conservation, institutional capacity building, and regional economic development. The standards established by the ICBG program emphasize the core principles of capacity building and community reciprocity. Bioprospecting activities such as those outlined in this chapter, in which poor communities in developing countries are cooperating with clear understanding and goodwill, can thus serve as a model for future similar agreements and initiatives.

This is not to say that the conceptual systems of developed countries work are transferable to developing countries, as the final two chapters of this section demonstrate. According to Hansen and Van Fleet,[7] indigenous knowledge, or TK, particularly that which involves a region's native flora and fauna (biodiversity), is not fully amenable to the legal constructs of intellectual property. Fundamentally, TK is cumulative, communal, and largely undocumented in the formal literature. Because of these characteristics, **TK often does not fulfill novelty requirements for establishing IP rights** or the condition that ownership of the intellectual property resides with an individual or individuals. Indeed, in the case of TK, it may be exceedingly difficult to identify the original individual inventors or authors, or even the current holders or curators of the knowledge. Finally, because TK is largely unrecorded but exists as "living" knowledge passed from individual to individual orally or through observation and apprenticeship, it is largely unavailable for consideration by IP offices of novelty within the complete repository of human knowledge.

But despite these difficulties in applying the criteria for intellectual property to TK, a number of forms of IP rights protection (primarily trade secrecy, geographical indications, plant variety protections, and patents) can be and have been used to establish ownership over elements of TK. However, the imperfect fit of TK into the definition of intellectual property has led to two interrelated dilemmas:

- In some cases, those who were not part of the indigenous community from which the TK originated may be able to use, and even to establish ownership over, elements of the TK without acknowledgment of (or recourse to) that indigenous community
- Those in indigenous communities who do hold TK may not be able to establish ownership, or even gain acknowledgment from others.

To address the first dilemma, anyone should make sure TK is disclosed, which will establish it as prior art. There are a variety of strategies to assist in establishing prior art status of TK. For the majority of TK, a defensive disclosure in the public domain (such as via a public registry) can prevent illegitimate IP claims over existing TK. For TK to which IP protections more easily apply, the TK holders may be able to themselves file applications. In addition, governments should require prior informed consent to be obtained from indigenous communities or national authorities when engaging in activities that could lead to the claiming of IP rights based on TK.

To address the second dilemma—that of maintaining control over TK—indigenous holders of TK can seek to use forms of IP protection. Hansen and Van Fleet discuss the advantages and disadvantages of the various options available. At least initially, most TK approximates a trade secret, and so it might easily be maintained within the original community as a trade secret. However, before the knowledge is more widely disseminated it may be necessary to use other forms of IP protection, including geographic indication, trademarks, plant variety protection, petty patents or utility models, or patents.

In the longer term, governments may create new forms of IP protection that accommodate the fundamental characteristics of TK (such as under the aegis of sui generis systems of plant variety protection as defined under the Trade-Related Aspects of Intellectual Property Agreement). In addressing the dilemma of control over TK, several issues outlined in the CBD ought to be worked out within national legal systems. **Of these issues, the foremost are conditions for granting/gaining access to genetic resources and any TK about them and requirements for equitable sharing of revenues or other benefits that might accrue from the development and use of TK-based technology in markets around the world.**

All of these approaches to preserving and protecting TK require a clear identification and attribution of specific TK claims. This can be a complex endeavor, but TK is important and often even essential to the survival of indigenous communities. It may also be an important source of life-giving technological innovation that could benefit millions around the world. The ultimate goal is to develop practical solutions within our legal frameworks that encourage indigenous communities both to sustain their traditions and to equitably share their knowledge with the wider world so that all may benefit.

Ammann[8] raises a different concern about how we think about food in developing countries and its impact on the developing world. He argues that **the commonly held distinction between organic and technologically intensive agriculture (focused on genetically modified organisms, or GMOs, or more specifically transgenic crops) has inhibited pragmatic approaches to creating agricultural management systems that build on local conditions, help alleviate poverty, respect local cultures and traditions, and draw upon a successful relationship with science.** This distinction between organic and technologically intensive agriculture is based on a deeper rift between systems of indigenous TK and western scientific knowledge, a rift that Amman contends is not only unproductive (hindering communication and exchange between the two) but artificial—reflecting differences in "worldviews, unfounded theories, or quasi-religious beliefs" held by respective proponents.

Still, the distinction between organic and technologically intensive agriculture is enormously significant. The designation of a technology as organic versus transgenic can attach very different regulatory requirements and offer different marketing opportunities for the technology, thus strongly influencing how and whether it is used and what its potential value is.

Ammann challenges the commonly held distinction between organic and transgenic technologies and proposes a series of tests of the definitions and principles advanced to define and distinguish the two. While they are different in some aspects, Amman finds none of the major distinguishing principles claimed by organic versus transgenic technology able to stand up to scrutiny. These include:

- the intrinsic genetic integrity of crop species genomes (crop species genes are not intrinsically more stable when considered transgenic or organic)
- the unnaturalness of transgenesis (transgenics are just as "natural" as organics)
- stability and predictability of progeny (organics and transgenics have stable and predictable inheritance patterns that are reproducible over time)
- unnaturalness of monocultures (irrespective of organic or transgenic status, growing all one type of either crop plant is not the natural state of the environment)
- erosion of biodiversity by transgenic technologies (transgenics have not been shown to decrease levels of biodiversity)
- systemic environmental superiority of organic versus transgenic crops (the overall conception that organics are superior to transgenics as a whole is not substantiated)

It is difficult, if not impossible, to consistently maintain a clear divide with respect to organic and biotechnology-based agricultural technology and methods. Yet, Ammann observes, "**power structures knowledge,**" **and interests on both sides are using and benefiting from a substantiation of**

the distinction between organic and transgenic agriculture.

Practical solutions to agricultural production—and practical solutions to medicine—could indeed benefit many if only we could manage to build bridges between TK and science-based knowledge systems and draw upon the best existing ideas and practices of both. ■

All chapters refer to: *Intellectual Property Management in Health and Agricultural Innovation: A Handbook of Best Practices.* 2007. A Krattiger, RT Mahoney, L Nelsen, JA Thomson, AB Bennett, K Satyanarayana, GD Graff, C Fernandez, and SP Kowalski (eds.). MIHR: Oxford, U.K., and PIPRA: Davis, U.S.A. Available online at www.ipHandbook.org. The online version contains for each chapter a detailed Editor's Summary, Implications, and Best Practices.

1 Chapter 16.1 by DS Karjala titled Biotechnology Patents and Indigenous Peoples, p. 1437.

2 Chapter 16.4 by C Costanza, L Christofersen, C Anderson, and JM Short titled Deal Making in Bioprospecting, p. 1495.

3 www.cbd.int/world/map.asp.

4 Chapter 16.2 by CG Thornström titled Access and Benefit Sharing: Understanding the Rules for Collection and Use of Biological Materials, p. 1461.

5 Chapter 16.3 by CG Thornström and L Björk titled Access and Benefit Sharing: Illustrated Procedures for the Collection and Importation of Biological Materials, p. 1469.

6 Chapter 16.5 by DD Soejarto, C Gyllenhaal, JA Tarzian Sorensen, HHS Fong, LT Xuan, LT Binh, NT Hiep, NV Hung, BM Vu, TQ Bich, BH Southavong, K Sydara, JM Pezzuto, and MC Riley titled Bioprospecting Arrangements: Cooperation between the North and the South, p. 1511.

7 Chapter 16.6 by SA Hansen and JW Van Fleet titled Issues and Options for Traditional Knowledge Holders in Protecting Their Intellectual Property, p. 1523.

8 Chapter 16.7 by K Ammann titled Reconciling Traditional Knowledge with Modern Agriculture: A Guide for Building Bridges, p. 1539.

FOR GOVERNMENT POLICYMAKERS

- ✔ Equity is a moral issue that has repercussions with respect to the distribution of benefits and environmental conservation. Thus, **equity is in the eye of the beholder.**
- ✔ The western system of IP rights, and particularly of patenting, is based on the premise that anything that is already known cannot be protected. Indigenous or traditional knowledge (TK) is often communal, has been disclosed, and has been passed on from previous generations. **The very nature of indigenous knowledge, therefore, does not meet some of the criteria for intellectual property protection** (such as novelty).
- ✔ In the longer term, **new forms of IP protection that are more amenable to the fundamental characteristics of TK** could be created by governments, such as under the aegis of sui generis systems of plant variety protection (PVP), as defined under the TRIPS Agreement.
- ✔ **Indigenous communities** often play a significant role as gatekeepers to a country's potential biodiversity wealth. They are the regional specialists with respect to the flora and fauna. Their knowledge can often exceed that of leading scientists.
- ✔ Patent laws per se do not "create" biopiracy. Rather, **biopiracy is a form of misappropriation**, unfair acquisition, and inequitable sharing of benefits with respect to biological resources.
- ✔ Policymakers ought to formulate methods for **equitable access to TK** held by indigenous societies and for compensating the TK's owners. However, this issue involves a delicate balance: access should be granted only via authorized permission, yet the price that is assessed for permission to bioprospect should not be so high that it dissuades companies and individuals from seeking access.
- ✔ Countries should consider implementing an **access and benefit sharing** (ABS) regime that balances equitable access to biological resources, as well as related TK, with opportunities arising from R&D expertise of potential foreign partners in development. Such policies should be grounded in, and consistent with, the Convention on Biological Diversity and the TRIPS Agreement.
- ✔ ABS regimes, including **the process for obtaining permits, should be transparent** and easily available to any scientist or institution that wishes to enter into biodiversity prospecting or collection activities. A complex system discourages foreign bioprospectors and may inhibit national researchers in their activities.
- ✔ The commonly held **distinction between organic and biotechnology-based agriculture** inhibits pragmatic approaches to creating agricultural management systems that build on local conditions, help alleviate poverty, respect local cultures and traditions, and benefit from a successful relationship with science. The world has much to gain by **reconciling organic and biotechnology-based agriculture** though realizing any gain will have to deal with the "**power structures of knowledge**," and overcome limitations imposed by those people who maintain the distinctions.

Given that IP management is heavily context specific, these Key Implications and Best Practices are intended as starting points to be adapted to specific needs and circumstances.

FOR SENIOR MANAGEMENT

(UNIVERSITY PRESIDENT, R&D MANAGER, ETC.)

- ✔ The technology transfer office should work with senior management to establish **policies and systems for accessing indigenous or traditional knowledge** (TK), bioprospecting activities, and benefit sharing in an equitable manner.

- ✔ Equity is a moral issue that has repercussions with respect to the distribution of benefits and environmental conservation. Thus, **equity is in the eye of the beholder.**

- ✔ Given the complexity of the health and agricultural industry and the enormous variety of applications and products that could be developed through the **biodiversity access agreement** (BAA), it is very difficult to know the profit margins for a company, product, or application ahead of time. Technology transfer, as well as information and data sharing, in the long run, may be more important than royalties.

- ✔ With adequate funds often lacking in public sector research centers, international donors should seriously consider loans or grants for training and equipment purchases. **Entering into bioprospecting activities, the public sector has much to gain by:**
  - having a clear institutional policy
  - building national scientific capabilities, and along with it, the possibility of adding value to biodiversity elements, which increase the negotiating strengths and benefit sharing stipulated in contract agreements
  - having internal capacity for negotiations, which includes adequate legal and counseling skills about the main aspects of commercial and environmental law

- ✔ Managers can identify which **nonmonetary benefits** companies could provide (such as capacity building,and technology transfer), that would be of greatest use to the institution. This approach will enable flexibility in benefit sharing and sustainability in the R&D relationships.

- ✔ Public sector institutions can provide important **intellectual and programmatic leadership in how cross-cutting agricultural research programs** can build bridges between TK and science and between organic agricultural and science-based agricultural practices. In so doing, they will help to advance the state of knowledge, the regulatory structure, and public perceptions of agricultural systems.

- ✔ The commonly held **distinction between organic and biotechnology-based agriculture** inhibits pragmatic approaches to creating agricultural management systems that build on local conditions, help alleviate poverty, respect local cultures and traditions, and benefit from a successful relationship with science. The world has much to gain by **reconciling organic and biotechnology-based agriculture** though realizing any gain will have to deal with the "*power structures of knowledge*," and overcome limitations imposed by those people who maintain the distinctions.

Given that IP management is heavily context specific, these Key Implications and Best Practices are intended as starting points to be adapted to specific needs and circumstances.

FOR SCIENTISTS

- ✔ **Scientists and anyone else accessing biodiversity must ask, and answer, the following questions prior to initiating collecting activities:** Under which conditions may I *enter* another sovereign state's territory in my scientific capacity? Under which conditions may I *collect* biological material and related information? Under which conditions may I *carry out or export* biological material and related information from that sovereign state's territory? Under which conditions may I make *further use* of collected biological material and related information?

- ✔ Scientists must be aware, not only of the biological and sociological value of indigenous or traditional knowledge and related genetic resources, but also of their **potential commercial value**. Hence, investigations and research ought to be conducted within guidelines set by the technology transfer office, for example, appropriate and timely disclosure of any potential inventions.

- ✔ **Interactions with foreign colleagues and collaborators** ought to be established according to institute or university policy guidelines, guidelines that are established to both preserve and reap the full value of these national natural resources.

- ✔ When working with colleagues from foreign countries, you should be aware that those colleagues may be authorized to make **collections of biological materials** only under specified circumstances. Before proceeding with joint activities, check with your institution's technology transfer office to make sure that all the requirements have been met.

- ✔ It is essential to understand the fundamental principles of the Convention on Biological Diversity (CBD) and Access Benefit Sharing (ABS) regimes. These exist to both **protect the resources of your country as well as to encourage collaborative projects in R&D that would foster a broad and equitable distribution of benefits** flowing from the development of the country's biological resources.

- ✔ The commonly held **distinction between organic and biotechnology-based agriculture** inhibits pragmatic approaches to creating agricultural management systems that build on local conditions, help alleviate poverty, respect local cultures and traditions, and benefit from a successful relationship with science. The world has much to gain by **reconciling organic and biotechnology-based agriculture** though realizing any gain will have to deal with the "**power structures of knowledge**," and overcome limitations imposed by those people who maintain the distinctions.

Given that IP management is heavily context specific, these Key Implications and Best Practices are intended as starting points to be adapted to specific needs and circumstances.

FOR TECHNOLOGY TRANSFER OFFICERS

- ✔ The three guiding principles for a successful relationship in bioprospecting and related endeavors are a **commitment to maintaining a fair, trusting, long-term relationship**; efficient and reasonable authorization; and the equitable sharing of benefits between a company and its collaborators in the host country.

- ✔ The western system of IP rights, and particularly of patenting, is based on the premise that anything that is already known cannot be protected. Indigenous knowledge is often communal, has been disclosed, and has been passed on from previous generations. **The very nature of indigenous knowledge, therefore, does not meet some of the criteria for intellectual property protection** (such as novelty).

- ✔ A **successful biodiversity access agreement** includes a clear definition and assignment of legal rights to all genetic resources involved; prior informed consent from all domestic parties affected by the bioprospecting (including landowners and managers); a clear statement of rights to patent and commercialize products derived from discoveries made; and terms of confidentiality. The BAA also establishes a noncompetitive relationship between the parties; trust that no transfer of proprietary technologies or technical capacity involved under the agreement will occur with respect to third parties; and that no exclusivity requirements exist.

- ✔ Patent laws per se do not "create" biopiracy. Rather, **biopiracy is a form of misappropriation**, unfair acquisition, and inequitable sharing of benefits with respect to biological resources.

- ✔ **Prior informed consent** is an important principle in bioprospecting. This should include informed consent in the case of collection and use of plant/genetic materials, as well as informed consent of individuals and their communities regarding traditional medicinal use or uses of a plant.

- ✔ When dealing with **foreign bioprospectors**, your office will function as the gateway and regulator of their activities. As such, technology transfer officers will provide oversight to negotiating agreements for equitable sharing of rewards, defining access, discussing possible patentability, and protecting the rights of the indigenous peoples who are the stewards of these resources.

- ✔ Negotiating access to your country's genetic resources, biodiversity, and TK will require a balanced, nuanced approach. **Equitable benefit sharing** must simultaneously ensure fair returns to your country, yet not inhibit the R&D initiatives of foreign partners. Solid agreements will benefit all parties: your country, your partnering organization, and the country or community that provides the resource. Extreme situations, such as an expectation of immediate windfall returns or wanton biopiracy by outsiders, will freeze the resources and ultimately lead to their demise.

- ✔ Both monetary and **nonmonetary benefits** may be attractive to the university or institute; both, therefore, should be considered. Nonmonetary benefits could include training opportunities for scientists and donation of equipment.

Given that IP management is heavily context specific, these Key Implications and Best Practices are intended as starting points to be adapted to specific needs and circumstances.

Putting Intellectual Property to Work: Experiences from around the World

"*By far the best proof is experience,*" wrote Francis Bacon. Given the experience of countries—both developing and developed—that have used intellectual property, IP (intellectual property) protection, and IP management to stimulate innovation, there is ample proof that good IP management has benefited multitudes of people around the world with new technologies, products, and services. Innovations in health and agriculture have greatly enriched lives. But does this experience apply to all countries? If the best proof is experience, then what can be said authoritatively about the effects of using IP systems wisely in developing countries?

The 28 case studies in this section of the *Handbook* (and the 21 case studies in the insert of this *Executive Guide* and more online) demonstrate that a great deal can be said. Developing countries already have a vast amount of experience with IP protection, and this experience proves that they can use intellectual property to their advantage. With more chapters than any other section, this portion of the *Handbook* amply reveals how developed and developing countries alike are deploying and adapting IP management to meet their needs. Tapping into the dynamism of product development partnerships (PDPs) and utilizing the potential of their universities, public sector institutions, and private companies, many **developing countries are quickly and creatively building on the experience of their own institutions, of neighboring countries, and of countries around the globe.**

EXPERIENCES FROM AROUND THE WORLD

Satyanarayana describes India's experience in the pharmaceutical sector. According to Satyanarayana,[1] during the past 50 years, **India** has made great strides in science through a series of policy initiatives promoting high-quality research. But especially since 2005, when India became fully compliant with the agreement on Trade-Related Aspects of Intellectual Property Rights (TRIPS), big changes have occurred. India's rigorous IP rights regime and professional IP management in both private sector companies and public sector research institutions are driving success. But this is only part of a larger coordinated attempt that includes increased public and private R&D expenditures, new policies governing traditional medicines, overhauled regulatory regimes for new drugs and biotechnologies, initiatives to emphasize and build on already competitive regions or technologies, and newly created governmental, research, and educational institutions.

In the pharmaceutical sector, the effects of these policies can be seen in:

- a shift in the Indian pharmaceutical industry from an approach based solely on the low-cost manufacture of generic drugs to

Krattiger A, RT Mahoney, L Nelsen, JA Thomson, AB Bennett, K Satyanarayana, GD Graff, C Fernandez and SP Kowalski. 2007. 17: Putting Intellectual Property to Work: Experiences from around the World. In *Executive Guide to Intellectual Property Management in Health and Agricultural Innovation: A Handbook of Best Practices* (Krattiger A, RT Mahoney, L Nelsen et al.). MIHR (Oxford, UK), PIPRA (Davis, USA), Oswaldo Cruz Foundation (Fiocruz, Rio de Janeiro, Brazil), and *bio*Developments-International Institute (Ithaca, USA). Available online at www.ipHandbook.org.

research-driven innovation of novel drugs for the global market
- the emergence of an entrepreneurial biotechnology sector in India
- the consideration by multinational pharmaceutical companies of investing in R&D and manufacturing operations in India

In agriculture, these effects are apparent in a rich pipeline of new innovations that promise to make India's agricultural sector more competitive and profitable. Besides a substantial allocation of funds for R&D by the government, two new initiatives were started in 2005: the National Agricultural Innovation Project (NAIP) and the Indo-U.S. Agricultural Knowledge Initiative (AKI).

India's transition from a protected economy to an open, global economic power has prompted the government to take a series of steps to address the new challenges of globalization, and the lessons it has learned apply broadly to many developing countries. Strengthening R&D, establishing policies to create and manage intellectual property, and fostering PDPs are all important steps for making important health products available for public distribution available in all countries.

According to Wolson,[2] technology transfer offices (TTOs) are a crucial part of IP management in **South Africa.** But a number of problems challenge nascent TTOs there: a weak flow of invention disclosures, skepticism or a lack of awareness among faculty about the TTO's role, low levels of research funding, high patenting costs, few experienced technology transfer practitioners, and unrealistic expectations about financial returns. Indeed, many there believe that the main motivation for undertaking technology transfer activities at a university is to generate income.

Solutions to these problems are being addressed organizationally by the Southern African Research & Innovation Management Association (SARIMA), legislatively by the Framework for Intellectual Property Rights from Publicly Financed Research (the Framework), and financially through the Innovation Fund. Established in 2002, SARIMA is a stakeholder organization providing a platform for those from government, academia, and industry with an interest in using research and innovation management to foster networking and promote common interests. The Framework is intended to bridge the "innovation chasm": the gap in South Africa between knowledge generators (in particular, universities and research institutions) and the market. It calls for a consistent approach to protecting intellectual property developed with public financing and draws heavily on the U.S. Bayh-Dole Act. Of course, as other countries have discovered, the Bayh-Dole Act cannot simply be imported. Its principles must be adapted to local frameworks and needs. In South Africa, for example, research funding comes mostly from external sources and requires a different structure for determining the use and ownership of project intellectual property.

TTOs in South Africa have already met with success. Some have been operating for several years and more are being launched. A vibrant stakeholder organization provides a platform for networking and professional development in the field, and links have been forged that strengthen international research collaborations and technology transfer partnerships. All of this has government support.

Other chapters in this section describe the experiences of **Brazil,**[3] **Chile,**[4] **China,**[5] the **European Union,**[6] and **Japan.**[7]

PUBLIC SECTOR INSTITUTIONS AND UNIVERSITIES

Salicrup and Rohrbaugh[8] provide more evidence of the ability of for-profit and nonprofit institutions in developing countries to bring new products to market that meet critical regional public health needs. The authors discuss the technology transfer and licensing approach of the U.S. **National Institutes of Health** (NIH). The institution's technology transfer experience has shown that many combinations of licensing strategies can be used to segment the world market to meet each region's needs. Even when patent protection is unavailable, unique biological materials (for example, an essential component of a vaccine) can be licensed for commercial use.

Institutions in developing countries have been found to be dependable licensees and partners. With careful review, a capable institution with commercialization capabilities may be found, and one should keep an open mind because, depending on the country, it may be a for-profit company, a nonprofit or government entity, or a semi-privatized company. NIH has several examples of different strategies involving various types of institutions that have reached the early stages of the commercialization process.

While discussions continue about IP capacity building in developing countries, some leading institutions are simply forging ahead and building their own capacity. The **State University of Campinas, or Unicamp**, one of the leading research universities in Brazil, is an example described by Ceron Di Giorgio.[9] A large university with a diversity of affiliated research institutes, Unicamp has moved up the patenting league tables in recent years to become the single largest patentor in Brazil. The university's current portfolio includes almost 50 granted, and 400 filed patents. Unicamp emphasizes chemistry, which accounts for close to half of its portfolio, and engineering, which accounts for a third. In addition, Unicamp conducts significant research in the life sciences (for example, a soy-based phytoestrogen for hormonal therapy licensed to a Brazilian pharmaceutical company).

These major advances in technology transfer at Unicamp are largely due to the efforts of its new technology transfer office, *Inova Unicamp*, founded in 2003. *Inova* began its operations by assessing all of the technologies being researched in Unicamp's many laboratories and institutes. It then aggressively pursued new patent applications and licensing deals for the most promising technologies. In the short space of two and a half years, the office signed 128 technology transfer agreements with both private industry and government agencies. It also saw ten start-up companies in the university's business incubator become self-sustaining.

What lies behind these successes in Brazil? New public policy. In particular, the work of *Inova* is directly informed by two pieces of legislation. A 1996 law gave the university ownership rights to employee inventions. A 2004 law on innovation, however, gives the university the option to either hand over title to the employee inventors, or share 5%–33% of any royalties with them. In addition, the government has instituted a number of sector-specific incentives to support innovation in Brazil, including tax deductions on royalty payments, R&D investments, and foreign IP filing fees, as well as subsidies to firms to help pay scientists' salaries.

The 2004 innovation law requires all government universities and R&D institutions to open an IP management or a technology transfer office. One major consequence of these policies will likely be increased patenting and licensing activities at universities throughout Brazil. Currently, Unicamp's rapid establishment of a functioning technology transfer office stands as a sterling example for other institutions in Brazil to emulate.

Other case studies in this section of the experiences and approaches of a range of institutions include: **Arizona State University**[10] in the United States, **Chinese Universities**,[11] the **Donald Danforth Plant Science Center**[12] in the United States, the **National Health Service** in England,[13] **Stanford University**'s Office of Technology Licensing,[14] the **University of California System**,[15] and the **University of California Agricultural Experiment Station**.[16]

PRODUCT DEVELOPMENT PARTNERSHIPS (PDPS)

Banerji and Pecoul[17] describe the **Drugs for Neglected Diseases Initiative (DND*i*)** that seeks to give patients in developing countries the opportunity to directly benefit from new products of drug R&D for diseases that lack a viable market. Only a tiny fraction (1.3%) of the drugs that came to market from 1975 to 2004 targeted tropical diseases (such as human African trypanosomiasis, Chagas' disease, leishmaniasis, helminthic infections, schistosomiasis, onchocerciasis, malaria, and tuberculosis) that all together make up 12% of the global disease burden and kill more than 35,000 people a day. The drugs that do exist are either inaccessible to patients or unbearably costly. DND*i* believes that drug research can exist in the public domain, and that patented products do not always benefit those who need them most.

As clearly articulated in its IP Policy statement, DND*i* is committed to managing intellectual property to pragmatically and effectively advance its mission of providing the most vulnerable populations in developing countries with equitable access to critically needed medicines. As the preamble of DND*i*'s IP policy states:

The DNDi IP approach will be pragmatic, and decisions regarding the possible acquisition of patents, ownership, and licensing terms will be made on a case-by-case basis. DNDi will put the needs of neglected patients first and will negotiate to obtain the best possible conditions for them. The DNDi's decisions regarding IP will contribute to ensuring access and encouraging further innovations.

DND*i* has led two successful campaigns to negotiate terms that allowed them to get important drugs to the world's neediest people at minimal cost. In the first case, DND*i* approached French pharmaceutical giant sanofi-aventis in 2003 to develop artesunate-amodiaquine, a fixed-dose combination therapy for chloroquine-resistant malaria. The negotiation process eventually led to a contract with very favorable terms for DND*i*; the drug was made available for production by generic manufacturers with no payment owed to either sanofi-aventis or DND*i*, and sanofi-aventis agreed to supply the drug at cost to the public sector, NGOs, and international organizations. In the second case, DND*i* successfully collaborated with the University of California, San Francisco's (UCSF) business development office to support research leading to treatments for the lethal human African sleeping sickness. While conventional wisdom holds that a university should always seek the largest possible return on research investment, DND*i* was able to convince university officials of the seriousness of its mission, and a compromise was reached that advances the effort to bring new treatments to persons suffering from this deadly and largely neglected disease.

In pursuing its humanitarian mission, DND*i* has learned that it is crucial to thoroughly familiarize all parties with the organization's aims and guiding principles. By the end of contract negotiations with UCSF, for example, decision makers expressed great personal satisfaction at helping to advance DND*i*'s work. Through similar efforts DND*i* hopes to have developed and made available, by 2014, six to eight field-relevant treatments.

Boadi and Bokanga[18] describe the building of public-private partnerships in Africa by the **African Agricultural Technology Foundation (AATF)**. AATF emerged from a Rockefeller Foundation initiative in the early 2000s following a wide-ranging and unprecedented consultation among African, European, and North American stakeholders who were, and are, actively seeking to improve food security and reduce poverty in sub-Saharan Africa. AATF recognizes that new and unique public-private partnerships (PDPs) are needed to remove many of the barriers that have prevented smallholder farmers in sub-Saharan Africa from gaining access to existing agricultural technologies. Focusing on the creation of these PDPs, it promotes efforts to create sustainable markets and seeks to dramatically improve access to agricultural technologies, materials, and know-how.

AATF has two unique characteristics: first, it is prepared to in-license technologies from the private sector, which it then sublicenses to its partners. This is no small issue and requires careful considerations of a range of issues, including liability. Second, AATF strongly focuses on downstream activities or, to put it more broadly, on technology stewardship. This includes facilitating access to local, national, and regional markets for products based on transferred technologies. The goals are to create more sustainable technology transfer mechanisms and to allow national institutions to more effectively absorb new technological concepts and adopt them for productive use.

But the fundamental raison d'être of AATF goes much deeper than "merely" IP management. As Gordon Conway, then president of the Rockefeller Foundation, put it in the AATF annual report of 2005:

We should examine the current system and ask ourselves, 'How can those who care about the fate of the small-scale farmer make technological options more available?' The rise of a sophisticated global

IP system covering many building block technologies has meant public researchers [in Africa] have little access to new ideas and tools in their field. Left to its own devices, the gap is likely to grow—with wealthy nations' farmers using techniques that are ever more sophisticated and poor farmers left with the same tools they have used for centuries.

Other case studies sharing PDP experiences describe **PATH,**[19] and **ICIPE,**[20] a nonprofit that partnered with **Africert Ltd** in transferring standards certification know-how, critical for the introduction of new products.

FOCUS ON SOLUTIONS: ACCELERATING PRODUCT DEVELOPMENT AND DELIVERY

Numerous partnership efforts are underway to accelerate access and delivery for agricultural and health products in developing countries. For example, in the tropics, where just about everyone eats eggplant, it is commonly infested by **eggplant fruit and shoot borer** (**EFSB**), which inflicts a 70% crop loss. Conventional efforts to breed for resistance have been unsuccessful, so farmers rely heavily on pesticides. These chemicals, however, are expensive, and the pest is becoming more and more resistant to them. Moreover, some pesticides damage the environment and/or are illegal.

Recently, a new solution to the problem of EFSB was developed in partnership with many organizations, writes Medakker and Vijayaraghavan,[21] including by MAHYCO, a private Indian company. It was the first company in India to develop a transgenic hybrid eggplant genetically engineered with a gene that provides resistance to EFSB. The gene (cry1Ac) is obtained from the bacterium *Bacillus thuringiensis* (Bt). A spore-forming bacterium, Bt produces crystal proteins (called Cry proteins) that are toxic to many species of insects, including EFSB. Cultivation of the hybrid eggplant reduces the need for pesticide applications.

This breakthrough was made possible when MAHYCO obtained the rights under license for the use of the Bt cry1Ac gene technology for insect pest management from the Monsanto Company. The license also allows for sublicensing of the technology on a royalty-free basis to a partnership of public institutes and agricultural universities in India, Bangladesh, and the Philippines. This consortium is developing a nonhybrid form of Bt eggplant for use by farmers in developing countries. The nonhybrid form will be less expensive, but the yield is higher for the hybrid technology. Therefore, more farmers might choose the hybrid technology.

Commercial release of the first transgenic Bt hybrids developed by MAHYCO is planned for India by the end of 2007, after the fulfillment of all regulatory requirements. The transgenic Bt open-pollinated varieties under development by the public-private partnership are expected to be commercialized about six months later. This approach to EFSB is an excellent example of how biotechnology applications can be concurrently commercialized for the market and subsidized for poorer market segments.

In health, a prominent example of improvement regarding access to innovations in health is the PATH Malaria Vaccine Initiative (MVI), a program funded by the Rockefeller Foundation that analyzed whether consolidating patents in the malaria vaccine field could streamline access by advancing and accelerating the development of vaccines. The project was designed to ensure **market access for the malaria vaccine candidates** that are most likely to receive regulatory approval and be developed as products. The study, described by Shotwell,[22] assessed the status of the relevant patents, determined their availability for licensing, and explored the potential of patent consolidation or technology trust to enhance access to the vaccine. Developing a broad-based technology trust for existing malaria antigen patents was not recommended. Instead, several other steps were recommended for consolidating available rights and improving access with regard to future patent families.

Before this study, MVI had identified some potentially obstructive IP issues for a malaria vaccine for developing-country markets. Public and academic institutions—institutions with missions that in many cases include some form of public benefit—hold many of the patents related to malaria antigens. As the study's findings reveal, with few exceptions the patents held by public and academic institutions have been assigned or exclusively licensed to private companies and, therefore, are currently unavailable for licensing from the original public institution patent holders.

While it may be possible to sublicense these malarial antigen patents from the current private holders of the technology, it is likely to be more difficult and costly; engaging the patent holders to contribute to a patent pool or clearinghouse also might be challenging. Moreover, a patent pool for a malaria vaccine might generate further obstacles: potential antitrust issues, real or perceived, might trigger scrutiny by the U.S. Department of Justice and the Federal Trade Commission. And while the concept of a technology trust or patent pool may be useful for patents filed in the future, even some of those would be under option for license by the private companies holding the current patents. Finally, the number of high-priority cases for any malaria antigen is small, as is the number of entities likely to seek access to any given patent family. This makes the expense of a patent pool even less justifiable. Taking all of these things into consideration means fewer missteps and faster progress towards a vaccine for malaria.

Other chapters in this section provide case studies of licensing experience related to the **Cohen-Boyer patents at Stanford University,**[23] IP issues related to **molecular pharming, specifically for plant-derived vaccines,**[24] **corn/maize breeding and the impact of biotechnology on the breeding and commercialization process,**[25] the **University of California's Strawberry Licensing Program**[26] (the most successful program in terms of the generation of licensing revenues of any U.S. university), the successful resolution of IP constraints that led to the introduction of **virus-resistant papayas,**[27] and a project on the **somatic embryogenesis of grapes in Chile.**[28]

CONCLUSIONS

If indeed the best proof is experience, then the case studies described here, in the *Handbook,* and in the insert of this *Executive Guide* do indeed speak for themselves. The experiences represented by these case studies provide all the evidence needed to spur further efforts to build upon the IP strengths of developing countries. Many forward-thinking people have seen the possibilities, and this section broadly maps out work that is already underway around the globe to make these possibilities into realities. Such experiences offer the most powerful proof of the benefits that can be obtained through creative IP management in developing countries and indeed around the world. ■

All chapters refer to: *Intellectual Property Management in Health and Agricultural Innovation: A Handbook of Best Practices.* 2007. A Krattiger, RT Mahoney, L Nelsen, JA Thomson, AB Bennett, K Satyanarayana, GD Graff, C Fernandez, and SP Kowalski (eds.). MIHR: Oxford, U.K., and PIPRA: Davis, U.S.A. Available online at www.ipHandbook.org. The online version contains for each chapter a detailed Editor's Summary, Implications, and Best Practices.

1 Chapter 17.5 by K Satyanarayana titled Current IP Management Issues for Health and Agriculture in India, p. 1605.

2 Chapter 17.7 by R Wolson titled Technology Transfer in South African Public Research Institutions, p. 1651.

3 Chapter 17.1 by CI Chamas, SM Paulino De Carvalho and S Salles-Filho titled Current Issues of IP Management in Health and Agriculture in Brazil, p. 1563.

4 Chapter 17.2 by C Fernandez, MR Moynihan titled A Model for the Collaborative Development of Agricultural Biotechnology Products in Chile, p. 1577.

5 Chapter 17.3 by Z Chen, W Gao and J Xu titled IP Rights in China: Spurring Invention Driving Innovation in Health and Agriculture, p. 1585.

6 Chapter 17.4 by A Blaya titled Experiences from the European Union: Managing IP Under the Sixth Framework Programme, p. 1593.

7 Chapter 17.6 by J Chapman and KN Watanabe titled Current Issues of IP Management for Health and Agriculture in Japan, p. 1621.

8 Chapter 17.12 by LA Salicrup ML Rohrbaugh titled Partnerships for Innovation and Global Health: NIH International Technology Transfer Activities, p. 1709.

9 Chapter 17.16 by R Ceron Di Giorgio titled From University to Industry: Technology Transfer at UNICAMP in Brazil, p. 1747.

10 Chapter 17.8 by PJ Slate and M Crow titled The New American University the Role of "Technology Translation": The Approach of Arizona State University, p. 1661.

11 Chapter 17.9 by H Guo titled IP Management at Chinese Universities, p. 1673.

12 Chapter 17.10 by KR Schubert titled Application and Examples of Best Practices in IP Management: The Donald Danforth Plant Science Center, p. 1683.

13 Chapter 17.11 by T Bates titled IP Management in the National Health Service in England, p. 1697.

14 Chapter 17.13 by N Page titled The Making of a Licensing Legend: Stanford University's Office of Technology Licensing, p. 1719.

15 Chapter 17.14 by AB Bennett M Carriere titled Technology Transfer at the University of California, p. 1729.

16 Chapter 17.15 by GD Graff and AB Bennett titled Intellectual Property and Technology Transfer by the University of California Agricultural Experiment Station, p. 1739.

17 Chapter 17.19 by J Banerji and B Pecoul titled Pragmatic and Principled: DNDi's Approach to IP Management, p. 1775.

18 Chapter 17.18 by RY Boadi and M Bokanga titled The African Agricultural Technology Foundation Approach to IP Management, p. 1765.

19 Chapter 17.17 by S Brooke, CM Harner-Jay, H Lasher and E Jacoby titled How Public-Private Partnerships Handle Intellectual Property: The PATH Experience, p. 1755.

20 Chapter 17.20 by P Munyi and R Nyagah titled From Science to Market: Transferring Standards Certification Know-How from ICIPE to Africert Ltd, p. 1783.

21 Chapter 17.25 by A Medakker and V Vijayaraghavan titled Successful Commercialization of Insect-Resistant Eggplant by a Public–Private Partnership: Reaching and Benefiting Resource-Poor Farmers, p. 1829.

22 Chapter 17.21 by SL Shotwell titled Patent Consolidation and Equitable Access: PATH's Malaria Vaccines, p. 1789.

23 Chapter 17.22 by MP Feldman, A Colaianni and C Kang Liu titled Lessons from the Commercialization of the Cohen-Boyer Patents: The Stanford University Licensing Program, p. 1797. The online version of the *Handbook* also lists one of the Cohen-Boyer licenses.

24 Chapter 17.23 by A Krattiger and RT Mahoney titled Specific IP Issues with Molecular Pharming: Case Study of Plant-Derived Vaccines, p. 1809.

25 Chapter 17.24 by V Gracen titled How IP and Plant Breeding Come Together: Corn as a Case Study for Breeders and Research Managers, p. 1819.

26 Chapter 17.27 by AB Bennett and M Carriere titled The University of California's Strawberry Licensing Program, p. 1833.

27 Chapter 17.27 by M Goldman titled The IP Management of the PRSV-Resistant Papayas Developed by Cornell University and the University of Hawaii and Commercialized in Hawaii, p. 1837.

28 Chapter 17.28 by C Fernandez titled Fundación Chile: Technology Transfer for Somatic Embryogenesis of Grapes, p. 1845.

Glossary

The definitions contained in the glossary are derived, in part, from McCarthy's Desk Encyclopedia of Intellectual Property.[1] In addition to this glossary, the reader is encouraged to refer, for expanded definitions and additional terms, to online intellectual property glossaries, including those found on the following Web sites:

- World Intellectual Property Organization: www.wipo.int/tk/en/glossary/index.html.
- U.S. Patent and Trademark Office: www.uspto.gov/main/glossary/index.html.
- U.S. Department of State: usinfo.state.gov/products/pubs/intelprp /glossary.htm.

assignment - A transfer of intellectual property (IP) rights. An assignment of a patent, for example, is a transfer of sufficient rights so that the recipient has title to the patent. An assignment can be a transfer of all rights of exclusivity in the patent, a transfer of an undivided portion (for example, a 50 percent interest), or a transfer of all rights within a specified location (for example, a certain area of the United States). Anything less is considered to be a license transfer, rather than a patent transfer.

Berne Convention - A major multinational copyright treaty, with nearly 150 members. There are five main points to the Berne Convention: (1) national treatment, that is, nondiscrimination with respect to foreign authors and copyright owners; (2) no formalities, that is, copyright is automatically granted and is not conditioned on formalities such as registration or notice; (3) minimum duration of copyright; (4) moral rights provided to authors under the national laws of member nations; and (5) copyright protection independent of whether such protection exists in the country of origin.

best mode - A condition for the grant of a patent, found in the patent specification. An inventor must describe and disclose the best method he or she knows for carrying out the invention.

claims - The section of the patent that defines an invention (the technology that is the exclusive property of the patentee for the duration of the patent) and is legally enforceable; that is, the claims set the metes and bounds of the patent rights. The patent specification must conclude with a claim, particularly pointing out and distinctly claiming the subject matter that the applicant regards as the invention or discovery. The claim or claims are interpreted as set forth in the specification: the terms and phrases used in the claims must be sufficiently described in the specification, that is, patent claims must read in the light of the specification. The specification discloses and the claims define the invention.

commercialization - The process of taking an invention or discovery to the marketplace. It involves working the idea into a business plan, consideration of protection options, and determining how to market and distribute the finished product.

compulsory license - A license granted by the state upon request to a third party that, through the license, is permitted to exploit a patented invention after the owner of the patent has refused to provide a voluntary license under acceptable conditions.

confidential disclosure agreement - See **confidentiality agreement.**

MIHR/PIPRA. 2007. Glossary. In *Executive Guide to Intellectual Property Management in Health and Agricultural Innovation: A Handbook of Best Practices* (Krattiger A, RT Mahoney, L Nelsen et al.). MIHR (Oxford, UK), PIPRA (Davis, USA), Oswaldo Cruz Foundation (Fiocruz, Rio de Janeiro, Brazil), and *bio*Developments-International Institute (Ithaca, USA). Available online at www.ipHandbook.org.

confidentiality agreement (nondisclosure agreement, confidential disclosure agreement) - A legal document through which intellectual property can be disclosed by one party to another wherein the latter party is permitted to use the information for certain purposes, and only those purposes, that are stated in the agreement and agrees not to disclose the information to others.

contributory infringement - An indirect infringement of IP rights in which people, or organizations, contribute to a direct act of infringement by another (in order to aid or abet the act of infringement), for example, knowingly selling an article that is used solely to practice a patented process or to manufacture a patented product.

copyright - An exclusive right conferred by the government on the creator of a work to bar others from reproducing, adapting, distributing to the public, performing in public, or publicly displaying said work. Copyright does not protect an abstract idea; it protects only the concrete expression of an idea. In order to obtain copyright protection, a work must have originality and some modicum of creativity.

cross licensing - A legal agreement in which two or more parties that have potentially conflicting patent claims, or other conflicting IP rights, reach an agreement to share the IP rights in question through a reciprocal licensing arrangement.

dependent claim
A claim in a patent that refers back to a previous claim and defines an invention that is narrower in scope than that in the previous claim. A dependent claim is written in such a way as to be more restricting than the technology defined in the previous claim (often an independent claim).

descriptive mark - A word, picture, or other symbol that describes some quality or trait of a product or service, such as the purpose, size, color, class of users, or end effect on users. A descriptive term is not considered to be inherently distinctive; to establish validity of a descriptive mark for registration or protection in court, proof of acquired distinctiveness of the mark is needed. This acquired distinctiveness confers secondary meaning. For example, "Kentucky Fried Chicken" a mark that originally was descriptive, subsequently acquired secondary meaning as a trademark for a distinctive type of commercial food product.

design patent - A government grant of exclusive rights in a novel, nonobvious, and ornamental industrial design. A design patent confers the right to exclude others from making, using, or selling designs that closely resemble the patented design. A design patent covers the ornamental aspects of a design; its functional aspects are covered by a utility patent. A design patent and a utility patent can cover different aspects of the same article.

differential pricing (tiered pricing) - The practice of setting different prices for different markets—typically higher prices in richer markets and lower prices in poorer markets.

disclosure of origin - A requirement imposed on patent applicants to disclose in patent applications the geographic origin of biological material on which the invention (subject of the patent application) is based.

divisional patent application - A patent application that is carved out of a parent application, such that the parent application is divided into one or more divisional patent applications. Divisional applications are entitled to the original filing-date priority of the parent application.

due diligence - Investigations undertaken to assess the ownership and scope of one or more IP rights that are being sold, licensed or used as collateral in a transaction. This is done in order to identify business and legal risks associated with the IP rights being analyzed.

duration - The term, or length of time that an IP right lasts. A U.S. utility patent on an invention, for example, has a duration of 20 years from the date on which the patent application was filed, as does a plant patent. The duration of a U.S. copyright is usually the life of the author plus 70 years (for works created after January 1, 1978). Protection of information as a trade secret lasts as long as the information remains secret. Duration of a trademark continues as long as it is used (as a source indicator) and properly maintained/protected.

examination - *See **patent examination**.*

exclusive license agreement - A legal document licensing intellectual property to another party for its exclusive use. Exclusively licensed patent rights cannot, within the scope or field of the exclusive license, be subsequently or simultaneously licensed to any other party.

field-of-use restriction - A provision in an IP license that restricts use of the licensed intellectual property by the licensee to only in a defined product or service market.

first to file - A rule under which patent priority is determined. The rule gives priority to the party that first files a patent application for an invention, rather than to the party that is first to invent. First to file is followed by almost every nation in the world except the United States. For trademarks, priority between conflicting applications to register a trademark is handled by publishing the application with the earliest filing date for possible opposition by the applicant with a later filing date. In the United States, ownership of a trademark is determined by who was first to use it, not by who was first to file an application for registration. However, under the intent-to-use system, an application for registration can be filed prior to actual use of a mark.

first to invent - A rule under which patent priority is determined by which inventor was the first to actually invent, rather than by who was the first to file a patent application. This is the rule followed in the United States. *Compare to **first to file**.*

freedom to operate - The ability to undertake research and/or commercial development of a product without infringing the unlicensed intellectual or tangible property rights of others.

functionality - That aspect of design that makes a product work better for its intended purpose, as opposed to making the product look better or to identify its commercial source.

Indigenous Cultural and IP Rights - Indigenous Cultural and IP Rights refers to the rights to a heritage, that its, to the objects, sites, knowledge, and methods of transmission of communities that have traditionally been defined by the social ownership of knowledge. This right privileges customary law over modern law. Heritage includes all aspects of culture (art, music, dance, literature, and so on), indigenous knowledge (medicinal, nutritional), and land management practices. There are numerous attempts today to give legal substance and scientific validity to indigenous knowledge. Article 29 of the Draft Declaration of the Rights of World Indigenous People states that "*[i]ndigenous people are entitled to the recognition of full ownership, control and protection of their cultural and intellectual property.*"

industrial property - Industrial property is a subset of intellectual property, referring to those types of intellectual property that have an industrial application. Specifically, it refers to patents, trademarks, designs, mask works, and plant breeders' rights.

infringement - An invasion of an exclusive right of intellectual property. Infringement of a utility patent includes making, using, or selling a patented product or process without permission. Infringement of a design patent involves fabrication of a design that, to the ordinary observer, is substantially the same as an existing design, where the resemblance is intended to induce the observer to purchase one thing supposing it to be another. Infringement of a trademark consists of the unauthorized use or imitation of a mark that is the property of another in order to deceive, confuse, or mislead others. Infringement of a copyright involves reproducing, adapting, distributing, performing in public, or displaying in public the copyrighted work of someone else.

intellectual property (IP) - Creative ideas and expressions of the human mind that have commercial value and are entitled to the legal protection of a property right. The major legal mechanisms for protecting intellectual property are copyrights, patents, and trademarks. IP rights enable owners to select who may access and use their intellectual property and to protect it from unauthorized use.

international patent application - *See **Patent Cooperation Treaty (PCT)**.*

intellectual property management - The means by which an institutionally owned IP portfolio is managed with regard to marketing, patenting, licensing, and administration.

invention - The creation of a new technical idea and of the physical embodiment of the idea or the means to accomplish it. To be patentable, an invention must be novel, must have utility, and would not have been obvious to those possessing ordinary skill in the particular art of the invention.

inventive step (nonobviousness) - A condition for patentability, which means that the invention would not be obvious to someone with knowledge and experience in the technological field of the invention. According to the European Patent Convention, "*An invention shall be considered as involving an inventive step if, having regard to the state of the art, it is not obvious to a person skilled in the art.*"

joint inventors - Two or more inventors of a single invention who work together during the inventive process.

know-how - Information that enables a person to accomplish a particular task or to operate a particular device or process. *Refer to **trade secret**.*

license - A grant of permission to use an IP right within a defined time, context, market line, or territory. There are important distinctions between exclusive licenses and nonexclusive licenses. An exclusive license is "exclusive" as to a defined scope, that is, the license might not be the only license granted for a particular IP asset, as there might be many possible fields and scopes of use that can also be subject to exclusive licensing. In giving an exclusive license, the licensor promises that he or she will not grant other licenses of the same rights within the same scope or field covered by the exclusive license. The owner of IP rights may also grant any number of nonexclusive licenses covering rights within a defined scope. A patent license is a transfer of rights that does not amount to an assignment of the patent. A trademark or service mark can be validly licensed only if the licensor controls the nature and quality of the goods or services sold by the licensee under the licensed mark. Under copyright law, an exclusive licensee is the owner of a particular right of copyright, and he or she may sue for infringement of the licensed right. There is never more than a single copyright in a work regardless of the owner's exclusive license of various rights to different persons.

licensee - A party obtaining rights under a license agreement.

licensor - A party granting rights under a license agreement.

license out - The process by which one person, company, or institution extends to another person, company, or institution permission to use the former's intellectual property.

license in - The process by which a person, company, or institution obtains permission to use the intellectual property owned by someone else.

material transfer agreement (MTA) - A contract between the owner of a tangible material and a party seeking the right to use the material for research or other assessment purposes. The material may be either patented or unpatented. Material transfer agreements tend to be shorter than license agreements. The purpose of an MTA is to document the transfer the material and outline the terms of use, including identification of the research or assessment project, terms of confidentiality, publication, and liability.

maintenance fees - Fees for maintaining in force a patent. The fees typically have to be paid at irregular intervals, depending on the jurisdiction, and significantly increase over time.

notice - A formal sign or notification attached to items that embody or reproduce an intellectual property asset—for example, the presence of the word *patent* or its abbreviation, *pat.*, together with the patent number, on a patented article made by a patent holder or his/her licensees. The formal statutory notice of U.S. trademark registration is the letter *R* inside a circle: ®, *Reg. U.S. Pat. & Tm. Off.*, or *Registered in U.S. Patent and Trademark Office*. Many firms use informal trademark notices, such as *Brand*, *TM*, *Trademark*, *SM*, or *Service Mark*, adjacent to words or other symbols considered to be protectable marks. Notice of copyright consists of the letter *C* in a circle symbol: © or the word *Copr.* or *Copyright*, the copyright owner's name, and the year of first publication.

nonassignable - A condition whereby a licensing agreement and/or the rights, obligations, and terms thereof may not be assigned to any party who is not a signatory to the agreement.

nondisclosure agreement
See **confidentiality agreement**.

nonexclusive license
A license under which rights are granted to the licensee but not exclusively to that licensee; the licensor reserves the right to give the same or similar rights to use the licensed materials to other parties.

nonobviousness
One of three conditions an invention must meet to be patentable. *See also* ***inventive step***.

nontransferable
The licensing agreement and/or the rights, obligations, and terms thereof that may not be sold, given, assigned, or otherwise conveyed to any party who is not a signatory to the agreement.

novelty
One of three conditions an invention must meet to be patentable.

obviousness
A condition of an invention that makes it ineligible to receive a valid patent; the condition of an invention whereby a person with ordinary skill in a field of technology can readily deduce it from publicly available information (prior art). *See also* ***ordinary skill in the art***.

ordinary skill in the art
The level of technical knowledge, experience, and expertise possessed by the ordinary engineer, scientist, or designer in a technology that is relevant to an invention.

Paris Convention
The main international treaty governing patents, trademarks, and unfair competition. The Convention is administered by the World Intellectual Property Organization (WIPO) and has four principal provisions: (1) national treatment for all seeking protection of IP rights, whether foreign or nationals; (2) minimum level of protection; (3) Convention priority, with a specified time (12 months for patents, six months for trademarks) for applications to be filed in other member nations; and (4) administrative framework within the Paris Union.

patent (U.S.)
A grant by the federal government to an inventor of the right to exclude others from making, using, or selling his or her invention. There are three kinds of patents in the United States: a standard utility patent on the functional aspects of products and processes; a design patent on the ornamental design of useful objects; and a plant patent on a new variety of a living plant. Patents do not protect ideas, only structures and methods that apply technological concepts. Each type of patent confers the right to exclude others from a precisely defined scope of technology, industrial design, or plant variety. In return for the right to exclude, an inventor must fully disclose the details of the invention to the public so that others can understand it and use it to further develop the technology. Once the patent expires, the public is entitled to make and use the invention and is entitled to a full and complete disclosure of how to do so.

patent application - A technical document that describes in detail an invention for which a patent is sought.

patent examination - A process of review of a patent application, undertaken by a patent examiner, to determine whether the application complies with all statutory requirements for patentability. The examination process reviews prior art to ensure novelty, along with determining compliance with other statutory requirements, rules, and matters of procedure and form.

Patent Cooperation Treaty (PCT) - An international treaty that provides a mechanism through which an applicant can file a single application that, when certain requirements have been fulfilled, may then be pursued as a regular national filing in any of the PCT member nations. There are currently more than 120 PCT member nations.

patent pooling - A patent pool is an agreement between two or more patent owners to license one or more of their patents to one another or to third parties. A patent pool allows interested parties to gather all the necessary tools to practice a certain technology.

patent searching - A process carried out by the patent examiner for checking the novelty of a patent application. The subsequent patent research report lists published items comprising both patent and nonpatent literature relevant to the subject of the invention.

plant breeders' rights - Plant breeder's rights are used to protect new varieties of plants by giving exclusive commercial rights to market a new variety or its reproductive material.

plant patent - In the United States, the Plant Patent Act of 1930 provides a grant of exclusive IP rights to applicants who have invented or discovered a new asexually propagated variety of plant. Tuberous plants are not covered by plant patents.

plant variety protection (PVP) - A form of patent-like protection for sexually propagated plants, as well as hybrids, tubers, and harvested plant parts. The Plant Variety Protection Act of 1970 is administered by the U.S. Department of Agriculture and not the U.S. Patent and Trademark Office (which does issue plant patents).

prior art - The existing body of technological information against which an invention is judged in order to determine whether it is novel and nonobvious and can thus be patented.

prior informed consent - The consent given by a party with respect to an activity after being fully informed of all material facts relating to that activity. The Convention for Biological Diversity requires that access to genetic resources shall be subject to the prior informed consent of the country providing the resources.

priority date - The date of the first filing of a patent application that describes an invention in detail. Priority date, as well as patentability, with respect to novelty of invention, is determined in light of any relevant prior art existing at the time of filing. In other words, depending on the specific jurisdiction, if the invention was known or published previous to the priority date, the applicant will be unable to obtain a patent.

provisional application - A provisional application is a document in patent actions that serves to establish an early priority date of an invention. A provisional application will not mature into a regular application, and does not form the basis of a grant of a patent. It is a document that precedes the complete application upon which the grant is based. A provisional application establishes a priority date for disclosure of the details of an invention and allows a period of up to 12 months for development and refinement of the invention before the patent claims take their final form in a complete, regular patent application.

process claim - A claim of a patent that covers the method by which an invention is performed by defining the steps to be followed. This differs from a product claim or an apparatus claim, which covers the structure of a product.

product-by-process claim - A patent claim through which a product is claimed by defining the process by which it is made. The product-by-process form of claim is most often used to define new chemical compounds, since many new chemicals, drugs, and pharmaceuticals can practicably be defined only by describing the process of making them.

public domain - The status of an invention, creative work, commercial symbol, or any other creation that is not protected by some form of IP right. Items that have been determined to be in the public domain are available for copying and use by anyone.

reduction to practice - The physical part of the inventive process that completes and ends the process of invention by demonstrating that the invention has a practical application. Reduction to practice can be carried out either by the actual construction of an apparatus, by performing the steps in a process, or by formally filing a patent application (constructive reduction to practice).

research tools - The term *research tool* includes the full range of tools that scientists may use in the laboratory, including cell lines, monoclonal antibodies, reagents, animal models, growth factors, combinatorial chemistry and DNA libraries, clones and cloning tools (such as PCR), methods, and laboratory equipment and machines.[2] There is concern about the patenting of research tools, because such patents may inhibit the free undertaking of research.

royalty - Income derived from the sale or use of a licensed product or process.

tiered pricing - *See **differential pricing**.*

trademark - (1) A word, slogan, design, picture, or other symbol used to identify and distinguish goods. (2) Any identifying symbol, including a word, design, or shape of a product or container, that qualifies for legal status as a trademark, service mark, collective mark, certification mark, trade name, or trade dress. Trademarks identify one seller's goods and distinguish them from goods sold by others. They signify that all goods bearing the mark come from, or are controlled by, a single source and are of an equal level of quality. And they advertise, promote, and generally assist in selling goods. A trademark is infringed by another if the second use causes confusion of source, affiliation, connection, or sponsorship.

trade secret - Business information that is the subject of reasonable efforts to preserve confidentiality and has value because it is not generally known in the corresponding trade. Such confidential information is protected against those who gain access to it through improper methods or by a breach of confidence. Misappropriation of a trade secret is a type of unfair competition.

traditional knowledge - Tradition-based creations, innovations, literary, artistic or scientific works, performances and designs originating from or associated with a particular people or territory.

Trade-Related Aspects of Intellectual Property Rights (TRIPS) - An international agreement that was initiated under the forerunner of the World Trade Organization (WTO), the General Agreement on Tariffs and Trade (GATT), under the Uruguay round of trade negotiations. The TRIPS Agreement is the most comprehensive multilateral agreement on Intellectual Property covering all IP instruments. It was the first IP rights accord to legitimize the patenting of living organisms. TRIPS provides the guidelines for the harmonization of IP rights laws under the WTO. All WTO member countries have substantive TRIPS obligations.

unfair competition - Commercial conduct that the law views as unjust, providing a civil claim against a person who has been injured by the conduct. Trademark infringement has long been considered to be unfair competition. Other recognized legal categories of unfair competition are false advertising, trade libel, misappropriation of a trade secret, infringement of the right of publicity, and misappropriation.

UPOV (the Convention of the International Union for the Protection of New Varieties of Plants) - An international treaty that guarantees to plant breeders in member nations national treatment and a right of priority. National plant variety protection statutes of member nations are brought into harmonization with the various UPOV provisions, for example, the requirements of distinctness, uniformity, stability, and novelty for new crop varieties.

utility - The usefulness of a patented invention. To be patentable an invention must operate and be capable of use, and it must perform some "useful" function for society. ■

---

1 McCarthy JT, RE Schechter and DJ Franklyn. 1995 and 2004. *McCarthy's Desk Encyclopedia of Intellectual Property*, 2nd and 3rd editions. The Bureau of National Affairs: Washington, DC.

2 From NIH Research Tools Guidelines. ott.od.nih.gov/policy/rt_guide.html.

Biographical Sketches of Authors and Members of the Board of Patrons

ARNTZEN, Charles J.

Charles Arntzen is the Florence Ely Nelson Presidential Endowed Chair at Arizona State University (ASU) in Tempe and a Regents Professor. He was Director of Research at the DuPont Company in Delaware from 1984 to 1988, followed by service as Deputy Chancellor for Agriculture, Dean, College of Agriculture and Life Sciences and Director, Texas Agricultural Experiment Station at the Texas A&M University System. In 1995 became President and CEO of Boyce Thompson Institute, a not-for-profit corporation affiliated with Cornell University. He was elected to the US National Academy of Sciences in 1983 and to the National Academy of India the following year. He currently serves as a member of the Council of Advisors on Science & Technology of President George W. Bush and the US Nanotechnology Advisory Board.

He is a fellow of The American Association for the Advancement of Science and received the Award for Superior Service from the U.S. Department of Agriculture for international project leadership in India. He served as chairman of the National Biotechnology Policy Board of the National Institutes of Health, as chairman of the National Research Council's Committee on Biobased Industrial Products, and on the National Research Council's Committee on Space Biology and Medicine. He served for eight years on the Editorial Board of Science. Dr. Arntzen currently serves on the Board of Directors and the Scientific Advisory Board of Advanced BioNutrition, Inc., and is on the Advisory Board of the Burrill & Company Agbio Capital Fund and The Nutraceuticals Fund. He also serves as a Distinguished Advisor on the Council for Biotechnology, and is a member of the Board of Directors of the National Center for Genome Resources in Sante Fe, New Mexico.

BEACHY, Roger

Roger Beachy is president of the Donald Danforth Plant Science Center in St. Louis, Missouri. He previously held academic positions at Washington University, St. Louis, and the Scripps Research Institute, La Jolla, California. His research includes projects to reduce virus infection in plants via biotechnology, and in studies of the control of gene expression in plants. Beachy is a member of the U.S. National Academy of Sciences and a Fellow of the Academy of Microbiology; he has received several awards for his work, including the Wolf Prize in Agriculture. The Danforth Center has committed significant efforts to research in developing countries, including through private-public partnerships, and Beachy is involved in a variety of efforts with regard to rationalizing regulations that control commercialization of agricultural biotechnology.

Beachy is President of the International Association of Plant Biotechnology. He belongs to numerous institutional boards, including the PNAS Editorial Board, the NRC Governing Board, the Board on Agriculture and Natural Resources (National Research Council of the National Academy of Sciences), Malaysia's International Advisory Panel, and the Governing Board of Directors of the International Crops Research Institute for the Semi-Arid Tropics (ICRISAT), and the Burrill and Company Board of Advisors.

BENNETT, Alan B.

Alan Bennett currently serves as the Associate Vice Chancellor for Research at U.C. Davis. He is responsible for technology transfer, strengthening research-based alliances with industry, and supporting technology-based economic development in the Sacramento/Davis region. He is the founding Executive Director of the Public Intellectual Property Resource for Agriculture (PIPRA), an organization consisting of 37 universities in nine countries that is dedicated to the collective management of intellectual property and supports broad commercial innovation and humanitarian uses of technology in agriculture. From 2000 to 2004, Dr. Bennett served as the Executive Director of the University of California Systemwide Office of Technology Transfer and Research Administration, where he was responsible for IP management and research policy for the University of California system; this task involved managing a portfolio of more than 5,000 cases, 700 active licenses, and revenue in excess of US$350 million for the four-year period. He earned B.S. and Ph.D. degrees in Plant Biology at U.C. Davis and Cornell University, respectively. He joined the U.C. Davis faculty in 1983. His research in plant molecular genetics has focused on cell-wall disassembly and fruit development. Dr. Bennett has published over 130 research papers in leading scientific journals, holds several utility patents related to crop quality traits, and is a regular speaker at universities, international symposia, and private companies. He is a Fellow of the

American Association for the Advancement of Science (AAAS) and of the California Council for Science and Technology (CCST).

BORLAUG, Norman E.
In 1970, Norman Borlaug won the Nobel Peace Prize for his lifelong work to feed a hungry world. His work, more than that of any other person, is credited with saving lives.

In 1944, Dr. Borlaug joined the Rockefeller Foundation's pioneering technical-assistance program in Mexico, at which he was a research scientist in charge of wheat improvement. For the next two decades, he worked to solve a series of wheat production problems in Mexico and to train a generation of young scientists.

With the establishment of the International Maize and Wheat Improvement Center (CIMMYT) in Mexico in 1966, Borlaug assumed leadership of the wheat program; he continues to serve as a consultant for it. The high-yielding, disease-resistant wheat cultivars he developed, along with improved management practices, transformed agricultural production in Mexico during the 1950s and in Asia and Latin America in the 1960s and 1970s. This transformation has come to be known as the Green Revolution.

In 1984, Dr. Borlaug joined Texas A&M University and was named Distinguished Professor of International Agriculture. Since 1986, he has also served as president of the Sasakawa Africa Association and leader of the Sasakawa-Global 2000 agricultural program in Sub-Saharan Africa, in partnership with former U.S. President Jimmy Carter and Yohei Sasakawa.

Borlaug has been awarded 58 honorary doctorate degrees, and is a member or fellow of the academies of science in 12 nations. The U.S. National Academies of Science awarded him the National Service Medal in 2002 and in 2004 President Bush bestowed upon Borlaug the U.S. National Medal of Science. He was the driving force behind the establishment of the World Food Prize in 1985 and serves as Chairman of its Council of Advisors.

CHEN, Zhang Liang
Zhang Liang Chen was born on February 3, 1961, in Fujian, China. He received his Ph.D. in 1987 from Washington University for his research in the Division of Biology and Biomedical Sciences in the field of plant molecular biology and his work in early transgenic plant research. He then returned to China as an associate professor. Two years later, he was a full professor at Beijing University. He has continued his research in transgenic plants and biosafety. He served as director of National Key Laboratory of Protein Engineering and Plant Genetic Engineering. In 1995, he became vice-president of research at Peking University. In 2002, he became the president of China Agricultural University. He and his research group have published over 190 international papers and seven book, and hold over eight patents.

Dr. Chen is also Chair of the Plant Biotech Committee of UNESCO, Consultant for the International Society for Plant Molecular Biology (ISPMB), and member of the Sino-Euro Administration Committee for Biotechnology Cooperation. He also serves as a member and Vice-Chairman of the Council of Scientific Advisers to the International Center for Genetic Engineering and Biotechnology (ICGEB) in Italy and India.

DIOUF, Jacques
Jacques Diouf is Director-General of the Food and Agriculture Organization of the United Nations (FAO). He has held this position since 1994 and is currently serving his third six-year term, which began in January 2006. Over the years, he has held numerous position including Ambassador for the Senegalese Permanent Mission to the United Nations, member of the Senegalese Parliament, and Secretary of State for Science and Technology for Senegal. In addition, he has represented Africa in the Consultative Group on International Agricultural Research and served on the Council of African Advisers of the World Bank. The recipient of numerous awards from countries in every corner of the world for his work in agricultural development, Dr. Diouf has undertaken field trips to various agricultural institutions around the world and has participated in major international meetings representing Senegal and the Central Bank for West African States. He has been a leading voice on issues related to agricultural development and the environment.

DRYDEN, Sam
Sam Dryden is internationally recognized as a successful investor and developer of life-sciences ventures. His particular expertise is in the application, scale-up and commercialization of early-stage technologies worldwide.

Sam is a Managing Director of Wolfensohn & Company, a corporate advisory and investment firm located in New York, where he focuses on private equity investments in biofuels and other alternative energies. He is also CEO of Emergent Genetics, LLC—a life-sciences investment holding company.

Until June 2006, Sam served as the Chair and Corporate CEO of Emergent Genetics, Inc.—a global leader in the development and marketing of biotechnology-enhanced seed products. Emergent Genetics' operations were based in Europe, the United States, Argentina, and Brazil and comprised one of the largest seed companies in India. The largest portion of the company's worth was acquired in April 2005 by Monsanto Company. Remaining operations were acquired in June 2006 by Syngenta AG.

Sam began his career as an analyst with the U.S. Department of Commerce, Bureau of Economic Analysis, where he was responsible for modeling and forecasting selected sectors of the U.S. economy. He was then employed by the Union Carbide Corporation from 1974 to 1980, with responsibilities for various aspects of new corporate ventures. These transactions involved extended assignments in Japan, Europe, and South America.

In 1980, Sam led the spinout of Union Carbide's biotechnologies and related business operations and subsequently cofounded Agrigenetics Corporation and served as its president and CEO. The company grew to become one of the world's largest seed enterprises and was acquired in 1985. It is now part of Dow AgroSciences. During this same period, he was also chairman of an affiliated partnership that managed and invested US$60 million in proprietary plant sciences research conducted in leading universities, as well as private and public research institutions worldwide.

Following the sale of Agrigenetics, Sam founded and became President of Big Stone Inc.—a private venture investment and development company focused on the life sciences. The firm participated in founding over a dozen companies in areas such as biopesticides, novel nucleic acid-based therapeutics and diagnostic products, transgenic animals, fermentation-based production of vitamins, pharmaceutical clinical trialing, environmental toxicological testing, and biotherapeutics. Sam also served as the nonexecutive chairman of Celgro Inc., independent venture of Celgene Corporation and a company focused on the development of novel, single-isomer, agricultural chemical compounds.

In addition to his for-profit activities, Sam has extensive pro bono involvement in efforts relating to food security and international economic development. Currently he is an advisor to the World Bank regarding rural development strategy. He is a member of the board of directors of the Global Crop Diversity Trust. Sam serves on the National Academies Panel on Science and Technology for Global Sustainability. In the past, he served on the steering committee for the Global Assessment on Agricultural Science and Technology, led by the World Bank. He was a member of the executive council, as well as chair of the Private Sector Committee of the Consultative Group on International Agricultural Research. He has been an advisor to the Rockefeller Foundation and a member of the Design Advisory Committee and Scientific Advisory Board of its African Agricultural Technology Foundation—an organization created for the advancement of African food security. In the mid-80's, Sam chaired a Rockefeller Brothers Fund development initiative to benefit developing country food security. He also served on the Board of the South/North Development Initiative—a private Rockefeller Family foundation for alleviation of rural poverty in less-developed countries through entrepreneurial development. He is a past member of the U.S. Government's Agricultural Sciences and Technology Review Board.

Sam is a member of the Council on Foreign Relations and serves on its Advisory Committee on Intellectual Property and American Competitiveness. He has also served on its study group analyzing trade issues between the United States and Europe surrounding genetically-modified foods.

He has written and lectured widely on the policy issues of food security, the evolving nature of global public goods and new mechanisms for public and private sector relations. In this regard, his travels have taken him on missions to most countries in Latin America, including Cuba, as well as Europe, Asia, Africa, and the Middle East.

Sam, a native of eastern Kentucky, received his BA in economics from Emory University in 1973.

FATHALLA, Mahmoud F.

Dr. Fathalla is a professor of Obstetrics and Gynecology and former Dean of the Medical School at Assiut University in Egypt and is currently the chairman of the World Health Organization (WHO) Global Advisory Committee on Health Research. He has served as the Director of the UNDP, UNFPA, World Bank, WHO Special Programme of Research, Development and Research Training in Human Reproduction and has served as a consultant to various international bodies such as the WHO, UNPF, IPPF, Population Council, and the Ford and Rockefeller foundations. He is the author of more than 150 scientific publications. Professor Fathalla has been an international campaigner for Safe Motherhood and a founder of the Safer Motherhood Initiative. His scientific interests include women's health, safe motherhood, reproductive health, ethics and human rights, and contraceptive research and development.

FERNÁNDEZ, Carlos

Carlos Fernández studied agronomy at Universidad de Chile. After working as an Assistant Professor at the Agronomy Faculty of the same university, he received a Ph.D. in Plant Physiology at the University of California, Davis. Upon graduation, he joined Monsanto Company, where he held various management positions that gave him responsibilities in several countries. He led the development of agricultural technologies in Latin American countries, first from the company headquarters in St. Louis and later from Sao Paulo, Brazil. Among other things, he contributed to the development of new applications for Roundup, the most successful herbicide in the world, and the development of nontillage systems for various crops. In Europe, he developed new products and actively participated in the design of the Roundup post-patent policy for Europe and Africa.

While working for Monsanto in California, he evaluated and contributed to the development and introduction of transgenic crops to the market. During his stay in California, he returned to the University of California, Davis, and earned an M.B.A. In 1999, he returned to Santiago, Chile, and began working at Fundación Chile, where he coordinated programs related to technology transfer, intellectual property, regulatory matters, and the development of transgenic crops. He contributed to the Cooperative Agreement between the University of California, Davis and Fundación Chile. In addition to his work at Fundación Chile, he serves as a consultant to the Food and Agricultural Organization of the UN and the Chilean Ministry of Economy. Some of his latest contributions as a consultant include two studies sponsored by the Ministry of Economy of Chile: "Comparative Analysis of Biotechnology Policies in N. Zealand, Canada, United States, Australia, Japan, China, Argentina, Brazil, Spain and Chile" and "Formulation of a Model for a Technology Transfer Office for Chile." He also contributed to a recent study, sponsored by UNDP, titled "Commercialization Impact on Agricultural Export Products Caused by the Introduction of GMO in Chile."

As of July 2006, Dr. Fernández is the Head of Strategic Studies and the technology transfer unit of the Foundation for Agriculture Innovation.

FREIRE, Maria

Dr. Maria C. Freire is CEO and President of The Global Alliance for TB Drug Development, a position she has held since 2001. During her service, the Alliance has built the largest pipeline of TB drugs in the world, advanced compounds into clinical testing, and pioneered precedent-setting agreements with industry.

From 1995 to 2001, Dr. Freire directed the Office of Technology Transfer at the NIH, where she was responsible of technology transfer policies and procedures for the Department of Health and Human Services and for patenting and licensing activities at the NIH and the FDA.

Dr. Freire is an internationally recognized expert in technology commercialization. She is a member of the NIH Advisory Board for Clinical Research, a Governor of the New York Academy of Sciences, and the Chair of the Working Group for New TB Drugs for the global Stop TB Partnership. Dr. Freire was selected as one of ten Commissioners of the World Health Organization's Commission on Intellectual Property Rights, Innovation and Public Health (CIPIH) and a member of Time magazine's Global Health Summit Board of Advisors.

Born in Lima, Peru, Dr. Freire trained at the Universidad Peruana Cayetano Heredia. She holds a Ph.D. in biophysics and completed post-graduate studies in immunology and virology at the University of Virginia and the University of Tennessee, respectively, and at the John F. Kennedy School of Government at Harvard University. She has received numerous national and international awards, including the Arthur S. Flemming Award, DHHS Secretary's Award for Distinguished Service, and the Bayh-Dole Award.

GRAFF, Gregory D.

Gregory D. Graff is an applied economist with expertise in the economics of innovation, entrepreneurship, intellectual property, and technology transfer, especially as they apply to the agricultural life sciences and biotechnology. He applies microeconomic and econometric tools to scientific, patent, regulatory, and commercial data, building uniquely thorough industry-level datasets to analyze the impacts of innovation and technology transactions on markets, industrial organization, and the political economy of science policy.

Dr. Graff currently manages research projects for the Public Sector Intellectual Property Resource for Agriculture (PIPRA), a consortium of 37 agricultural research universities and institutes that is hosted by the University of California. PIPRA uses an innovative model of collaborative intellectual property management to mobilize its members' technologies for the purpose of genetically improving "orphan" crops. Dr. Graff has taught as a university lecturer at both U.C. Berkeley and U.C. Davis and has recently published articles in *The Review of Economics and Statistics, World Development, California Management Review,* and *Nature Biotechnology* as well as chapters in several books. Dr. Graff has a Ph.D. in agricultural and resource economics from U.C. Berkeley (2002), an M.A. in economics from Ohio State University (1995), and a B.S. in biology from Cornell University (1992).

HENNESSEY, William O.

William Hennessey is Professor of Law at the Franklin Pierce Law Center. He directed Pierce Law's graduate programs in intellectual property and summer from 1986 until 2003. A noted IP expert, author, and lecturer, he recently directed the fourth annual Pierce Law Intellectual Property Summer Institute at Tsinghua University School of Law in Beijing, China. He co-authored a legal casebook on international IP law and policy.

Professor Hennessey has served as a legal advisor to the governments of Indonesia and the People's Republic of China and has served as a consultant to the World Bank, Asian Development Bank, United Nations Development Programme, U.S. Agency for International Development, U.S. Department of State, and the U.S. Patent and Trademark Office. He has also served as consultant for the World Intellectual Property Organization in many countries on various issues concerning IP protection and economic development.

IDRIS, Kamil

Dr. Kamil Idris has been Director General of the World Intellectual Property Organization (WIPO) since November 1997. He is head of the International Union for the Protection of New Varieties of Plants (UPOV). He was formally re-appointed to a second six-year term as Director General of WIPO on May 27, 2003. His mandate will end on November 30, 2009. Formerly, Kamil Idris was a member of the International Law Commission from 1992 to 1996 and from 2000 to 2001.

Kamil Idris holds a Bachelor of Law (LL.B.) from Khartoum University, Sudan; a Bachelor of Arts in Philosophy, Political Science and Economic Theories from Cairo University, Egypt; a master's in International Law and International Affairs from Ohio University, United States; and a Doctorate in International Law from the Graduate Institute of International Studies, University of Geneva, Switzerland.

JAHN, Molly

Molly Jahn holds degrees from Swarthmore College, M.I.T., and Cornell University, and pursued postdoctoral work at U.C. Berkeley. At Cornell, Molly focused her research on plant breeding, genetics, genomics and molecular biology and on the development of improved crop germplasm. Her group at Cornell has produced a number of globally successful crop varieties currently grown commercially on six continents. Molly has worked extensively internationally in Latin America, Asia and Africa to link crop breeding objectives to outcomes that improve human welfare, such as nutritional status and income. Molly was recently named a Fellow of the AAAS and was elected to the Board of Directors of The World Vegetable Center, the international research center for vegetables. On August 1, 2006, she was named the twelfth dean of the College of Agricultural and Life Sciences at the University of Wisconsin - Madison.

KHUSH, Gurdev Singh

Dr. Khush was born in a small village in Punjab. After receiving his education at the Punjab Agricultural University and the University of California, Davis, Dr. Khush, in 1967, joined the International Rice Research Institute in the Philippines where he served as the Head of Plant Breeding, Genetics, and Biochemistry Division until 2002. As a result of wide-scale adoption of his high-yielding varieties, rice production increased 135% between 1967 and 2000, to feed an estimated one billion additional consumers. His contributions to rice genetics and biotechnology are equally well recognized. He has written three books, more than 80 book chapters and 160 research papers.

Dr. Khush has served as consultant to rice breeding programs of 15 countries as well as The Rockefeller Foundation, the Third World Academy of Sciences, Italy, and the International Science Foundation, Sweden. He is now serving as a member of Scientific Advisory Committee (overseas) to the Department of Biotechnology, Government of India and member of Science Council, an advisory body to Chinese Academy of Agricultural Sciences, Beijing.

For his monumental contributions to the World Food Security, Dr. Khush has been honored with numerous awards and honors such as the Japan Prize (1987), World Food Prize (1996), Rank Prize (1998), Wolf Prize (2000), International Scientific and Technological Cooperation Award from the Government of China (2001), and Padma Shriaward from the president of India. He is one of five Indian scientists who have been elected to membership of Royal Society (FRS) as well as the U.S. National Academy of Sciences. Dr Khush has received Doctor of Science, *honoris causa*, degrees from nine universities including from University of Cambridge in England and Ohio State University.

Commenting on his life work, Dr. Cantrell, Director of the International Rice Research Institute said, "While Dr. Khush's name may have passed the lips of many, his life's work has passed the lips of almost half of humanity."

KOWALSKI, Stanley P.

Stanley P. Kowalski was born and grew up in a working-class neighborhood in Pittsburgh, Pennsylvania, where he attended Catholic primary and public high school. He matriculated at the Pennsylvania State University, and later at the University of Pittsburgh, earning B.S. degrees in horticulture and biology, with emphases in genetics and biochemistry. Later, he earned a Ph.D. in plant breeding from Cornell University. Dr. Kowalski's experience as a research scientist has included studies of plant nutrition at the Pennsylvania State University, wheat breeding at the University of Nebraska, purification and characterization of DNA polymerases at the University of Rochester, biochemical characterization of insect resistance in potatoes at Cornell University, lipid-mediated signal transduction at the National University of Singapore, plant genome mapping at Texas A&M University, glycolipid biosynthesis at Cornell University, and a study of the biochemical/genetic basis of plant/insect interactions at the U.S.D.A. Beltsville Agricultural Research Center. He has been long interested in international development, due both to his exposure to the dynamic international programs at Cornell and the influence of Professor Norman Borlaug, whose office was located directly across the hall from Dr. Kowalski's laboratory at Texas A&M University.

The second phase of Dr. Kowalski's career has been defined by a transition from research to international work. He received a foreign language area studies scholarship and completed Cornell's one-year intensive Chinese-language program (Chinese FALCON). Subsequently, he worked for the International Service for the Acquisition of Agri-Biotech Applications (ISAAA) in the intellectual property/technology transfer initiative, during which time he conducted the preliminary freedom-to-operate analysis of *Golden*Rice. After working at ISAAA, he earned a J.D. with an emphasis in intellectual property at the Franklin Pierce Law center. He has published numerous research and legal articles.

KRATTIGER, Anatole

Anatole Krattiger, a Swiss citizen, began his career as a farmer, lived in many parts of the world, and is currently a research professor at the Biodesign Institute at Arizona State University (ASU). As adjunct professor at the Sandra Day O'Connor College of Law at ASU, he co-teaches a course on innovation management and controversies in health and agri-biotechnology. He is an Adjunct Professor at Cornell University where he co-teaches a course on IP management in the life sciences. He founded, and serves as Chairman of, *bio*Developments-International Institute, a nonprofit organization that brings people together to jointly develop solutions to problems that extend beyond geographic and cultural frontiers. He recently served as Executive to the Humanitarian Board for *Golden*Rice, a position that required him to work on licensing, technology transfer, and regulatory issues; he also served as Director of Research at MIHR in the U.K. during its formative years. In the early 1990s, he contributed to the international establishment of ISAAA, a global agri-biotechnology broker developing public-private partnerships in agriculture; he served as executive director of ISAAA until 2000. He also briefly worked on biodiversity-policy issues at the International Academy of the Environment in Geneva, Switzerland, and as a scientist in biotechnology at CIMMYT in Mexico.

Dr. Krattiger is a member of the Advisory Council on Intellectual Property of the Franklin Pierce Law Center in Concord, New Hampshire, and a member of the board of the Black Sea Biotechnology Association. He is editor-in-chief of *Innovation Strategy Today* and a member of the editorial boards of the *International Journal of Biotechnology* and the *International Journal of Technology Transfer and Commercialization*. He was a Distinguished Advisor to the Council for Biotechnology Information in Washington, D.C., until the Council merged with BIO. He holds a diploma in farming, a bachelor's degree in agronomy from the Swiss Agricultural College, a master's degree in plant breeding, and a Ph.D. in biochemistry and genetics from the University of Cambridge, U.K.

MAHONEY, Richard T.

Richard T. Mahoney is Director, Vaccine Access, for the Pediatric Dengue Vaccine Initiative, a program of the International Vaccine Institute (IVI) in Korea. Previously, he was Research Professor in the School of Life Sciences and in the Biodesign Institute of Arizona State University. As a consultant to the Rockefeller Foundation, he played a lead role in the consultative process that led to the formation of MIHR. Previously, he was responsible for institutional development in the establishment and launching of the IVI in Seoul, Korea. In this role, he was responsible for cultivating relations with vaccine manufacturers and managing intellectual property, among other things. Dr. Mahoney has had a long career in public health and is known for his work with the International Task Force on Hepatitis B Immunization, accomplished while he was with the Program for Appropriate Technology in Health (PATH). Before co-founding and joining PATH, he was a Program Officer in Population with the Ford Foundation. He oversaw the development and implementation of IP management policies for the Ford Foundation, PATH, and IVI. Prof. Mahoney continues to write on policy and economic research.

MANGENA, Mosibudi

Mosibudi Mangena is the Minister of Science and Technology in South Africa and President of the Azanian People's Organisation (AZAPO). He was born in Tzaneen, matriculated from Hebron Training College in 1969, and received an M.Sc. degree in Applied Mathematics from the University of South Africa (called the University of Azania on the AZAPO website). He joined the South African Students' Organisation (SASO) and was elected to the Student's Representative Council at the University of Zululand in 1971. Moving back to Pretoria, he became chairperson of the SASO Pretoria branch in 1972. He chaired the Botswana region of the Black Consciousness Movement of Azania (BCMA) in 1981 and the BCMA central committee from 1982 to 1994. He returned from exile in 1994 and became leader of Azapo. He was appointed Deputy Minister of Education in South Africa by Nelson Mandela in 2001, and became Minister of Science and Technology in 2004.

MASHELKAR, R.A.

Dr. R.A. Mashelkar is presently the President of the Indian National Science Academy (INSA) and President of Global Research Alliance (GRA), a network of publicly funded R&D institutes from five continents with over 60,000 scientists. Prior to this, for over eleven years Dr. Mashelkar served as the Director General of Council of Scientific and Industrial Research (CSIR), an organization with thirty-eight laboratories and about 20,000 employees. His leadership transformed CSIR into a user-focused, performance-driven organization, a process of transformation that has been recently heralded as one of the ten most significant achievements of Indian Science and Technology in the 20th century.

Dr. Mashelkar is only the third Indian engineer to have been elected as a Fellow to the Royal Society (FRS), London, in the 20th century. He was elected Foreign Associate of the National Academy of Science (U.S.) in 2005, and was only the eighth Indian since 1863 to be elected. He was elected a Foreign Fellow of the U.S. National Academy of Engineering (2003), Fellow of the Royal Academy of Engineering (U.K.) in 1996, and Fellow of the World Academy of Art & Science (U.S.) in 2000. Twenty-six universities have honored him with honorary doctorates, including the universities of London, Salford, Pretoria, Wisconsin, and Delhi. He is currently the President of the Materials Research Society of India.

In post-liberalized India, Dr. Mashelkar has played a critical role in shaping the country's S&T policies. He was a member of the Scientific Advisory Council to the Prime Minister and also of the Scientific Advisory Committee to the Cabinet set up by successive governments.

Dr. Mashelkar has won more than 50 awards and medals, including the S.S. Bhatnagar Prize (1982), the Pandit Jawaharlal Nehru Technology Award (1991), the G.D. Birla Scientific Research Award (1993), the Material Scientist of Year Award (2000), the IMC Juran Quality Medal (2002), the HRD Excellence Award (2002), the Lal Bahadur Shastri National Award for Excellence in Public Administration and Management Sciences (2002), the World Federation of Engineering Organizations (WFEO) Medal of Engineering Excellence by WFEO, Paris (2003), the Lifetime Achievement Award by the Indian Science Congress (2004), the Science Medal by the Academy of Science for the Developing World (2005), and the Ashutosh Mookherjee Memorial Award by the Indian Science Congress (2005), among others.

The President of India honored Dr. Mashelkar with the Padmashri (1991) and with the Padmabhushan (2000), which are two of the highest civilian honors in India, in recognition of his contribution to nation building.

MCCALLA, Alex F.

Alex is Professor of Agricultural and Resource Economics, Emeritus, at the University of California, Davis. He was born in Alberta, Canada, and received his first two degrees from the University of Alberta before moving on to the University of Minnesota where he received his doctorate in Agricultural Economics in 1966. Throughout his academic career he was associated with the University of California-Davis where he served as Dean of the College of Agricultural and Environmental Sciences and Associate Director of the California Agricultural Experiment Station (1970–1975) and Founding Dean, Graduate School of Management (1979–1981).

Dr. McCalla is best known for his research in international trade where he has published extensively. The quality of his research and communication skills has been recognized by the American Agricultural Economics Association, which presented him with its Quality of Communication Award in 1979 and its Quality of Research Discovery Award in 1982. He was elected Fellow of the American Agricultural Economics Association in 1988, Fellow of the Canadian Agricultural Economics Society in 2000, and a Distinguished Scholar of the Western Agricultural Economics Association in 2004. He was a founding member and co-convener of the International Agricultural Trade Research Consortium. He served as the Chair of the Technical Advisory Committee (TAC) of the Consultative Group on International Agricultural Research (CGIAR) from 1988 to 1994.

He elected early retirement from the University of California in June 1994 and was appointed Director of the Agriculture and Natural Resources Department of the World Bank in Washington, D.C., effective September 12, 1994. During his tenure he led a major effort to revitalize the World Bank's commitment to Rural Development. He was appointed Director of Rural Development in July 1997, following a Bank reorganization. He retired from the World Bank December 31, 1999.

In June 1998 he was awarded the Degree of Doctor of Science, *honoris causa,* by McGill University in Montreal, Canada. On December 28, 1999, he was awarded the Doctor's Degree of Honor by the Georgian State Agrarian University. In September of 2004 he received the Distinguished Alumni Award from the University of Alberta.

He served as Chair of the Board of Trustees of CIMMYT, the International Maize and Wheat Improvement Center with Headquarters in Mexico, (2001–2005) and is a member of the Board of Directors of the Danforth Plant Science Center in St. Louis.

MOREL, Carlos

Dr. Carlos M. Morel is currently the Director of the Centre for Technological Development in Health (CDTS), a new unit being implemented at the Oswaldo Cruz Foundation (FIOCRUZ, Rio de Janeiro, Brazil) to stimulate health product innovation.

A molecular biologist and medical doctor by training, Dr. Morel received his M.D. from the Medical Faculty of the Federal University of Pernambuco. He completed his graduate studies at the Biophysics Institute of the Federal University of Rio de Janeiro and at the Molecular Biology Department of the Swiss Cancer Institute in Lausanne (ISREC), Switzerland. His research has been in the field of molecular parasitology, and he has collaborated with various international organizations and research programs working on neglected diseases and capacity building.

Dr. Morel was previously a Professor at Brasilia University (UnB, Brasilia, Brazil) and President of FIOCRU. He was also Director of the UNICEF/UNDP/World Bank/WHO Special Programme for Research and Training in Tropical Diseases (TDR) at the World Health Organization in Geneva, where he established close working relationships with product-development public private partnerships and global ventures committed to public health. He participated actively in the establishment of the Medicines for Malaria Venture (MMV), the Global Alliance for TB Drug Development (GATB), the Drugs for Neglected Diseases Initiative (DNDi), and the Foundation for Innovative New Diagnostics (FIND).

A member of the Brazilian Academy of Sciences and an Honorary Fellow of the Royal Society of Tropical Medicine and Hygiene in London, Dr. Morel holds the National Order of Scientific Merit (Brazil) and Doctor *Honoris Causa* from the Federal University of Pernambuco (Brazil). He has been a member of the MIHR Board of Trustees since its founding.

NELSEN, Lita

Lita Nelsen is the Director of the Technology Licensing Office at the Massachusetts Institute of Technology, where she has been since 1986. Every year, the office manages over 400 new inventions originating from M.I.T., the Whitehead Institute, and Lincoln Laboratory. Typically, the office negotiates over 100 licenses and starts up over 20 new companies each year. Ms. Nelsen earned her B.S. and M.S. degrees in Chemical Engineering from M.I.T., as well as an M.S. in Management from M.I.T. as a Sloan Fellow. Prior to joining the M.I.T. Technology Licensing Office, Ms. Nelsen spent 20 years in industry, primarily in the fields of membrane separations, medical devices, and biotechnology; she worked at such companies as Amicon, Millipore, Arthur D. Little, Inc., and Applied Biotechnology. Ms. Nelsen was the 1992 President of the Association of University Technology Managers. She serves on the board of the Mount Auburn Hospital and the Scientific Advisory Board of the Children's Hospital Oakland Research Foundation. She also serves as the intellectual property advisor to the International AIDS Vaccine Initiative and is a founding and current board member of MIHR. Ms. Nelsen is widely published in the fields of technology transfer and university/industry collaborations. She was a CMI Fellow at Cambridge MIT Institute (at the University of Cambridge), where she studied the role of university/industry/government partnerships in technology transfer and local economic development. She is a co-founder of Praxis, the U.K. University Technology Transfer Training Programme.

POTRYKUS, Ingo

Ingo Potrykus is the engine behind the *Golden*Rice Project and the Humanitarian Board. Together with Peter Beyer, he was one of the inventors of the *Golden*Rice technology. Since his retirement as a professor in 1999, far from settling down, he has devoted enormous efforts to bringing biofortified *Golden*Rice to those who need it.

Prof. Potrykus was born in 1933 in Hirschberg, Silesia, Germany. He has been married since 1960, and has three children and eight grandchildren. In 1968, he earned a Ph.D. in Plant Genetics at the Max-Planck-Institute for Plant Breeding Research, Cologne, Germany.

He conducted research in botany at the University of Basel, Switzerland, and was an Assistant Professor at the Institute of Plant Physiology, Stuttgart-Hohenheim from 1970 to 1974. From 1974 to 1976, he was Research Group Leader at the Max-Planck-Institute for Genetics, Ladenburg-Heidelberg, and then, until 1986, at the Friedrich Miescher-Institute, Basel, Switzerland. From 1986 until his academic retirement in 1999, he was Full Professor in Plant Sciences at the Swiss Federal Institute of Technology (ETH), Zurich.

Since 1974, his research has focused on plant-science-based contributions to food security in developing countries, where he was involved in the development and application of genetic engineering technology for "food security" crops such as rice (*Oryza sativa*), wheat (*Triticum aestivum*), sorghum (*Sorghum bicolor*), and cassava (*Manihot esculenta*). Focusing on problems in the areas of disease and pest resistance that were difficult to solve with traditional techniques, he worked to improve food quality and yield, improved exploitation of natural resources, and improved biosafety. This work was performed by an international team of 60 coworkers, on average, that was financed from competitive grants and core funding. The *Golden*Rice project, initiated in 1991 as Ph.D. project, was possible only because of that core funding. Details of the *Golden*Rice project can be found in approximately 340 publications in refereed journals and 30 international patents.

Professor Potrykus's teaching activities have included lectures and courses in basic and advanced plant biology and plant biotechnology in Biology, Agronomy, Pharmacy, Forestry, and Environmental Sciences departments, as well as International Training Courses such as EMBO. His numerous awards include: the KUMHO (ISPMB) Science International Award in Plant Molecular Biology and Biotechnology in 2000, the American Society of Plant Biologists (ASPB) Leadership in Science Public Service Award in 2001, the Crop Science of America (CSSA) Klepper Endowment Lectureship in 2001, the

CSSA President's Award in 2002, and the European Culture Award in Science in 2002. He received an Honorary Doctorate from the Swedish University of Agricultural Sciences in 2002. He is a member of *Academia Europaea*, the World Technology Network, the Swiss Academy of Technical Sciences, and the Hungarian Academy of Sciences.

REDDY, K. Anji

Dr. K. Anji Reddy is the founder of Dr. Reddy's Laboratories. He spent his early life in Tadepalli village, Guntur, Andhra Pradesh. His father was a peasant and grew turmeric. After completing his schooling at the local high school, Dr. Reddy earned a degree in science in 1958 from the A.C. College in Guntur City. He also earned a BSc in pharmaceuticals and fine chemicals from Bombay University. Later he obtained a PhD in chemical engineering from the National Chemical Laboratory in Pune. Dr. Reddy started his career working for a state-owned company, then called Indian Drugs and Pharmaceuticals Ltd. He was the founder and managing director of Uniloids Ltd., where he worked from 1976 to 1980, and Standard Organics Ltd., where he worked from 1980 to 1984. In 1984, Dr. Reddy laid the foundation for Dr. Reddy's Laboratories. The company established new standards in the Indian pharmaceutical industry and transformed the Indian bulk-drug dependency of the mid-80s into a self-sufficient industry in the mid-90s. By then, the Indian pharmaceutical industry had developed into an export-oriented industry and has remained such ever since.

In 1993, Dr. Reddy's Laboratories emerged as India's first drug-discovering company, and in April 2001, it became the first non-Japanese, Asian pharmaceutical company listed on the New York Stock Exchange

Dr. Reddy is an active member of the Prime Minister's Council on Trade and Industry. Dr. Reddy founded and has donated generously to Dr. Reddy's Foundation for Human and Social Development, which works to support development. During his long career, Dr. Reddy has been the recipient of many awards and civic honors. These include the Padma Shri Award, Sir P. C. Ray Award, FAPA-Ishidate Award for Pharmaceutical Research (1998), Businessman of the Year (2001), and the Achiever of the Year (2000).

RODIN, Judith

Judith Rodin has served as president of the Rockefeller Foundation since March 2005. Trained as a research psychologist, Dr. Rodin was previously the president of the University of Pennsylvania, and earlier the provost of Yale University. The Rockefeller Foundation was established in 1913 by John D. Rockefeller, Sr. to "promote the well-being" of humanity by addressing the root causes of serious problems. The Foundation works globally to expand opportunities for poor and vulnerable people and to help ensure that the benefits of globalization are shared more equitably.

Judith Rodin was born and raised in Philadelphia, Pennsylvania. She graduated from the University of Pennsylvania, and received her Ph.D. from Columbia University. A pioneer in the behavioral medicine movement, she taught at New York University before embarking on 22 years on the faculty at Yale, where she ultimately held appointments in both the School of Arts and Sciences and the School of Medicine. Named president at Penn in 1994, she was the first woman to serve as president of an Ivy League institution.

Dr. Rodin serves on a number of leading nonprofit boards, as well as on the boards of AMR Corporation, Citigroup, and Comcast Corporation. She is the author of more than 200 academic articles and chapters and has written or co-written 11 books. She served on President Clinton's Committee of Advisors on Science and Technology. A member of a number of leading academic societies, including the Institute of Medicine of the National Academy of Sciences, she has received nine honorary doctorate degrees.

SALIM, Emil

Emil Salim is on the faculty of economics at the University of Indonesia. Previously, he was the State Minister for Population and Environment from 1978 to 1993. He currently serves as a member of many international and national committees, including the United Nations High Level Advisory Board on Sustainable Development. He serves as Chairman of the National Economic Board, an economic expert team to President Abdurachman Wahid. He was a member of the economic expert team to President Suharto on debt and development issues of the nonaligned countries, and a member of the Indonesian Peoples' Assembly. In addition, he was Co-chairman of the World Commission on Forestry and Sustainable Development.

Dr. Salim also serves as Chairman of the Board of Trustees for a number of leading Indonesian environmental organizations, including the Indonesian Biodiversity Foundation, the Foundation for Sustainable Development, and the Indonesian Ecolabelling Institute. He received his master's degree and his doctorate in economics from the University of California, Berkeley, in the United States.

SASSON, Albert

Professor Albert Sasson, a Moroccan, is a world-renowned international consultant in biotechnology. He has authored more than 200 publications concerning his research and popularization activities in soil microbiology, algology, and agrobiology. He has published books and contributed to publications on biology teaching, environment and development issues, biotechnologies, and food and nutrition. *Biotechnologies in Developing Countries* is one of his outstanding publications.

Professor Sasson is a prolific speaker, with invaluable information and insight in the areas of cloning, genetically modified foods, the use of biotechnology in agriculture and its possible impact on man and the environment, and ethical and legal issues related to biotechnology. He has expert knowledge of how biotechnology can reduce poverty and the successes and failures of its application worldwide.

After a career as a university dean, he joined UNESCO in 1974, where he served as Special Advisor to the UN for over 27 years. Since January 2000, Prof. Sasson has been senior consultant to UNESCO, Moroccan institutions, and the company Publicis Dialog (Paris). He provides special advice to governments worldwide on the development of national policies on biotechnology, and is an advocate for the adaptation of technologies by the third world for their social and economic development.

Professor Albert Sasson is a man with a passion for science, especially for discoveries in the life sciences. He is truly fascinated with the application of science to food, agriculture, medicine, pharmaceuticals, energy, the environment, and bio-remediation.

SATYANARAYANA, Kanikaram

Kanikaram Satyanarayana holds a doctorate degree in biosciences. After a brief postdoctoral stint, he joined the Council of Scientific and Industrial Research in New Delhi. In 1980, he moved to the Indian Council of Medical Research (ICMR). He is involved in science and technology policy and evaluation, and is Chief of the Intellectual Property Rights Unit. For over twenty years, he has worked extensively in the areas of science and technology evaluation and science policy issues; he was instrumental in the formulation of Indian national policies in these areas.

In 1996, Dr. Satyanarayana published the first guidelines for promoting industry-academia partnerships in medical research in *Contract Research, Consultancy and Technology Transfer policy of the ICMR*. These guidelines are currently being revised to be in agreement with the new WTO and IPR regimes. He has organized several training workshops on WTO and IP rights issues for the benefit of scientists at ICMR institutes, medical colleges, and other institutes. Some of these training workshops were conducted with international funding (WHO). He set up the Intellectual Property Rights Unit at the ICMR in 1999 and brought out the *Intellectual Property Rights Policy of ICMR* in 2002. He is a member of several national committees on intellectual property and has participated in several national and international conferences on such topics as globalization, the impact of TRIPS on public health, access to health care in developing countries, and so on. An active researcher, he has obtained competitive grants from various agencies in India and the World Health Organization. He has also published several papers in national and international journals. He is closely associated with the U.K.-based Centre for the Management of intellectual property in Health R&D (MIHR) and has contributed to their Manual for Technology Transfer Managers. Currently, he is the only member of the International Editorial Board of the second edition of this *Handbook* who is from a developing country. He is a founder and Secretary of the Society for Technology Management, India, and is currently a Senior Deputy Director-General and Chief of the Intellectual Property Rights Unit at the ICMR.

SEKI, Akinori

Akinori Seki is president of the Sasakawa Peace Foundation (SPF), an organization committed to fostering international understanding, exchange, and cooperation. Seki studied at the Gakushuuin University of Economics and received his Ph.D. from the London School of Business.

He worked for many years for the Marubeni Corporation, where he became General Manager (Strategies and Coordination) and Deputy Executive Officer (Corporate Strategies Department). He also lived in Africa briefly as President of Gambia Fisheries' Co. Ltd. He joined the SPF in 1999, initially as Program Director, before becoming Chief Operating Officer, then Executive Director, and now President.

He has served as an advisor to many organizations, including the Myanmar Economic and Management Institute, the United Nations Industrial Development Organization (UNIDO), and the University of Cambodia, and he was a committee member of KEIDANREN and of the study group for Indo-China. He serves on the Board of Directors of the Bellagio Forum and is Member of the Advisory Committee, UNIDO (Tokyo Office). He is an Honorary Professor of Tafaccur University, in the Republic of Azerbaijan.

SERAGELDIN, Ismail

Ismail Serageldin is Director of the Library of Alexandria and also chairs the Boards of Directors for each of the Biblioteca Alexandria's affiliated research institutes and museums. He is also a Distinguished Professor at Wageningen University in the Netherlands. He serves as Chair and Member of a number of advisory committees for academic, research, scientific and international institutions and civil society efforts, including the Institut d'Egypte (Egyptian Academy of Science), TWAS (Third World Academy of Sciences), the Indian National Academy of Agricultural Sciences, and the European Academy of Sciences and Arts. He is former Chairman of the Consultative Group on International Agricultural Research (CGIAR, 1994-2000), Founder and former Chairman of the Global Water Partnership (GWP, 1996-2000) and the Consultative Group to Assist the Poorest (CGAP), a microfinance program (1995-2000). Serageldin has also served in a number of capacities at the World Bank, including as Vice President for Environmentally and Socially Sustainable Development (1992-1998), and for Special Programs (1998-2000). He has published over 50 books and monographs and over 200 papers on a variety of topics, including biotechnology, rural development, sustainability, and the value of science to society. He holds a Bachelor of Science degree in engineering from Cairo University and a Master's and Ph.D. from Harvard University. He has received 19 honorary doctorates.

SHEVELUKHA, Victor S.

Victor Shevelukha was born in 1929, currently lives in Moscow, and is head of the Agricultural Biotechnology Department, Russian State Agrarian University, Moscow. He is a member of the V.I. Lenin All-Union Academy of Agricultural Sciences (VASKhNIL), the Russian academy of Agricultural Sciences, the International Academy of Agrarian Education, the Slavonic Academy, the Agrarian Academy of the Belarus Republic, the International Academy of Informational Sciences, and the Academy of Natural Sciences, among other public academies.

Victor has authored more than 400 scientific works, including 10 monographs and manuals on plant production, plant breeding, seed production, agricultural biotechnology, plant physiology, and agricultural economic policy. He has advised 45 Ph.D. students and 12 doctors of sciences and is currently Chairman of the Scientific Council in RSAU-MAAS, which confers doctorate degrees in the fields of genetics, biotechnology, plant breeding, and seed production.

He worked as a senior agronomist at MAAT's training farm, Druzhba, in the Yaroslavl region (1955-1957); as a secretary of the Ryazantcev CPSU district committee, Yaroslavl region (1957-1959); as the head of agricultural department, Yaroslavl CPSU regional committee; as the first vice-chairman of Yaroslavl regional executive committee (1959-1964); as senior lecturer, associate professor, professor, and head of Crop Science Department at the Belarus Agricultural Academy (1964-1973); the director of Belarus Research Institute for Arable Farming (1973-1974); a secretary of the Central Committee, Belarussia Communist Party (1974-1979); a Deputy Minister of Agriculture of the USSR; a member of Collegium in the USSR Ministry of Agriculture (1979-1983); academic-secretary of Plant Production and Breeding Department, V.I. Lenin All-Union Academy of Agricultural Sciences and Russian Academy of Agricultural Sciences (1983-1994); a deputy of the State Duma, Federal Assembly of the Russian Federation; and vice-chairman of the Committee for Education & Science, the State Duma (1994-2000).

Prof. Shevelukha is also a member of both the Russian Federation Union of Writers and the Russian Federation Union of journalists. He has written and published 10 volumes of fiction and sociopolitical journalism.

Finally, Victor has been awarded the K.A. Timiryazev and V.I. Vernadsky gold medals, orders and medals of the USSR, Russia, and foreign countries, and honorary deeds and titles from the State Duma (Russian Parliament), the Federal Assembly of the Russian Federation, the Ministry of Agriculture, and the Ministry of Education and Science.

SWAMINATHAN, M. S.

Professor M. S. Swaminathan has been acclaimed by *TIME* magazine as one of the twenty most influential Asians of the 20th century, one of the only three from India, the other two being Mahatma Gandhi and Rabindranath Tagore. He has been described by the United Nations Environment Programme as "the Father of Economic Ecology," and by Javier Perez de Cuellar, Secretary General of the United Nations, as "a living legend who will go into the annals of history as a world scientist of rare distinction." He was Chairman of the UN Science Advisory Committee, set up in 1980 to take follow-up action on the Vienna Plan of Action. He has also served as Independent Chairman of the FAO Council and President of the International Union for the Conservation of Nature and Natural Resources. He is the current President of the Pugwash Conferences on Science and World Affairs.

A plant geneticist by training, Professor Swaminathan's contributions to the agricultural renaissance of India have led to his being widely referred to as the scientific leader of the green revolution movement. His advocacy of sustainable agriculture leading to an "evergreen revolution" has made him an acknowledged world leader in the field of sustainable food security. The International Association of Women and Development conferred on him their first international award for his significant contributions to promoting the knowledge, skill, and technological empowerment of women in agriculture, and for his pioneering role in mainstreaming gender considerations in agriculture and rural development. Professor Swaminathan was awarded the Ramon Magsaysay Award for Community Leadership in 1971, the Albert Einstein World Science Award in 1986, and the first World Food Prize in 1987.

Professor Swaminathan is a Fellow of many of the leading scientific academies of India and the world, including the Royal Society of London and the U.S. National Academy of Sciences. He has received 55 honorary doctorate degrees from universities around the world. He currently holds the UNESCO Chair in Ecotechnology at the M. S. Swaminathan Research Foundation in Chennai (Madras), India, and was Chairman of the National Commission on Agriculture, Food, and Nutrition Security of India until October 2006.

THOMSON, Jennifer A.

Jennifer Ann Thomson is Professor of Microbiology in the Department of Molecular and Cell Biology at the University of Cape Town, South Africa (UCT). Previously, she had held the positions of Head of the Department of Microbiology at UCT, the Director of the Laboratory for Molecular and Cell Biology at the Council for Scientific and Industrial Research, and Associate Professor in the Department of Genetics at the University of the Witwatersrand in Johannesburg. Her research involves the development of genetically modified maize that is resistant to the Maize streak virus (endemic to Africa) and tolerant to drought. She received an honorary doctorate from the Sorbonne University, Paris in 2005, and the UNESCO/L'Oreal award for Women in Science in 2004. She is Chair of the Board of the African Agricultural Technology Foundation, based in Nairobi, Kenya. She is a Director of the South African Pebble Bed Modular Reactor (Pty) Ltd. She has published a book, *Genes for Africa: Genetically Modified Crops in the Developing World.*

VAN MONTAGU, Baron Marc

Baron Marc Van Montagu is an Emeritus Professor at Ghent University, and founder and Chairman of the Board of IPBO, the Institute for Plant Biotechnology for Developing Countries. He received a Ph.D. in organic chemistry/biochemistry from Ghent University in 1965, and served as the Director of the Department of Genetics at the Flanders Interuniversity Institute for Biotechnology, before joining the faculty at Ghent University in 1999.

Dr. Van Montagu has made pioneering contributions to plant gene discovery, including the discovery of the gene transfer mechanism between *Agrobacterium* and plants, which was central to the development of transgenic plants. His work at the Lab of Genetics, Ghent University, produced two spin-off biotech companies, Plant Genetic Systems (PGS) and Crop Design. His research at PGS led to the construction of the first herbicide tolerant plants, as well as the construction of the first plants producing the Bt (*Bacillus thuringensis*) insecticide. His was listed among the top 100 living contributors to biotechnology by *The Scientist* magazine and, until 2004, was the most cited scientist in the field of Plant and Animal Science.

XUAN, Vo-Tong

Dr. Vo-Tong Xuan is a distinguished agricultural scientist, an outstanding educator, a low-profile institution builder, and a national and international leader in agricultural development.

As a scientist, he is widely recognized for his expertise in the management of saline and acid-sulphate soils and other problem soils in Vietnam. He is an expert in rice production and in rice-based farming systems, as well as in agricultural diversification in the Mekong Delta. His technical expertise and strong farmer-focused leadership in the Mekong Delta greatly increased rice productivity and contributed to the emergence of Vietnam as the third-largest rice exporting country in the world. Xuan has authored and co-authored six books and more than 100 technical papers about agricultural, rural development, and sustainable food security.

As an educator, he emphasized scientific as well as down-to-earth hands-on-training in the University of Cantho, at which he served as Chairman of the Departments of Bio-Agronomy and Agronomy, and Assistant Dean of Agriculture. He rose to the rank of Vice Rector of the University of Cantho and, in 2000, was elected President of Angiang University, a position he still holds.

As an institution builder, Xuan developed and strengthened the Mekong Delta Farming Systems Research and Development Institute and served as its Director from 1983 to 2001. He also served as FAO Project Coordinator for the establishment of Agricultural Service Centers for Small Farmers. He organized the Vietnam Farming Systems R & D Network and has been serving as its Coordinator since 1991.

As a national leader in agriculture, Dr. Xuan was appointed member of the National Council on Science and Technology, the National Council on Education, the National Council on Professorial Titles Advisory Council of the Vietnam Chamber of Commerce and Industry, the Steering Committee of the Vietnam-Holland Research Program on Rural Development, and the Consultants' Group to the Prime Minister.

As an international leader in agriculture, he is widely recognized for his integrated approaches to agricultural development and deep concern for efficient and effective use of natural resources, sustainability, and environmental issues, as well as, for food security problems of developing countries. He is a strong advocate of the farming system approach in agricultural development.

He has served in key positions in the following international organizations: Member, Board of Governors, Asian Institute of Management in Manila; Member, Board of Trustees of IRRI; Member, Board of Trustees of The Rockefeller Foundation; Member, Board of Trustees of the International Potato Center at Lima, Peru; Member, FAO's Advisory Committee on Farmer-Centered Agricultural Resource Management Program; Member, Technical Advisory Committee of the CGIAR; Member, Policy Advisory Council, Australian Centre for International Agricultural Research; Member, Advisory Council of the Asian Development Research Forum.

Dr. Xuan served as international consultant, lecturer of IFAD, FAO, DANIDA, SIDA, and IDRC-Singapore since the 1980's.

He received from the Prime Minister of Canada a certificate of recognition for his "dedication and contribution to the world of sciences." The Ministry of Agriculture, Fisheries and Forestry of the Republic of France, awarded him the "*Chevalier de l'Ordre du Merite Agricole Medal.*" He was elected the 2002 Nikkei Asia Prize for Regional Growth; Most Distinguished Alumnus of the University of the Philippines College of Agriculture Alumni Association; Ramon Magsaysay Award for Government Service; and the 2005 ASTD Derek Tribe Award. Other awards include: the People's Teacher Award, Vietnam Farmers' Federation Medal "For the Cause of the Farmers' State Award as "Hero of the Working Class," Outstanding Scientific Achievement Award from the Prime Minister, Most Distinguished Alumnus Award from the University of the Philippines at Los Banos.

YUTHAVONG, Yongyuth

Yongyuth Yuthavong is a scientist. His interests lie in antimalarial drug development and broad issues of science, technology, and public policy. In 1962, Professor Yuthavong was awarded a Thai government scholarship to study in the United Kingdom. He obtained a bachelor's degree in chemistry with first-class honors from the University of London in 1966 and a doctoral degree in organic chemistry from the University of Oxford in 1969. He spent many years at Mahidol University in Thailand, where, beginning in 1983, he served on the faculty as a professor of biochemistry. He was actively engaged in the establishment in 1992 of the National Science and Technology Development Agency (NSTDA) of Thailand (1992) and became the agency's first president, serving two terms. In 1998, he returned to the research career at the NSTDA's National Centre for Genetic Engineering and Biotechnology (BIOTEC), where he had once served as its director. Currently he heads a research group in BIOTEC, where he is working on the development of new antimalarials with grants from the Medicines for Malaria Venture and the Wellcome Trust.

In 2004, Yuthavong received the Nikkei Asia Prize for Science, Technology, and Innovation from Nihon Keizai Shimbun, Japan, for his work on antimalarial drug targets. The same year, Thailand's National Identity Board named him Person of the Year. In 1984, he received the Outstanding Scientist of Thailand Award from the Foundation for Promotion of Science and Technology. He has received honorary doctorates from Prince of Songkla University and Mahidol University. In 2006, Bangkok's The Nation newspaper named him one of Thailand's 35 most influential people over the last 35 years.

Yuthavong is past chairman of the Foundation of Thai Academy of Science and Technology and the Foundation for Promotion of Talents in Science and Technology. Today, he serves as a member of five state university councils. He is a member of the National Education Board and the National Research Council Committee on Chemistry and Pharmaceutical Sciences. He has coauthored 115 research articles published in international journals and 16 book chapters and books on biomedical science, policy, and general issues of science and technology.

Yuthavong was appointed Thailand's minister of Science and Technology in 2006. ■

Readers are encouraged to consult the online version (www.iphandbook.org) for fuller indexing, a robust search engine, and a "Web log" presentation of key content areas with comments.

Index